The Changing World and Africa

Xinfeng Li · Chunying An
Editors

The Changing World and Africa

SSAP

palgrave
macmillan

Editors
Xinfeng Li
Institute of West-Asian and African Studies
Chinese Academy of Social Sciences
Beijing, China

Chunying An
Institute of West-Asian and African Studies
Chinese Academy of Sciences
Beijing, China

ISBN 978-981-16-4982-0 ISBN 978-981-16-4983-7 (eBook)
https://doi.org/10.1007/978-981-16-4983-7

Jointly published with Social Sciences Academic Press
The print edition is not for sale in China (Mainland). Customers from China (Mainland) please order the print book from: Social Sciences Academic Press.

Translation from the Chinese language edition: 《变化中的世界与非洲》 by Xinfeng Li, and Chunying An, © Social Sciences Academic Press 2021. Published by Social Sciences Academic Press. All Rights Reserved.

This Palgrave Macmillan imprint is published by the registered company Springer Nature Singapore Pte Ltd.
The registered company address is: 152 Beach Road, #21-01/04 Gateway East, Singapore 189721, Singapore

Foreword

Africa is both an ancient continent that gave birth to the earliest human civilizations and a new continent, which is now witnessing an annual economic growth of 5% and a stabilizing political environment. Africa is now marching toward stability, independence, development, and prosperity despite its miserable history of having been invaded, plundered, enslaved, and exploited. As a continent with rich natural resources, an abundant young labor force, and potentially huge consumer market, Africa has been attractive to foreign forces with various purposes throughout time. The gradually stabilizing development of Africa, particularly the development of China–Africa relations, has caused different countries to adjust their strategies and policies regarding Africa and to strive to achieve their own policy objectives through new exchanges and interactions with Africa.

Organized by the China-Africa Institute (CAI), many scholars have contributed to this book who strive to provide an in-depth and comprehensive introduction to many countries' relations with or policies regarding Africa. The 18 research articles contained here include an analysis of new strategies and policies regarding Africa by traditional Western colonial powers, the introduction of newly industrialized countries' (such as China, India, Turkey, and Brazil's) African policies, an interpretation of both the history and current situation of African policies by Latin American countries (such as Cuba), and an explanation of Russia's "return to Africa" policies. Some of these articles provide a comprehensive review and analysis of historical and contemporary developments among various countries' relationships with Africa from the perspectives of political relations, economic and trade cooperation, development aid, and cultural exchanges, while other articles provide in-depth insights from the perspectives of ruling parties, heads of state, or policies covering different fields. From these articles, we may better understand the common backdrop of national political adjustments regarding Africa, different starting points, the dynamics and behavioral approaches underlying these policy adjustments, and even mutual influences among such policy adjustments.

Due to globalization, the "special bilateral relations" between Africa and traditional colonial powers are being sidelined by a now more widely accepted multilateralism. Not only have many international multilateral institutions (such as the World Health Organization, World Food Programme, United Nations Industrial Development Organization, and many multilateral banks) financed African development, many countries have also actively engaged in African affairs and development. Instead of being simply a sphere of influence or a "protectorate" of former colonial empires, Africa is now rebuilding its relations with the world amid opportunities and challenges.

In the context of globalization, the African economy has not only undergone its own rapid development, but has also become one of the driving forces supporting global development. From 1995 to 2017, the average annual growth rate of the African economy was around 3–5%. With the advent of the new millennium, Africa has maintained an average annual economic growth rate of 5%, exceeding that of the rest of the world. Meanwhile, as a continent with rich natural resources, Africa has the world's third largest oil reserves and the fourth largest natural gas reserve, which, together with the labor force dividend brought about by its annual population growth of 3%, contribute to the emergence of a growing "Africa Craze" in the world. The relevant countries have not only promoted the significance of African affairs in their foreign relations, but have also initiated the adjustment of strategies and policies regarding Africa. This has triggered a new round of resource reallocation and even competition for influence. In its strategy adjustment regarding Africa, for example, the US regards China as its direct competitor and as such treats Africa as the staging ground for a China–US game. The French media also aim "to deal with China's rising influence in Africa;"[1] the European Union believes that "Africa has become a game field of great powers".[2]

Most African countries are neither willing to see powerful nations play a leading role in Africa, nor are they willing to choose sides among those same major powers, and as a result refuse to go along with strategies such as that deployed by the US. However, guided by the principle of cooperation, some countries have adopted a pragmatic approach in their policy adjustments regarding Africa, doing so by weakening the traditional superiority of Western countries over Africa and developing relationships with Africa through the expansion of investments and employment as well as integrating security cooperation with development cooperation. Some countries have begun to draw lessons from others' experiences and practices, putting forward concepts such as an "equal partnership with shared benefits" (such as with the EU and UK) or "respecting the sovereignty of Africa countries" (such as with Germany and Italy). At the same time, most countries have begun to pay closer attention to people-to-people and non-governmental exchanges and lay more emphasis on the building of soft power.

Distinct characteristics can be found in the African policies held by different countries and organizations. By preaching its "America First" policy, the Trump government of the US targets China as the object of its hedge diplomacy in Africa. In order to prevent an influx of Africa immigrants and refugees, the EU hopes to solve the "security" problems of Africa through investment and development. Due to different national resource endowments, the emphasis and approach of their policies regarding Africa may vary significantly. For example, France pays attention to sports exchanges and cuisine cooperation. Meanwhile, India has launched the ministry of yoga and traditional medicine in Africa and appointed the first yoga minister in history. Cuba, on the other hand, has developed advantages in medical aid.

Among the European countries, the UK has put forward the diplomatic concept of a "global Britain" after Brexit, and has subsequently tried to profit from Africa's rich resources, huge market, and demographic dividends by exploiting its historical role in African countries. Realizing

[1] Emmanuel Macron à la conquête de l'Afrique de l'Est, *LeParisien*, 11 Mars 2019.

[2] Louis Michel, "Europe-Africa: The Indispensable Partnership", http://www.lse.ac.uk/collections/LSEPublicLecturesAndEvents/pdf/20080117_Michel.pdf, accessed 2018-11-20.

that today's globalized Africa is no longer a private backyard for France, French President Emmanuel Macron proposed the establishment of a more realistic relationship without a special France–Africa relationship, hoping to essentially neutralize the security challenges posed by African immigrants and refugees through the promotion of economic development and employment. Germany formulated the African Marshall Plan, which put forward slogans promoting such ideas as a "Compact with Africa," "Reform Partnership," and shifting from "for Africa" to "with Africa." In its dealings with Africa, Germany went beyond development assistance by using the three pillars of trade and employment, peace and security, and democracy and the rule of law, to encourage and support private enterprises to invest in Africa. It also explicitly committed to respecting the diversity of African countries and prioritized African development as basis for further Germany–Africa cooperation. Italy targeted Sub-Saharan Africa, the home countries of African immigrants, and transitional countries in North Africa as its main partners for cooperation, hoping to ease the pressure of immigrants and security through developmental cooperation. Portugal has strengthened its economic and social ties with African Portuguese-speaking countries through two platforms: the Community of Portuguese-Speaking Countries (CPLP) and the Forum for Economic and Trade Cooperation between China and Portuguese-Speaking Countries (Macao).

In the Asia-Pacific region, Japan continues, on one hand, to prioritize short-term interests, while on the other hand, carries out a more balanced diplomacy. In 1993, the Tokyo International Conference on African Development (TICAD) established a regular meeting mechanism that initiated a new period of attaching importance to its diplomatic strategies regarding Africa. It then put forward the concept of the "Asia-Africa Growth Corridor," gradually strengthening its strategic influence over Africa and forging a long-term strategic African policy. As for India, the "neighborhood first" policy of the Modi government prioritized Africa in its foreign relations. Against the backdrop of strategic cooperation with China, India has mobilized various resources, including strengthening relations with the West and Japan and integrating strategic demands for regional leadership in its various policies. These policies have mainly been established in order to deal with the Belt and Road Initiative (BRI) of China. South Korea, meanwhile, has carried out a policy of strengthening the construction of a fulcrum country and sharing the development model. Under the background of "new contacts," Australia has deepened its existing political and economic contacts and cultural exchanges with African countries.

In the Middle East, Turkey has completed three aspects of its diplomatic system on the African continent over the course of twenty years: soft power, economic diplomacy, and security system. With the aggressive humanitarian diplomacy embodied in its policies regarding Africa, Turkey has become the second largest humanitarian aid country in Africa after the US. This has not only enhanced its political and economic ties with African countries, but also greatly improved its influence and voice in regional affairs. Saudi Arabia, relying on its huge oil revenues, has also become a major aid provider to Africa. It has rallied the support of African countries in the Arab–Israeli conflict, prevented them from being affected by political turmoil in the "Arab Spring," and supported the development of the Arab nation.

In South America, Brazil and Cuba have both attached great importance to African affairs because of the integration of a large number of Black people within their populations. For Brazil, the expansion of export markets, creating jobs, and increasing foreign exchange earnings are important considerations for adjusting its relations with Africa. In 2008, relying on various platforms for South-South cooperation (such as the community of Portuguese-speaking countries) and diversified policies, Brazil expanded its economic ties with Africa, strengthened its institutional presence in Africa, and won the explicit support of African countries on the issue of Brazil's bid to be added as a permanent member of the UN Security Council. African countries also provided important international support for Cuba. In its African policy, Cuba not only focused on strengthening solidarity with the third world by opposing colonialism and supporting national independence, but also paid more attention to practical results. Medical and military diplomacy are important contents, characteristics, and advantages of existing Cuba–Africa relations. Cuban overseas guerrillas are also known as "medical guerrillas." Cuba identified Angola and South Africa as important links in the development of Cuba–Africa relations, and maintained the continuity and independence of its African policy with the help of multilateral organizations such as the Non-Aligned Movement (NAM).

The recent "Africa Craze" has forced Russia to "return" to Africa. Judging from the available statistics, Russia's total cooperation with Africa is not particularly significant, but its influence extends over almost all of Africa. Russia attaches great importance to the building of multilateral mechanisms on a political level, the exploitation and processing of energy resources on an economic level, has inherited the tradition of military industry and trade, and has increased the training of African students and cadres in its aid. Africa is becoming a new growth point in Russia's regional diplomacy.

Under the guidance of Chairman Mao Zedong and Premier Zhou Enlai, and supported by the efforts of various Chinese leaders and practitioners at different times, relations between modern-day China and Africa have evolved into multiple close cooperative ties. Since 2019, China has become Africa's largest trading partner for ten consecutive years, and has embraced closer cooperative relations in the fields of politics, the economy, security, culture, and education. In 2015, Chinese President Xi Jinping proposed that the new strategic partnership between China and Africa should be upgraded to a comprehensive strategic partnership, strengthening "five pillars"[3] and carrying out "ten major cooperation projects."[4] Different from the competitive thinking promoted by some countries, the principles of "equal cooperation and mutual benefits" have topped the national agenda since the debut of China–Africa relations. This approach was embodied in the construction of the "five connectivities" (policy connectivity, infrastructure connectivity, trade connectivity, financial connectivity, and people-to-people connectivity) put forward by the Belt and Road Initiative (BRI). With the rapid development of China–Africa relations, the BRI has become a powerful engine to support the building of a China–Africa community with a shared future.

China has been playing an increasingly more obvious role in Africa's relations with the outside world. As China and African countries are experiencing rapid economic growth, continued cooperation between China and Africa is having an impact on the distributive patterns of global resources and wealth. Therefore, many other countries are likely to take

[3] At the 2015 FOCAC Summit in Johannesburg, South Africa, President Xi proposed that equality and mutual trust in politics, win-win cooperation in the economy, mutually enriching cultural exchanges, mutual assistance in security and solidarity, and coordination in international affairs should be the "five pillars" of China-Africa cooperation, http://img02.sogoucdn.com/app/a/200698/efe30b257d500fe29af0dbdabd6d8d02_1579932571_1100x4149.

[4] China-Africa cooperation plans for industrialization, agricultural modernization, infrastructure, finance, green development, trade and investment facilitation, poverty reduction, public health, people-to-people exchanges, peace and security, https://baike.sogou.com/v141970527.htm;jsessionid=4C370CCFFF693D9B25356C3B70DE1468.

China as a frame of reference or comparison when formulating policies regarding Africa, producing various policies across themes such as competition, cooperation, reference, or confrontation. This is an issue that China needs to pay close attention to and deeply understand when handling its relations with Africa.

In his letter congratulating the inauguration of the Chinese Academy of African Studies in April 2019, President Xi proposed to "enhance the traditional friendship and mutual understanding between Chinese and African peoples and to draw on both sides' academic resources to provide good ideas and advice for China–Africa cooperation, as well as the cooperation between the two sides and other parties, to contribute to the development of the China–Africa relationship and the building of a community with a shared future for humanity."[5] I believe that this book will live up to President Xi's expectations and contribute to the deepening of a mutual understanding between the Chinese and African peoples.

Beijing, China
Autumn 2020

Hong Zhou

[5] Xinfeng Li: The Congratulatory Letter of President Xi Jinping: Foundation and Soul for Building and Prosperity of the China-Africa Institute, West and Asia, Vol 2, 2020, p. 7.

The original version of the book was revised: Affiliations of the chapter authors has been updated. The correction to the book is available at https://doi.org/10.1007/978-981-16-4983-7_23

Contents

Part III China and Africa Relations

About the Editors

Xinfeng Li is General-Director of the Institute of West-Asian and African Studies of the Chinese Academy of Social Sciences (CASS), Executive President of the China-Africa Institute, a Senior Research Fellow, Professor, and Doctoral Supervisor. Li has published more than 20 academic papers and works, including *China in Africa* (Chinese and English versions) and *Zheng He and Africa*. As a former chief reporter of *People's Daily* in South Africa, he has had a lasting impact on the African continent. He has earned more than ten national awards, including the China News Award (2006 & 2017), the first prize of the CASS Award for the Excellence in Countermeasure Studies (2012 & 2016), and the China-Africa Joint Exchange Program Research Award (2013 & 2016).

Chunying An is Senior Editor of the Institute of West-Asian and African Studies of the Chinese Academy of Social Sciences (CASS) and Secretary General of the China Society for African Studies. Her major fields of study cover African economies, poverty reduction and sustainable development in Africa, and China–Africa relations. Major publications include the following: *A Study on the Issues of Poverty and Anti-poverty in Africa* (monograph, 2010), *Experience and Lessons for Africa from China's Success in Economic Development* (paper, 2016), and *China-Africa Poverty Reduction Cooperation and Experience Sharing* (think tank report, 2018).

List of Figures

List of Tables

Chapter 21

PART I

Africa and Its Relationships with Developed Countries

CHAPTER 1

The Historical Logic and Pragmatic Choices Behind the US Strategy in Africa

Peng Wang

In December 2018, the Trump administration released its "New Africa Strategy" as an important component of its overall foreign strategy. The "New Africa Strategy" calls for the US to strengthen its economic and trade ties with Africa, enhance security by resisting violence and terrorism, and more "efficiently and effectively" aid the coordination of US foreign relations with Africa.

It has been more than a year since the Trump administration's Africa strategy was proposed. During this period, China–US relations have moved in a more confrontational than cooperative direction.[1] Then-US

[1]Wang Wen and Wang Peng, 2019. "20 Years of G20 Mechanisms: Evolution, Dilemma, and China's Response," *Contemporary International Relations*, no. 5, pp. 1–9.

This work is supported by the Major Program of National Social Science Foundation of China (18ZDA205).

P. Wang (✉)
Renmin University of China, Beijing, China
e-mail: wxy03@ssap.cn

X. Li and C. An (eds.), *The Changing World and Africa*,
https://doi.org/10.1007/978-981-16-4983-7_1

Secretary of State Mike Pompeo visited Africa in February 2020 and repeatedly publicly accused China and African countries of ordinary interactions, and even publicly warned African countries to "beware of Chinese infiltration," as well as interfered with China–Africa Belt and Road cooperative initiatives. During the outbreak of the global COVID-19 pandemic in early 2020, some American politicians and media even undermined China–Africa relations by making false remarks such as "China is spreading the virus" and "China is discriminating against Africans in China."[2]

In general, the US strategy toward Africa under the Trump administration is driven by both historical factors and contemporary features. Accordingly, this paper briefly traces the US Africa strategy in the 70 years after World War II and summarizes its historical logic and power base. On this basis, it discusses the policy guidelines and provides a preliminary assessment of the Trump administration's Africa strategy, as well as the US–Africa strategy from the perspective of Sino-American strategic competition.

1 Historical Logic and Power Resources of the US Strategy in Africa

From the outbreak of the Cold War to Trump's rise to power, the 70 years (1947–2017) of United States' Africa strategy can be broadly divided into two periods. The first 40 years or so makes up the Cold War era (1947–1990), which is mainly characterized by global superpowers competing to support their proxies and compete for spheres of influence in the African continent in the context of a US-Soviet rivalry. The period between 1991 and 2017 can be regarded as the post-Cold War era, which can also be separated by the "9/11" terrorist attack in 2001—the phase before "9/11" (1991–2001) was the US unipolar moment, during which the US was trying to claim a Cold War "victory" as the unique remaining superpower in Africa and to establish a new order to its own advantage by strengthening the US–Africa economic and trade investment connectivity; in the phase after "9/11" (2001–2017), the American counterterrorism war has been top of the US agenda in Africa.

[2] Paravicini, Giulia, 2020. "Pompeo Takes Veiled Swipe at China on Final Leg of Africa Trip," *Reuters*, February 19, https://www.reuters.com/article/us-usa-pompeo-ethiopia-idUSKBN20D1PB.

US Strategy in Africa During the Cold War (1947–1990): A Bipolar Model of Hegemony

After the end of the Second World War, a wave of anti-colonial and national liberation movements rose in Africa in the mid-1950s, with most African countries gaining independence in the 1960s. At the same time, the US, with its economic and military superiority, tried to establish a global hegemonic order encompassing the African continent. Africa's rich natural resources (such as ores, energy, precious metals, gems, and uranium) attracted the intervention of superpowers such as the US, the Soviet Union, and former European colonial sovereign powers, while tribal conflicts and regional wars within and among African countries also provided opportunities for outside forces to intervene. During the Cold War, Africa became the front line of many of the bloody conflicts and proxy wars between the US and the Soviet Union.

The entry of Soviet power into Africa was, to a large extent, a card played in response to the "containment" blockade imposed by the US on the Soviet Union in Eurasia. At the beginning of the Cold War, the US besieged the Soviet Union on three fronts: the western front (NATO confronted the Warsaw Pact in Eastern Europe), the Middle Eastern front (from the US–Israeli alliance in 1948 to the establishment of the Baghdad Pact in 1955, marking the initial formation of the US strategic deployment in the Middle East), and the eastern front (the US–Japanese and US–Korean alliances and the Southeast Asia Pact established in 1955), forcing the Soviet Union to try to use military force through an "indirect approach" to break the US blockade by expanding its sphere of influence in Africa, where US influence was somewhat weaker.[3] The opportunity for Soviet involvement in African affairs was the Suez incident of 1956—the Second Middle East War. By intervening in Egypt's disputes with Israel, Britain, and France, the Soviet Union successfully extended its strategic reach into North Africa.

In response to the strategic expansion of the Soviet Union, the US intensified its strategic deployment to Africa starting in the 1960s.[4] However, under an anti-imperialist and anti-colonialist ideology, it was

[3] Lefebvre, Jeffrey, 1998. "Moscow's Policy Towards Africa During and After the Cold War," translated by Zhu Jingfang et al., *West Asia and Africa*, no. 3.

[4] Scott, Thomas, 1998. *Africa and the End of the Cold War: An Overview of Impacts*. London: Wellington House, pp. 7–8.

easier for the Soviet Union to win the affection and support of the revolutionary countries in Africa than for the US. On the one hand, the US does not support colonialism ideologically and supports self-determination and the independence of African nations. However, a number of US Western European/NATO allies still maintain large colonies in Asia, Africa, and Latin America, so the US must support these Western European colonial suzerain states from the strategic point of view of supporting the allies and resisting the Soviet Union's expansion into Africa. This includes defending the British and French spheres of influence in West Africa and the interests of the Belgian colonial authorities in Africa. The US, on the other hand, has had its own national, bureaucratic/sectoral, and commercial interests. The takeover of a number of French military bases and enterprises in North Africa by various USS departments and commercial organizations has not only humiliated the USin front of the emerging African countries, which considered America the same aggressive force as the old Western European colonizers, but has also created divisions within the Western camp and accelerated the centrifugal tendencies of France, Belgium and other Western European powers.

Influenced by the African independence movement, by the end of the 1960s, 41 newly independent states had emerged in Africa. The first summit of the Non-Aligned Movement in 1961 and the establishment of the Group of 77 in 1964 strengthened the voice of the emerging African countries in the international arena and forced the US and the Soviet Union to reconsider the role of these African countries. As a result, both America and the Soviet Union had to adopt more flexible and pragmatic alliance strategies with African countries.

The chronic problems of the African countries themselves, however, such as their relatively homogeneous economic structures leading to a high degree of dependence on the former Western European suzerainty for markets and raw materials, left the newly independent African countries with little room for maneuvering in their external dealings and dependent on foreign aid for their economies. At the same time, the cruelty of proxy wars and the intensification of confrontation among some African countries have made it difficult for any country on the continent to truly choose "non-alignment" and stay out of confrontations and regional disputes among superpowers. The combination of these factors has made it impossible for most of the emerging African countries to escape from external ideological propaganda or the war of the great power struggle for hegemony.

The US Africa Strategy in the Post-Cold War Era (1991–2017): Unipolar Dominance

The strategic dominance of the US in Africa did not begin after the end of the Cold War, but rather, much earlier, in the late 1970s. At that time, the Soviet Union's influence in Africa began to decline due to its own economic, military, and overall national power limitations, as well as its ill-advised hegemonic strategy of excessive intervention in African countries. By the early 1980s, most African countries had begun to alienate themselves from the Soviet Union, even moving toward a policy of confrontation; and the Soviet Union itself had become trapped in the war in Afghanistan. The US seized this opportunity to fill the vacuum created by the withdrawal of the Soviet Union, which collapsed on December 25, 1991. After that, the African continent entered an era of unipolar domination by the US.

In the post-Cold War era, the United States' strategies and policies toward Africa have varied during the different presidential terms of office. During the senior George Bush Administration era, the US inherited the Cold War tradition and continued to promote the "democratization" of a pro-American Africa while seizing the Soviet Union's position. The Clinton era was more focused on the development of economic and trade ties between the US and Africa. During the George W. Bush era, after the September 11 incident, the war on terror prioritized US–Africa security cooperation and energy cooperation. After Obama came to power, he made four visits to Africa and issued related strategies and policies, which to some extent raised the importance of Africa in the diplomatic sequence of the US government. At the same time, Obama's policies integrated the post-Cold War US Africa strategy, centering on economic and trade cooperation, counterterrorism, and the so-called "democratization transformation" as the three core areas of the US Africa policy. In Obama's second term, with the overall improvement of China–Africa cooperation, the US perceived its influence in Africa as being threatened by China, and therefore, in line with its global "Pivot Asia" strategic adjustment, it also made significant adjustments to its Africa strategy, highlighting the growing strength of its competition with China. The above features have been inherited to varying degrees by Trump's New Africa Strategy.

In general, Africa has always been a relatively minor and marginal part of the United States' global geostrategic map, both during the Cold War and in the post-Cold War era. It is difficult for the US to design a stable and coherent strategy for Africa that is guided by core bilateral interests, as it did in the careful delineation of US–European/transatlantic relations. On the contrary, throughout the Cold War and post-Cold War eras, the US strategy toward Africa has always shown a considerable degree of reactivity: the Cold War period was directed against Soviet power expansion; the post-Cold War period shifted to a counterterrorism focus immediately after the Clinton era, which itself focused on economic and trade ties; from Obama's second term onward, there was a gradual shift toward targeting China's growing influence in Africa, and the Trump era has only reinforced this turn.

Since the US has always shown a high degree of "reactivity" in its strategic orientation toward Africa, it inevitably demonstrates a high degree of pragmatism at the strategic level, i.e., in order to serve and respond to the reorientation of US global strategy (whether Eurocentric during the Cold War or Asia-Pacific/Indo-Pacific-centric in recent years) on the African continent. The US strategic toolkit includes the following approach: selecting priority countries and then using a combination of economic and trade ties (including market access, investment in Africa, etc.) and military power (including arms sales, military aid, support for specific military factions, etc.) as a means of bringing them together in order to counter the perceived "threat." This has been the core logic of the United States' strategy in Africa for more than 70 years after the Cold War, with the economy and the military serving as the main components of the means of implementation and the foundation of power.

2 Policy Guidelines and a Preliminary Assessment of the Trump Administration's New Africa Strategy

On December 13, 2018, then-US National Security Adviser John Bolton delivered a speech at the Heritage Foundation to formally lay out President Trump's "New Africa Strategy."[5] The document "President Donald

[5] Bolton, John R. 2018. "Remarks by National Security Advisor Ambassador John R. Bolton on The Trump Administration's New Africa Strategy." The White House, December 13, https://www.whitehouse.gov/briefings-statements/remarks-national-security-advisor-ambassador-john-r-bolton-trump-administrations-new-africa-strategy/, accessed on April 12, 2020.

J. Trump's Africa Strategy Advances Prosperity, Security, and Stability" was released on the White House homepage on the same day.[6]

Compared with Bolton's speech, the Trump Africa Strategy is more concise and does not explicitly criticize the so-called "external threats" and "great power competition" of countries such as China and Russia. But by referencing the actual implementation of US Africa policy in the year and a half since December 13, 2018, as well as the words and deeds of Trump, Pompeo, and other figures in Africa and China–Africa cooperative affairs, we can still judge that the hostility advocated by Bolton's speech does not seem to have been materially reduced. Therefore, when discussing the concrete outline of the implementation of the US strategy for Africa, this paper will refer to Bolton's speech and the Trump Africa Strategy and other important texts in order to use them as the main basis for evaluating the implementation of the US strategy for Africa.

At the beginning of his speech, Bolton pointed out the three core US interests at the heart of the new US Africa strategy: trade, security, and foreign aid. Accordingly, the three main objectives of "prosperity, security, and stability" were also at the heart of Trump's New Africa Strategy.

First, in the field of economics, Bolton's speech advocated the strengthening of mutually beneficial trade and commercial ties between the US and African countries (US trade and commercial ties). Under the overall framework of the Trump administration's Africa strategy, the "Prosper Africa Initiative" has been launched as the main policy focus in the economic field. Among its aims, "investment in Africa" and developing a "business environment" are central to the initiative, which aims to enhance US economic and trade relations with Africa and create jobs in the US. Indeed, "the Trump administration's economic and trade policy toward Africa does not go beyond the post-Cold War direction and framework of US policy toward Africa." This is justified through the promotion of US trade and investment in Africa and by linking US investment in Africa to the political and institutional reforms in African countries themselves, which has been the dominant direction of US policy toward Africa since the Cold War, pursued by multiple successive administrations.[7]

[6] 2018. "President Donald J. Trump's Africa Strategy Advances Prosperity, Security, and Stability." The White House, December 13, https://www.whitehouse.gov/briefings-statements/president-donald-j-trumps-africa-strategy-advances-prosperity-security-stability/.

[7] Zhao Chenguang, 2019. "He U.S. 'New Africa Strategy': Changes and Constants." *International Studies*, no. 5, pp. 115–126.

Trump's Africa strategy has two distinctive features in economic matters. On the one hand, it follows the principle of "America first." Therefore, in addition to expanding market access for US exports, the US also places particular emphasis on improving opportunities for American workers and businesses, thereby supporting American jobs. The speech also repeatedly mentioned concepts such as "Americans' hard-earned taxpayer dollars." On the other hand, it seems to be related to the recent criticism of the so-called "China's rule-breaking" by the US and the "order/rule-based international order" in the "Indo-Pacific" region. In Africa, it emphasized "the rule of law" and criticized Russia, for example, for "disregarding the rule of law" in promoting relations with African countries," helping African partner countries "strengthen the rule of law" and "provide law enforcement services for their citizens," requiring all African countries receiving US aid to "promote the rule of law."

Second, in the area of security, Trump's New Africa Strategy confronts radical Islamic terrorism and violent conflict. However, reviewing the Trump administration's actions over the past four years, there still seems to be some distance between the actual investment in US–Africa security cooperation and the aforementioned goals. Immediately after taking office, Trump proposed a ten percent reduction in the US military presence in Africa, as well as in UN support for African peacekeeping operations. At the same time, in response to the cuts in US troop presence and peacekeeping/terrorism support, Trump has argued that African countries must take ownership of their own counterterrorism operations. This approach is logically consistent with his advocacy of more responsible obligations for allies in the framework of the NATO, US–Japan, and US–South Korea alliances.

Third, in the case of US aid to Africa, the emphasis is on "ensur[ing] that US taxpayer dollars for aid are used efficiently and effectively." In the presentation that followed, Bolton described at length the two "threat sources" that currently challenge US national security and economic interests on the African continent: one is the competition between the great powers represented by China and Russia, which falls within the traditional security context, but with a substantial economic and institutional rule component; the other is the non-traditional security threats represented by the ideology of terrorism. On this basis, Bolton further elaborated on the specific means of implementation.

A comparison of the US African strategy in the Obama era can provide a deeper understanding of the essential features of the Trump administration's African strategy.[8] Barack Obama was the most frequent sitting president in US history to visit sub-Saharan Africa: the first visit was to Ghana in 2009, the second to Senegal, South Africa, and Tanzania in 2013, the third to South Africa for the funeral of former President Nelson Mandela in 2013, and the fourth in late July 2015. Obama's four trips to Africa not only expressed the United States' concerns about Africa's strategic importance, but also implicitly vilified China–Africa relations, a move that illustrates a trend of increased strategic competition between China and the US in Africa.

From the earlier US Africa Policy for the twenty-first century released in 2009[9] to the complete US Strategy for Sub-Saharan Africa released by the White House for the first time on June 14, 2012,[10] Obama's Africa strategy underwent a gradual transformation.[11] Generally speaking, he inherited the US' previous overall strategic arrangement for Africa in the areas of security, economy, and trade, and also attached equal importance to the integrated use of military and economic (aid) means. What is different from the previous George W. Bush administration is that the importance of Africa in the Obama administration's diplomatic sequence has been elevated, such as the declaration of a "partnership" with Africa and a more flexible and pragmatic policy on the Sudan issue. In addition to the Obama administration continuing to promote cooperation, good governance, and providing medical assistance to fight epidemics in Africa, it has also more effectively expanded public diplomacy. All of these measures are aimed at strengthening the US' influence on Africa and its ability to draw on local resources, while promoting counterterrorism operations in Africa to defend the security of US citizens and US troops at home and abroad. And in the latter part of the Obama administration, with China's rapidly expanding influence in Africa, the US has taken a

[8] Zhongwei Liu, 2015. "Obama's Fourth Visit to Africa and the Trends of U.S. Policy Towards Africa." *Contemporary World*, no. 9, pp. 69–71.

[9] Carter III, Phillip, 2009. "U.S. Policy in Africa in the 21st Century." U.S. Department of State, February 9, https://2009-2017.state.gov/p/af/rls/rm/2009/117326.htm.

[10] 2012. "U.S. Strategy Toward Sub-Saharan Africa." The White House, June, https://obamawhitehouse.archives.gov/sites/default/files/docs/africa_strategy_2.pdf.

[11] Wang Lei and Sun Hong, 2014. "Analysis of the Obama Administration's Strategy for Africa." *International Studies Reference*, no. 11, pp. 15–18.

variety of measures to offset the growing influence of other powers, such as China, in Africa and ultimately to serve the US global footprint.

In contrast, if Obama's Africa strategy is based on being primarily "constructive" with "confrontational" as a supplement, then Trump's Africa strategy is, to a large extent, embodied in "confrontational" and only supplemented by being "constructive." Trump's strategy is more purposeful, targeted, exclusive, and focused than Obama's, and the above-mentioned characteristics have mainly served to exclude and suppress China and Russia's growing presence in Africa. In other words, the Trump administration's Africa strategy is stronger than the Obama Africa strategy in terms of purpose, targeting, exclusivity, and focus, and the above-mentioned characteristics are mainly used to exclude and suppress the growing influence of China and Russia in Africa and to combat terrorism. On the "constructive" level, such as strengthening economic and trade ties between the US and Africa, more efficient and effective use of US aid to Africa, etc., are not only the purpose of the economic and trade field, but also subordinate to and serve the aforementioned "confrontational" goals to a considerable extent. These features are concentrated in the targets and specific means of Trump's Africa strategy.

3 The US Africa Strategy from the Perspective of China–US Competition

As mentioned above, Trump's Africa strategy is, to a considerable extent, characterized by China's growing influence in Africa, particularly by the Forum on China–Africa Cooperation (FOCAC) and the China–Africa Belt and Road initiative. Thus, in the following section, this paper will trace the overall trend of changes in the US perception of China and US–China relations in recent years, and then assess the impact of these factors on the U.S.' African policy.

General Trends in US Perceptions of China and US–China Relations

Over the past 40 years of reform and opening up, China's rapid economic development (especially after the 2008 global financial crisis), has accelerated the rise of China's comprehensive strengths while closing the gap between China and the US, and, in doing so, has exerted a certain degree of systemic pressure on the US. At the same time, since the 18th CPC National Congress, China has become more proactive and enterprising on the international stage. The Chinese government and people have become more confident in China's roads, theories, institutions, and culture, which has caused frustration among those US policymakers who hoped to see a "peaceful evolution" of China's political system through "engagement."[12] At the international level, with the help of cooperative platforms such as the Belt and Road Initiative and the Asian Investment Bank, China's development path has, to a certain extent, contrasted with and competed with the "American model" on a global scale, thus changing the mindset of the US policy elite. Under the combined effect of these objective/material and subjective/cognitive dimensions, Sino-American relations have begun to undergo a qualitative change from "more cooperative than competitive" to "more competitive than cooperative."

To begin with, the accelerating rise of China's relative power has led to a corresponding rise in the US' desire to counter China. From a classical structural realist perspective, the factor that has most fundamentally made Sino-US relations "more cooperative than competitive" in the past is that China's overall national strength has long been far inferior to that of the US. However, since the 2008 global financial crisis, Western countries such as the US have demonstrated only a weak recovery, while China has continued to maintain high economic growth. As a result, the power gap between China and the US has been rapidly narrowing.[13] Particularly at the East Asia/Asia Pacific level, some scholars have identified the formation of a "China–US bipolar pattern in the Asia-Pacific."[14] If there is still a certain degree of controversy in the Chinese academic community about the trend of China–US bipolarity, then the US government

[12] Campbell, Kurt M. and Ely Ratner, 2018. "The China Reckoning: How Beijing Defied American Expectations." *Foreign Affairs*, March/April, https://www.foreignaffairs.com/articles/united-states/2018-02-13/china-reckoning.

[13] For related debatable articles, see Li Daokui, 2012. "The Rise of Great Powers in Turbulent Times: How China Copes with the Financial Crisis in the World." Citic Press, pp. 412; Xiao Feng, 2016. "Will World Multipolarity Go to 'China and the United States'?" *Contemporary World*, no. 10, pp. 72–75; Yuan Peng, 2012. "China and the United States: A New Bipolar Confrontation?" *World Knowledge*, no. 2, pp. 48–49; Ling Shengli, 2015. "Chinese and Foreign Scholars on China-US Interaction and the New Patterns in the Asia-Pacific." *Peace and Development*, no. 6, pp. 42–52.

[14] Yan Xuetong, 2015. "The Asia-Pacific Has Formed a Sino-US Bipolar Pattern." *International Herald Tribune*, February 5, Xinhua, http://www.xinhuanet.com/herald/2015-02/05/c_133972187.htm.

and the strategic and academic communities' consensus on the accelerated narrowing of the Sino-US power gap is clearly accelerating.[15] From the Obama administration's "Pivot Asia" in 2011 to the US Asia-Pacific rebalancing strategy proposed by US Defence Secretary Leon Panetta in 2012, it is clear that the US government and strategic and academic circles are accelerating the formation of a consensus on the rapidly narrowing of the Sino-US power gap. The National Security Strategy Report and National Defence Strategy Report released after Trump took office clearly depicts this trend and the path of its development and evolution.[16] In his discussion of the "Three Currents of Change in US Policy toward China," Cheng Li pointed out that "in the official U.S. Strategic Security Report, the National Defence Report, and Trump's State of the Union address, there is a description of China as one of the adversaries, and it is ranked ahead of Russia in terms of the need to strengthen the containment of China in the security field."[17]

In addition to the security sector, the economic and trade ties that have long played the role of a sort of "ballast" for China–US relations are now being transformed from "problem solvers" to "problems themselves." At the government level, this change is mainly reflected in Trump's "securitization" of economics and trade, networks, and other issues. In a speech at the APEC summit, Trump made it clear that "economic security is not just related to the security of the country; economic security is national security. It's a matter of life and death—for our national strength."[18] US Commerce Secretary Wilbur Ross also said on another occasion that, "economic security is military security. Without economic

[15] Green, Mike, 2016. "The Legacy of Obama's "Pivot" to Asia." *Foreign Policy*, September 3, http://foreignpolicy.com/2016/09/03/the-legacy-of-obamas-pivot-to-asia/.

[16] 2015. "FACT SHEET: Advancing the Rebalance to Asia and the Pacific." The White House, November 16, https://obamawhitehouse.archives.gov/the-press-office/2015/11/16/fact-sheet-advancing-rebalance-asia-and-pacific; Lieberthal, Kenneth, 2011. "The American Pivot to Asia: Why President Obama's Turn to the East is Easier Said Than Done." *Foreign Policy*, December 21, http://foreignpolicy.com/2011/12/21/the-american-pivot-to-asia/.

[17] Teng Jianqun, 2018. "Trump's 'America First' Security Strategy and the China-US Game." *Pacific Journal*, no. 1, pp. 18–24.

[18] "The United States has been reminded time and time again in recent years that economic security is not merely related to national security. Economic security is national security. It is vital – (applause) – to our national strength." Trump, Donald, 2017. "Remarks by President Trump at APEC CEO Summit | Da Nang, Vietnam." The White

security, there is no military security."[19] US policy elites believe that "Made in China 2025" will fundamentally challenge US technological and economic superiority, thus putting the US at an extreme disadvantage in its competition with China. The Trump administration's crackdown on Chinese companies such as ZTE and Huawei is a footnote to the above statement.

Second, the successful practice of China's development path at home and abroad has reinforced the perception of China's threat among US policy elites, which is an important psychological factor in the US decision to counteract/rebalance China.

The factors that have aroused the fears of US politicians and academics about China's strategy come not only from the narrowing of the power gap between the two countries, but also partly from their negative perception of China's behavior and their "pessimistic" expectations about the direction of China's future development. In recent years, China has become more self-confident in its own path, theories, system, and culture, but this trend has an assertive connotation in some Western countries and media.[20] On the domestic front, the vitality and attractiveness of China's development path has "disappointed" some of the US policy elites who expected China to "evolve" peacefully on its own. They used to insist that China's reform and opening-up would inevitably produce a growing middle class, which—according to Western theories of "democratization"—would inevitably lead to the "democratization" of the country's politics (reflecting Western standards of democracy). They are also willing to support the "integration" of China. Thus, despite their deep preconceptions and hostility toward the Chinese political system, the expectation that "China will definitely democratize" has

House, November 10, https://www.whitehouse.gov/briefings-statements/remarks-president-trump-apec-ceo-summit-da-nang-vietnam/.

[19] Lovelace Jr, Berkeley. 2018. "Commerce Secretary Wilbur Ross on Auto Import Probe: 'Economic Security is Military Security'." CNBC, May 24, https://www.cnbc.com/2018/05/24/wilbur-ross-on-auto-import-probe-economic-security-is-military-security.html.

[20] Christensen, Thomas J., 2011. "The Advantages of an Assertive China: Responding to Beijing's Abrasive Diplomacy." *Foreign Affairs*, February 21, https://www.foreignaffairs.com/articles/east-asia/2011-02-21/advantages-assertive-china; Gilsinan, Kathy, 2015. "Cliché of the Moment: 'China's Increasing Assertiveness'." *The Atlantic*, September, https://www.theatlantic.com/international/archive/2015/09/south-china-sea-assertiveness/407203/.

led them to support contacts and exchanges with China and to advocate China's "integration" into the US-led Western international community.[21] However, China's development practice has not been what they expected it to be.

If the success of China's developmental path at home is disappointing to "democratizing" policy wonks in the US, China's diplomacy after the 18th CPC National Congress has shifted from "biding its time" to what may more properly be characterized as "energetic and vigorous." "The US is increasingly aware of the so-called "threat." Given these circumstances, those voices advocating toughness on, prevention of, and containment of China in the fields of security, economy, and trade are rising, and American political and military circles have concurrently begun to adjust their perceptions and threat judgments of China.[22] This trend is also reflected in the US view of China–Africa cooperation. For example, Fu Ying and others conducted in-depth research on the "pessimists," "optimists," and those labeled as "cautious/anxious" in US policy circles toward China, analyzing US policy circles' perception of China. "Three kinds of disappointment with China" also form the consensus that: "First, the U.S.' expectation that 'China's modernization will inevitably bring about changes in the political system' has not happened, and China's version of 'Gorbachev' did not emerge; rather, China's successes and increased self-confidence have made its path more irreversible. Second, after integrating into the international system, China has not become a follower subservient to the will and needs of American interests, but has taken a more proactive stance to participate in designing and shaping the direction of international and regional affairs."

It is on the basis of the aforementioned situation that the latest consensus in the US Government is undoubtedly evident in such authoritative official documents as the National Security Strategy Report and the

[21] Jie Chen and Chunlong Lu, 2011. "Democratization and the Middle Class in China: The Middle Class's Attitudes toward Democracy." Political Research Quarterly, vol. 64, no. 3, p. 705; Minxin Pei, 2016. "Transition in China? More Likely than You Think." *Journal of Democracy*, vol. 27, no. 4, pp. 5–20.

[22] Mehta, Aaron, 2018. "National Defense Strategy Released with Clear Priority: Stay Ahead of Russia and China." Defensenews, January 19, https://www.defensenews.com/breaking-news/2018/01/19/national-defense-strategy-released-with-clear-priority-stay-ahead-of-russia-and-china/.

National Defense Strategy Report.[23] The warnings issued on different occasions by the US government on China–Africa cooperation and the speculation on China's investment in Africa as "threatening the sovereignty of African countries" and "harming the independence and autonomy of African countries" are all products of this concern and anxiety.

The "China Factor" in the US Strategy Toward Africa, Based on China–Africa "Belt and Road" Cooperation

Since being launched in late 2013, the Belt and Road Initiative has developed into a cooperative platform for China to participate in global governance and provide international public goods, making Africa one of the key regions in the construction of the Belt and Road. By the end of October 2019, China had signed 197 cooperation documents with 137 countries and 30 international organizations aimed at building the Belt and Road. This included 44 African countries and the AU, accounting for more than 80 percent of all African countries.[24] At the same time, the United States' perception of China's "Belt and Road" initiative in Africa has also gradually changed; this shifting perception can be divided into three rough stages.

[23] See Evan, Zachary, 2020. "'Be Wary of Authoritarian Regimes': Pompeo Warns African Allies Not to Rely on China." *National Review*, February 20, https://www.nationalreview.com/news/be-wary-of-authoritarian-regimes-pompeo-warns-african-allies-not-to-rely-on-china/. 2020. "Pompeo Warns Africa About China." *The Washington Post*, February 19, https://www.washingtonpost.com/world/africa/promising-true-liberation-pompeo-contrasts-us-role-in-africa-with-chinas/2020/02/19/4aa50280-52ef-11ea-80ce-37a8d4266c09_story.html. Rogin, Josh, 2018. "University Rejects Chinese Communist Party-linked Influence Efforts on Campus." *The Washington Post*, January 15, https://www.washingtonpost.com/opinions/global-opinions/university-rejects-chinese-communist-party-linked-influence-efforts-on-campus/2018/01/14/c454b54e-f7de-11e7-beb6-c8d48830c54d_story.html.

[24] List of countries that have signed cooperation documents with China to build the Belt and Road, China Belt and Road Network, April 12, 2019, https://www.yidaiyilu.gov.cn/gbjg/gbgk/77073.htm.

The U.S.'s Belittling "wait-and-See" Approach to the Belt and Road Cooperation Between China and Africa (September 2013–December 2015)
In the two years between Xi Jinping's September 7, 2013 proposal of the Silk Road Economic Belt in Kazakhstan, and December 2015, when China and South Africa signed the Belt and Road Memorandum of Understanding, the US has generally viewed China's Belt and Road Initiative with skepticism and contempt, as a mere diplomatic rhetoric or propaganda line of China's ruling party, rather than as a resort to action. As a result, research on China–Africa Belt and Road cooperation has also been largely absent from mainstream policy discussions in the US during this period.

The US Began to Notice and Perceive the "threat" from China–Africa Belt and Road Cooperation (December 2015–September 2018)
South Africa, as a BRICS member country and the first African country to formally engage in the Belt and Road Initiative, signed a memorandum of understanding with China in December 2015, affirming that the two sides would jointly build the Silk Road Economic Belt and the twenty-first Century Maritime Silk Road. From South Africa's formal accession to the Belt and Road Network in December 2015 to the successful convening of the Beijing Summit on the Forum for China–Africa Cooperation in September 2018, over the past three years China has signed intergovernmental understandings with 37 African countries and the African Union to build the Belt and Road.[25] The memorandum was signed by 70% of the 53 countries that attended the Beijing Summit on the Forum for China–Africa Cooperation.

The substantial progress made in the construction of the Belt and Road in Africa has attracted the attention of the US. American strategists have begun to attach much greater importance to the Belt and Road Initiative as a result. The US believes that China's "new economic diplomacy," symbolized by the Belt and Road and the Asian Infrastructure Investment Bank (AIIB), is challenging the international economic rules and

[25] This is a Brief History of the Belt and Road, Xinhua, April, 26, 2019, http://www.xinhuanet.com/world/2019-04/26/c_1124418156.htm.

order and poses a "challenge" to the US and its regional alliance system.[26] Accordingly, a growing number of US think tanks are working on a range of issues around the Belt and Road Initiative, such as The Center for Strategic and International Studies (CSIS), a US-based think tank in security studies. Thus, in-depth and systematic research has been conducted and countermeasures have been proposed on the basis of this think tank research. For example, experts from CSIS have argued that the Belt and Road Initiative focuses on infrastructure development (roads, railroads, bridges, ports, and other infrastructure), which has been neglected by the US for the past several decades. Therefore, calls have been made for more profound, fundamental research into China's Belt and Road infrastructure capacity and the infrastructural needs of relevant partner countries, in order to develop appropriate counter measures for American national interests.[27]

US Confrontation and Countermeasures Against China–Africa Belt and Road Cooperation (December 2018–Present)
On December 13th of the same year, just three months after the Forum on China–Africa Cooperation (FOCAC) in September 2018, John Bolton, then the US Assistant for National Security Affairs, gave a speech at the Heritage Foundation, a conservative think tank, in which he outlined the Trump administration's New Africa Strategy. From the text of his speech, this strategy, as mentioned earlier, focuses on the United States' policy toward Africa in three main areas: trade and commerce, counterterrorism, and aid, while Trump's personal Africa strategy heavily targets China–Africa relations.

First, the US has positioned China as a major strategic competitor, seeking to offset China's growing influence in Africa and assert its hegemony. In December 2017, the Trump administration released its first National Security Strategy Report (NSSR), which explicitly identifies China as a "strategic competitor." In terms of guiding principles, the "New Africa Strategy" is highly consistent with the NSSR: it not only

[26] China's Belt and Road Initiative: Five Years Later, CSIS, January 25, 2018, https://www.csis.org/analysis/chinas-belt-and-road-initiative-five-years-later-0.

[27] Goodman, Matthew P. and Jonathan E. Hillman, 2019. "China's Second Belt and Road Forum," *CSIS*, April 24, https://www.csis.org/analysis/chinas-second-belt-and-road-forum.

blames current US setbacks in Africa or rejection of the US by host countries on China's "designs," but also blames China's use of debt, bribery, and other so-called "strategic rivalries" as a means of "protecting its own interests." The US seeks a strategic competitive advantage in Africa by "illegitimate" means, thus excluding the "legitimate" presence of the US. The US believes that China's promotion of the Belt and Road construction in Africa has reduced the share of US investment in Africa, negatively affected US military operations in Africa, and jeopardized the United States' national interests, while also causing damage to the political development, democratic processes, judicial justice, and natural environments of local countries in Africa. The US is of the view that the significant increase in China's influence in Africa has upset the original "balance of power" in the region, and that the US must therefore take rebalancing measures in order to respond effectively to competition from China.

Second, the US has taken concrete action aimed at competing with China in Africa using a developmental model. The US believes that the "China model"—with its state-directed initiatives, in which the government plays an active role and state-owned enterprises are important market actors—represents a major challenge for the US, especially its neoliberal model of economic governance in Africa. From the US point of view, if African countries gradually accept the Chinese approach through their BRI cooperation and abandon the United States' model for development, the strategic and economic presence of the US in Africa will be in danger.

For reasons of all the above considerations, the US has proposed the Prosper Africa Initiative within the framework of the New Africa Strategy, which is designed to encourage African leaders to opt for so-called "high-quality, transparent, and sustainable" US-funded projects, replacing Chinese aid with modern and powerful development tools. The Trump administration's policy toward Africa has sought to highlight the differences between the "American model" and China's approach to Africa development with the aim of curbing China's growing influence in Africa. The Prosper Africa Initiative claims to focus on investment and business environment in Africa, in order to develop US economic and trade relations with the locals. The US is attempting to use this Initiative to complement its previous proposals, such as the African Growth and Opportunity Act (AGOA), to combine US investment and assistance with political transformation and democratization in Africa, and to increase its political and security influence.

It is noteworthy that the Trump administration has also criticized the previous US government's aid to Africa as "inefficient and ineffective," arguing that the previous aid policy neither helps African countries achieve development nor safeguards US interests in Africa. In light of this, the Trump administration is trying to cut relevant funds, such as the US African Development Foundation's 2018 budget (cut by 60%) and the President's Emergency Plan for AIDS Relief (PEPFAR). PEPFAR was cut by 17 percent. At the same time, a number of aid programs that were put in place during the Obama era have been halted by the Trump administration.

On the other hand, the US has stepped up its efforts to smear China's Belt and Road Initiative with the so-called "debt trap theory," "geopolitical tool theory," and other weapons of discourse in an attempt to drive a wedge between China and Africa by creating mistrust and contradictions. For example, the US compares the Belt and Road Initiative to a "constricting belt" and a "one-way" road with no benefit for the locals. It also cited the cases of foreign debt among countries such as Zambia and Djibouti and used false accusations that China was using the Belt and Road to seek strategic advantages in Africa.

Third, the US has tried to reduce its investment in Africa in the field of security. Since taking office, Trump has repeatedly criticized the failed US security policies and aid operations in Africa during his predecessor's presidency. As a self-claimed "improvement" on Obama, Trump not only proposed to reduce the US troop presence in Africa, but also to reduce the levels of support provided to the United Nations peacekeeping mission in Africa. In the area of counterterrorism, unlike under Obama, when the US proactively united African countries in order to carry out counterterrorism operations, the Trump administration has systematically reduced the intensity and scale of US troops' direct involvement in counterterrorism operations in Africa, while simultaneously trying to promote African countries' ownership of counterterrorism operations. Two major reasons may explain this change: one is that the Trump administration perceives major powers such as China and Russia as a greater challenge than terrorist and extremist groups to the US; the other is the Trump administration prefers to use a bilateral approach to bring key African countries together to safeguard United States' strategic interests.

4 Conclusions and Reflections

In summary, since the end of the Second World War, the United States' strategy for Africa has undergone a certain degree of evolution and adjustment at different times. Three core elements have remained unchanged: first, the marginal position of the African continent in the United States' global strategy has not changed, and President Trump's own emphasis on Africa has declined even more than that of his predecessor, Barack Obama. Second, at the level of great power rivalries, the comparative advantage of the US in Africa has not changed. Due to the huge size of the US, which is still the world's most important financial donor to Africa, it still has greater influence on Africa than anywhere else in the world. Especially at the military level, the U.S.'s own strength, military deployment, and security network is woven into Africa with the help of the EU, France, and other important allies make the U.S.'s advantage in the security field obvious. Third, the US strategy in Africa has always shown a certain "reactivity," or what might be known as a "stimulus-response" mode. That is, in response to so-called "external threats"—whether it be the Soviet Union during the Cold War, terrorism after September 11, or the so-called "China threat" that has persisted since Obama's second term in office, there seems to be very little endogenous/endogenous motivation and incentive on the part of the US to cooperate with Africa in mutual beneficial development. As a result, the US strategy toward Africa tends to be more "tactical" than "strategic," and is therefore more "destructive and confrontational" than the "constructive and cooperative" approach pursued by other players. In such a state, African countries are used by the US as "chess pieces" and battlegrounds against other major powers.

However, with the decline in the United States' own strength, changes in US domestic society and public opinion, and new changes in the international landscape, the United States' strategy in Africa has undergone a certain degree of adaptation and transmutation: first, existing world powers have attached greater importance to Africa. The Forum on China–Africa Cooperation (FOCAC) and China–Africa Belt and Road cooperation have gradually expanded China's influence in Africa. At the same time, the Valdai International Debating Club in Russia released its report on Russia's return to Africa, strategies, and prospects in October 2019, and the European Union announced on March 9, 2020 the launch of a comprehensive strategy for its relations with Africa, focusing on issues such as security, economic growth, climate change, digital transformation, and migration. These phenomena mark the increased attention and pragmatic cooperation with Africa by world major powers such as China, Russia, the EU, and the US. Although these players do not necessarily mean to challenge the U.S.'s existing influence in Africa, they will objectively bring African countries more cooperation options, thus potentially forcing the US to rethink its own reactive, confrontational, exclusionary Africa strategy.

Secondly, Trump's overall political principle of "America First" constrains the development of any degree of practical cooperation between the US and Africa in terms of both economy and security. Even if the focus is on the exclusion of China and the boycott of China's influence in Africa, the "America First" principle will hinder the realization of its own strategic purpose. The reason is because during the Cold War, the US and the Soviet Union competed with each other to finance the proxy wars fought by their allies. In Obama's era, the US—through the form of concessions trying to form "TPP" and other regional cooperation organizations—directly excluded China, thus requiring the US to pay allies to provide benefits in exchange for its strategic support. However, all these measures are against Trump's "America first" principle, which attempts to reconcile all economic and trade partners to sign new "fair" and "reciprocal" agreements, rather than "take free rides" or "take advantage of the US."

Third, the Trump administration uses "uncertainty/strategic ambiguity" as a major tactic for business and political struggles, which presents difficulties in terms of predicting the behavior of opponents, while also pushing up transaction costs and undermining the U.S.' international strategic credibility. In the long run, this hinders sustainable US–Africa relations over time.

CHAPTER 2

The Concept of "Global Britain" and the UK's Policy Toward Africa Beyond Brexit

Jingkun Li

With more than 400 years' involvement in Africa, the UK is history's largest empire and one of the foremost European countries engaged in imperialism in Africa. In terms of its historical liaisons and practical interests, Africa has always occupied a significant place in the UK's foreign policies. Simultaneously, changes in international relations, as well as in the power of the UK itself, have prompted an ongoing adjustment to the UK's African policy. Since the end of the Second World War the role of Africa in the UK's diplomatic decisions has declined, especially after most African countries gained their independence and the UK's accession to the then-European Community. It was not until the Labour Party, led by Tony Blair, took power in 1997 before Africa began to once again attract unprecedented attention from the UK government.

J. Li (✉)
Chinese Academy of Social Sciences, Beijing, China
e-mail: wxy03@ssap.cn

X. Li and C. An (eds.), *The Changing World and Africa*,
https://doi.org/10.1007/978-981-16-4983-7_2

However, after the outbreak of the financial crisis in 2008 and especially after the Conservative became the ruling party in the UK in 2010, domestic affairs—including contentions about the UK's EU membership—distracted the UK government's attention from Africa. In the year 2016, the UK held a referendum on its EU membership that turned out to be a victory for the "leave" camp, leading to the UK's formal departure from the EU on January 31, 2020. It is in this context that the UK is redefining its diplomatic strategy and has coined the concept of "Global Britain," primarily in order to maintain and strengthen the UK's role as a global power. Africa, one of the UK's strategic partners, has become more visible within the new framework of the UK's foreign policy, especially in areas such as trade, investment, and security, the core concerns of the UK. Opportunities have been brought about for a possible deepening of the relationship between the UK and Africa by the UK's exit from the EU; however, the UK's future African policy will still be confronted with a number of challenges and uncertainties.

1 "Global Britain:" Reinventing a Diplomatic Rhetoric in the Context of Brexit

Over the past few years, Brexit has exerted significant influence on the UK's internal and external policies and has become one of the pivotal factors determining the future direction of the UK's diplomacy toward the outside world. These policies serve as a response to the new circumstances after Brexit, aiming to reshape the UK into a "global power" and to remove all concerns and doubts about whether the UK will retreat into isolationism.

Development of the "Global Britain" Concept

No official document has yet been promulgated that defines what exactly constitutes a "Global Britain." This concept was first spelled out by then-Foreign Secretary Boris Johnson on a trip to the UN in July 2016. Although he did not specifically mention this exact concept, he pointed out that, far from withdrawing into isolationism, the UK wanted to play a greater role on the world stage, and that to that end, "Brexit means us being more outward looking, more engaged, more energetic, more enthusiastic and committed on the world stage than ever before."[1] At the Conservative Conference held in October 2016, former Prime Minister Theresa May unequivocally set forth the UK government's vision of a "truly Global Britain," which did not equal "a vote to turn in on ourselves, to cut ourselves off from the world." On the contrary, after throwing off the binding of European integration, the UK would have the freedom to "look beyond the continent of Europe and to the economic and diplomatic opportunities of the wider world."[2] In December 2016, Boris Johnson delivered his first major policy speech at Chatham House on the topic: "Beyond Brexit: a Global Britain," promising that the UK would not become a nation "hauling up the drawbridge or slamming the door" and that it will "be more outward looking and more engaged with the world than ever before."[3] Delivering a speech during his visit to Malaysia in April 2017, Liam Fox, Secretary of State for International Trade, further expounded on what "Global Britain" signified and described the UK after Brexit as "a nation that is outward looking and internationalist, rejecting insularity and continuing to play a prominent role in global affairs."[4] In a speech on the Britain–Africa partnership in June 2018, Minister for Africa Harriett Baldwin interpreted "Global Britain" as "open, inclusive, and outward facing, committed to playing a leading role on the world stage." Further, "leaving the European Union does not mean stepping away from our global responsibilities — quite the opposite."[5] Boris Johnson, after succeeding Theresa May as Prime Minister in July 2019, repeatedly stressed the value of opening up the

[1] 2016. "UK's Johnson Says Britain Wants Greater Role on Global Stage." *Daily Mail* online, July 22, http://www.dailymail.co.uk/wires/ap/article-3704159/UKs-Johnson-says-Britain-wants-greater-role-global-stage.html, accessed on September 23, 2018.

[2] 2016. "Theresa May—Her Full Brexit Speech to Conservative Conference." *The Independent*, October 2, http://www.independent.co.uk/news/uk/politics/theresa-may-conference-speech-article-50-brexit-eu-a7341926.html, accessed on October 2, 2018.

[3] "Beyond Brexit: A Global Britain," HM Government, December 2, 2016, https://www.gov.uk/government/speeches/beyond-brexit-a-global-britain, accessed on March 4, 2020.

[4] Fox, Liam, 2017. "Malaysia and Britain: Partners in a Post-Brexit World." HM Government, April 6, https://www.gov.uk/government/speeches/malaysia-and-britain-partners-in-a-post-brexit-world, accessed on November 5, 2018.

[5] 2018. "Minister Baldwin Speech on UK-Africa Relations." HM Government, June 26, https://www.gov.uk/government/speeches/minister-baldwin-speech-on-uk-africa-relations, accessed on September 28, 2018.

UK's policy to the outside world, in particular free trade. At the same time, the Labour party—the largest opposition party in the UK—has gradually accepted the idea of "Global Britain" and developed its own objectives and policy approaches from the perspectives of diplomacy, defense, and development, which have been written into its election manifesto.[6] In sum, the two biggest parties in the UK have reached a consensus that takes "Global Britain" as the starting point and basis for the UK's foreign strategy as a whole.

Contents of "Global Britain"

The Memorandum from the UK Foreign and Commonwealth Office to the House of Commons contains a brief summary of the context, objectives, and main contents of "Global Britain,"[7] which makes it clear that Britain has always assumed a leading role in responding to global challenges and in making the most of opportunities for the country. As an active global player, its strategic policy direction has not changed, which means that it will continue to shore up its international relevance and influence, maintain a cohesive global vision, and engage in global cooperation across a variety of fields. However, the pace of change in an ever-more-challenging global environment will inevitably significantly impact the way in which the UK is prepared to project its influence and protect its national interests. According to this Memorandum and selected speeches by UK government officials (including the Prime Minister), it can be inferred that "Global Britain" is composed mainly of the following points.

First, as for its position on the world stage, the UK still views itself as a "global power," retaining "global influence" and taking "global responsibility." Despite the incompatibility of such a vision with the practical reality that the UK's global influence has undeniably waned, it is beyond doubt that the UK is still a force to be reckoned with in a variety of areas.

[6] "A Global Britain: Labour Will Take All Necessary Measures to Protect the Security of Our Citizens and Country." The Labour, https://labour.org.uk/issue/a-global-britain/, accessed on October 5, 2018.

[7] 2018. "Appendix: Memorandum from the Foreign and Commonwealth Office." UK Parliament, March 12, https://publications.parliament.uk/pa/cm201719/cmselect/cmfaff/780/78008.htm#_idTextAnchor035, accessed on March 4, 2020.

Second, the UK will take on economic diplomacy as a focal point for its future external relations in order to circumvent the risk of possible difficulties with its trade and investments,[8] the key to which relies on promoting free trade and furthering trade relations with countries both within and outside Europe. This is not only a continuation of the UK's foreign policy tradition valuing free trade above all else, but also one of the leverages by which the UK will exert influences on the outside world. Former Secretary for Trade Liam Fox was confident in saying that the UK would become a champion of free trade after it left the EU.[9] In March 2020, one month after the UK's departure from the EU, in a speech addressed to the WTO General Council, Secretary of State for International Trade Elizabeth Truss outlined, "a bold new era" in which the UK "is and always will be an open, liberal, outward looking nation," one that is responsible for "championing the cause of free, fair, rules-based trade" and which "wants to turn the rise in protectionist measures around."[10] In order to promote "trade diplomacy," a Department of International Trade has been set up within the UK government after the Brexit referendum to provide institutional guarantees for its trade policy.

Third, by taking advantage of its strengths in security and military affairs, the UK will play a key role in global security, especially in terms of combating terrorism and extremism, especially since security and defense are the fields in which the UK is likely to have a voice in the world. Driven mainly by the UK, a clause has been inserted into the 2017 Hamburg G20 Leaders' Statement that deals with providing financial support to actions taken against terrorism. At the February 2018 Munich Security Conference, the then-UK Prime Minister Theresa May reasserted that as internal and external security matters became increasingly entwined, the security and prosperity of the UK was correspondingly increasingly bound to global security and prosperity. Furthermore, the UK would invest in

[8] Cui Hongjian, 2017. "Changes in the UK's Foreign Policies in the Process of Brexit." In *Annual Report on Development of the United Kingdom (2016–2017)*, edited by Wang Zhanpeng, 26. China: Social Sciences Academic Press.

[9] 2016, "Liam Fox: Britain Will Be Great Again After Brexit." The Telegraph, September 24, https://www.telegraph.co.uk/news/2016/09/24/liam-fox-britain-will-be-great-again-after-brexit/, accessed on October 24, 2018.

[10] 2020, "Elizabeth Truss Outlines Bold New Era for Trade." HM Government, March 3, https://www.gov.uk/government/speeches/elizabeth-truss-outlines-bold-new-era-for-trade, accessed on March 4, 2020.

global security knowing this was how it could best protect its people at home and abroad.[11] It follows from the above discussions that, in order to maintain its status as a global power, the UK will become more actively engaged in its military and peacekeeping operations all over the world, and in particular, it will encourage the international community to take concrete steps toward tackling terrorism and extremism.

Finally, the UK will continue its contribution to building and consolidating a rules-based international order and manage to assume a leadership role in the multilateral order. Foreign aid is one of the most important instruments by which the UK endeavors to construct and reform the international order and wield influence over regional and global affairs. The UK is one of the largest donors of overseas aid and one of the few countries who have hit the 0.7% ODA/GNI aid spending target set down by the UN. Previously, as an EU member, the UK implemented its foreign aid policy principally through the EU's collective actions. Despite the weakening—and even loss of—the "scale effect" compared with when the UK was part of the EU, it has, nevertheless, gained greater autonomy after Brexit in determining the targeted countries for foreign aid according to its own will, thus facilitating a multilateral order to develop in favor of its national interests.[12]

Thus far, the idea of "Global Britain" is a response by the UK to the multiple challenges it faced after it lost the EU as its foremost forum, with the objective of maintaining its global visibility and influence. Contrary to the opinions of some observers, this author believes that instead of becoming isolated after Brexit, the UK will enhance its engagement in foreign policy, endeavoring to consolidate and expand its relations with partners apart from its traditional allies such as the US and Europe, as illustrated by the shift in its recent policies toward Africa.

[11] 2018. "PM Speech at Munich Security Conference." HM Government, February 17, https://www.gov.uk/government/speeches/pm-speech-at-munich-security-conference-17-february-2018, accessed on November 17, 2018.

[12] Li Jingkun. 2017. "Prospects of the Middle East Policy of the UK and EU in the Context of Brexit." *West Asia and Africa*, no. 5, p. 35.

2 Reshaping the UK's Partnership with Africa Within the Framework of "Global Britain"

As the largest-ever colonizing state in Africa, the UK has been redefining its African policy in accordance with both the changing international situation and its own relative strength. Since the end of the Second World War, the UK's African policy, after experiencing ups and downs, has returned to a normal trajectory at last. After the 2016 Brexit referendum, and especially after coining the term "Global Britain" as the primary concept underlying its future diplomatic strategy, the UK has refocused its attention on Africa as an indispensable strategic partners, and now seeks to reshape its African engagement.

Context and Reasons for the UK's Adjustment of Its African Policy

The UK's exit from the EU on January 31, 2020, may lead to the declining importance of the UK in both Europe and the world. In order to preserve and, if possible, enhance its international status and influence, the UK has turned much of its focus to Africa and managed to reestablish alliances with the latter, a "neglected" partner in the past. It is believed that Africa possesses enormous opportunities from which the UK could greatly benefit.

The first of these opportunities is the so-called "demographic dividend." The population of Africa accounts for 16% of the world's human population and is, at the same time, expected to increase at a speed faster than that in the rest of the world. The population in Africa is estimated to reach 2 billion by 2050, representing a quarter of the world's total, with a large middle class whose size is expected to be equal to that of the total population in Europe and who will become the target consumers of British companies. The second opportunity for the UK is that Africa has some of the world's fastest-growing economies, the GDP of which is expected to reach US$30 trillion by the year 2050.[13] However, the inflow of FDI into Africa and the merchandise trade volume of Arica account for only 3% of the world's total FDI and merchandise trade,

[13] "Written Evidence—Foreign and Commonwealth Office (FPW0027)." UK Parliament, http://data.parliament.uk/writtenevidence/committeeevidence.svc/evidencedocument/international-relations-committee/foreign-policy-in-changed-world-conditions/written/79900.html, accessed on September 28, 2018.

respectively, which means that an enormous expansion of trade with and investment in Africa still exists. Thirdly, the weight of Africa in global affairs cannot be underestimated. By combining their powers and taking a common position—for example, in supporting the Paris climate change agreement—African countries may play a critical role on the world stage. Therefore, the UK could not disregard Africa's support in future multilateral cooperative ventures, especially as regards issues such as climate change. Fourthly, one of the core concerns of both the UK and Africa is joint opposition to terrorism in order to guarantee the national security of both the UK and African countries. After the 2005 London underground bombings, terrorist attacks have been named as the number one threat to the UK's national security, with the number of both terrorists and terrorist attacks on the rise since 2017 and more than 2000 terrorists having been arrested in 2010–2017.[14] As one of the most unstable regions in the world, Africa is rather susceptible to conflicts and issues such as terrorism and extremism, a direct source of threat to the UK's national security. To the UK, having just withdrawn from the EU, national security and domestic stability are not only at the top of the government's agenda, but a crucial test of the newly elected Conservatives. In view of this, the UK has to cooperate with Africa in the area of security. Finally, Africa is rich in oil and natural resources, most of which are of great importance to the UK's economy.

It is in such a context that the UK is planning to renew its partnership with Africa. Just as mentioned by former PM Theresa May before her 2018 trip to Africa, "Africa stands right on the cusp of playing a transformative role in the global economy, and as long-standing partners this trip is a unique opportunity at a unique time for the UK to set out our ambition to work even closer together." The central message her trip meant to convey was "to renew partnership between the UK and Africa, which will seek to maximize shared opportunities and tackle common challenges."[15]

[14] Ibid.

[15] 2018, "Theresa May to Lead Ambitious Three-Nation Trip to Africa." HM Government, August 26, https://www.gov.uk/government/news/theresa-may-to-lead-ambitious-three-nation-trip-to-africa, accessed on November 26, 2018.

High-Profile Visits and Institutional Guarantees

After the 2016 Brexit referendum, the UK has turned away from the past approach that alienated African countries and instead launched a series of high-level official visits, expressing its willingness to enhance its contacts with Africa. These visits included—to name just a few—former Chancellor of the Exchequer Philip Hammond's visit to South Africa at the end of 2016; former Secretary for International Trade Liam Fox's trip to Ethiopia, South Africa, Mozambique, and Uganda in 2017; Boris Johnson, then in his role as Foreign Secretary, to Ethiopia, Gambia, Kenya, Libya, Nigeria, Uganda, and Somali in 2017, and who also attended the leaders conference between the African Union and the European Union that same year; Priti Patel, then-Secretary for International Development and her successor, Penny Mordaunt, who paid visits to a number of African countries in 2016–2017. In addition, other UK officials in charge of African affairs from the Department of Foreign Affairs and Commonwealth, the Department of International Development, and the Department of Defence have paid a number of visits to Africa. Amongst these visits, Rory Stewart, then-Joint Secretary for African affairs of the Department for Foreign Affairs and Commonwealth and the Department of International Development, having taken office in June 2017, undertook dozens of visits to various African countries in the span of several short months.

Then-PM Theresa May's visit to South Africa, Nigeria, and Kenya in August 2018 represented the climax of a series of official visits by UK government officials. This was her first trip to Africa after she succeeded David Cameron as Prime Minister, and she is the first British prime minister to have visited Africa since 2013. It is generally believed that this visit was an important signal of renewed British interest in engaging with Africa.[16] Accompanying Theresa May was a delegation composed of 29 representatives from UK business across a wide range of fields, with a set of commercial agreements having been reached, which acted as starting point to deepen bilateral trade relations. In addition to trade, Theresa May promised to increase the UK's investment in Africa and strengthen

[16] Vines, Alex and Elizabeth Donnelly, 2018, "Theresa May's Focus on UK-Africa Ties Is Long Overdue." The Chatham House, August 29. https://www.chathamhouse.org/expert/comment/theresa-may-s-focus-uk-africa-ties-long-overdue, accessed on November 29, 2018.

security cooperation. The PM's visit was viewed as a milestone in terms of upgrading the UK's visibility and engagement in Africa. As a follow-up, Foreign Secretary Jeremy Hunt visited Senegal, Ghana, Niger, Ethiopia, and Kenya between April 20 and May 3, 2019, in order to consolidate the UK's comprehensive partnerships with African countries, especially Francophone countries.

In addition to these high-level visits, the UK government is dedicated to increasing the number of its delegations and offices in Africa in order to reinforce the existing institutional contacts between the two sides and provide an institutional guarantee for its presence in Africa. At present, the UK has 37 embassies and high commissions,[17] 19 offices under the British Council, and 16 agencies under the Department of International Development in Africa. The UK government has made public its plan to expand its presence in Africa alongside its desire for the UK to have more offices across Africa than any other European country.[18] In August 2017, the UK diplomatic staff in Abuja, Nigeria moved to a new building, a move which Boris Johnson, the then-Foreign Secretary, claimed "is a symbol of the UK's long-term commitment to Nigeria,"[19] at the opening ceremony of the new complex. In 2018, the UK opened new embassies in Lesotho and Eswatini, upgraded the office in Mauritania to an embassy, and planned to open embassies in Francophone countries such as Niger and Djibouti, aiming to expand its diplomatic visibility in the Sahel.

However, despite the above-mentioned steps taken by the UK government to refresh its relationships with Africa, it must be recognized that the UK does not prioritize Africa as much as some other regions and countries in its "Global Britain" foreign strategy. In answering an inquiry from the Foreign Affairs Committee of the House of Commons, the UK government highlighted the three regions it prioritizes in its future agenda: North America (the US in particular); Europe and its neighbors; and the

[17] The official name of the UK's embassies in the Commonwealth members is High Commission, with High Commissioner as its head.

[18] 2018. "Minister Baldwin Speech on UK-Africa Relations." TRALAC, June 26, https://www.tralac.org/news/article/13191-minister-baldwin-speech-on-uk-africa-relations.html, accessed on October 26, 2018.

[19] 2017. "UK Foreign Secretary Boris Johnson Opens Country's New Embassy in Abuja." *Premium Times*, August 31, https://premiumtimesng.com/news/top-news/242028-uk-foreign-secretary-boris-johnson-opens-countrys-new-embassy-abjua.html, accessed on October 31, 2018.

Indo-Pacific, due to their centrality to the global economy and politics, which is pivotal to the success of "Global Britain."[20] Although the UK government recognized that it is necessary to strengthen its support of UK–Africa relations, it is equally true that Africa is in no way a priority area for the UK's external strategy. In addition, the positions and weight of different African countries in the UK's foreign strategy is quite imbalanced, with members of the Commonwealth attracting more attention than the others.

3 Priorities and Orientations of the UK's Future African Policy

In March 2019, Secretary of State for African Affairs Harriet Baldwin, at the inquiry at the House of Commons, identified the five areas that the UK government prioritizes in its relations with Africa: prosperity and economic growth, long-term security and stability, the Sahel, demographic and climate change, and natural resource management.[21] At the same time, some other authors hold the following areas to be more pressing areas of the focus for the UK's future African policy, namely, market access, foreign direct investment, foreign aid, and security and partnerships.[22] In summary, the UK's future African policy will still center around trade, investment, and foreign aid and security, but its particular area of focus within each of these areas will be adjusted to some extent.

[20] "Written Evidence—Foreign and Commonwealth Office (FPW0027)." UK Parliament, http://data.parliament.uk/writtenevidence/committeeevidence.svc/evidencedocument/international-relations-committee/foreign-policy-in-changed-world-conditions/written/79900.html, accessed on September 28, 2018.

[21] 2019. "Oral Evidence: The Work of the Minister of State for Africa, HC 900." Foreign Affairs Committee of House of Commons, March 12, http://data.parliament.uk/writtenevidence/committeeevidence.svc/evidencedocument/foreign-affairs-committee/the-work-of-the-minister-of-state-for-africa/oral/97929.pdf, accessed on March 4, 2020.

[22] Kohnert, Dirk, 2018. "More Equitable Britain-African Relations Post-Brexit: Doomed to Fail?" *Africa Spectrum*, vol. 53, no. 2, pp. 119–130.

Trade

As discussed above, trade diplomacy is not only the key to constructing a "Global Britain," but is also one of the core fields in which the UK is rebuilding its relations with Africa. In December 2017, for example, UK Trade Policy Minister Greg Hands set out his vision for the UK-Africa trade relationship. He pointed out that the UK government wanted a thriving trade relationship with Africa that benefited all, since "the government places tremendous importance on strengthening our economic ties with the continent." Further, he claimed that "as we leave the EU, we will build and strengthen ties between British and African businesses and help the UK into Africa's trading partner of choice."[23]

After the UK leaves the EU, it will need to renegotiate trade agreements with all other countries. It appears likely that the UK will keep its previous agreements with various African countries as intact as possible, so as to maintain the stability of existing bilateral trade relationships and, at the same time, signify that the UK will continue to provide the same preferential treatments to African countries as already existed within the framework of the EU. In October 2017, the UK government issued a white paper titled, "Preparing for our Future UK Trade Policy,"[24] that committed to granting preferential treatment to the developing countries at conditions no less favorable than those provided for in the EU's existing agreements and which would help the least developed countries enter the UK's market. In September 2018, the UK enacted a Taxation (Cross-Border Trade) Act, promising to establish legal guarantees for granting preferential trade arrangements to the developing countries and tariff-free treatment to the least developed countries after its exit from the EU, as well as to substantially cut down on customs duties to the other 25 developing countries. This will be done in order to minimize the adverse impacts of Brexit on the present trade arrangements between the UK and African countries. The UK government made additional promises to open its market to the poorest African countries, maintain its existing "duty-free

[23] 2017. "Trade Policy Minister Sets Out Future UK-Africa Trading Relationship." HM Government, December 6, https://www.gov.uk/government/news/trade-policy-minister-sets-out-future-uk-africa-trading-relationship, accessed on November 16, 2018.

[24] 2017. "Preparing for Our Future UK Trade Policy." Department for International Trade, October, https://assets.publishing.service.gov.uk/government/uploads/system/uploads/attachment_data/file/654714/Preparing_for_our_future_UK_trade_policy_Report_Web_Accessible.pdf, accessed on October 27, 2018.

and quota-free" policy toward them, and to support the African countries in developing their economies through trade integration, particularly with an eye toward establishing a free trade zone.[25] On the other hand, the UK's determination to enhance its partnership with Africa could be stem from its newly established trade mechanisms. In June 2018, the UK government named Emma Wade-Smith as the first HM Trade Commissioner for Africa, alongside commissioners for Asia-Pacific, China, East Europe and Central Asia, Europe, Latin America, Middle East, North America, and South Asia. This position highlights the fact that "trade is at the heart of our approach to Africa,"[26] as indicated by International Trade Secretary Lam Fox at Wade-Smith's appointment ceremony.

Some aspects of the above commitments have already been realized. The UK has concluded trade deals with eleven African countries that will last until the end of February 2020 and which replicate the trade pattern that already exists within the European Economic Partnership Agreement. Among these are the January 2019 ESA-UK Economic Partnership Agreement (EPA) (covering Madagascar, the Seychelles, Mauritius, and Zimbabwe); a trade and political continuity agreement with Tunisia enacted in September 2019; trade agreements with Mozambique, South Africa, Botswana, Lesotho, Namibia, and Swaziland in October 2019, with the latter five being members of the South African Customs Union. Under the framework of these trade agreements, the UK market remains open to these African countries, a situation which is intended to remove doubts and concerns about the UK's trade policy after Brexit and which will become a model for future trade relations between the UK and other African countries.

However, a series of challenges and uncertainties still lie ahead, which may present obstacles to future trade relations between the UK and Africa. First, in 2017, bilateral trade between the UK and Africa stood at only £31 billion, a marginal rise over that of 2016 (£ 28.7 billion) and representing only a small portion of the total trade values of each side. The UK's goods exported to Africa accounted for 2.6% of the total

[25] Elliot, Kimberly Ann, 2020. "A Post-Brexit Trade Policy for Development and a More Integrated Africa." Centre for Global Development, January. https://www.cgdev.org/sites/default/files/Elliot-Trade-Africa-Post-Brexit.pdf, accessed on March 4, 2020.

[26] 2018. "HM Trade Commissioner for Africa Appointed." HM Government, June 20, https://www.gov.uk/government/news/hm-trade-commissioner-for-africa-appointed, accessed on October 21, 2018.

UK exports and Africa's to the UK for 3.6% of the total. Second, in spite of the fact that the UK has established trade relations with each African country, there are huge gaps between the different countries, with only five countries whose exports to the UK exceed 20% of their total export value. These are the Seychelles (33%), the Gambia (29%), Mauritius (28%), South Africa (27%), and Kenya (27%).[27] Third, although a diversification of the commodity trade between the UK and Africa has emerged in recent years, the trade between the two sides is still dominated by a pattern in which the UK exports manufactured goods to Africa and imports raw materials from Africa. Finally, the UK is faced with serious challenges and competition from other international players such as the EU and the US, which may constrain its efforts at developing future trade ties with Africa.

Investment

While the UK has long been the biggest global investor in Africa, it has recently been surpassed not only by the US and France, but also by the Netherlands, whose investment in Africa has consistently risen in recent years. It is now the fourth largest investor in Africa.[28] In terms of the number of investment projects in Africa, it ranks third, after the US and France. In the year 2018, the UK's investment stock in Africa reached £38.7 billion, representing 6% of the total inflowing investment into Africa and a 14% increase over that of 2017, concentrated particularly on the mining sector.[29]

A number of events have signaled the UK government's determination to encourage and expand its investments, especially private investments, into Africa. During her visit to Africa, Theresa May announced that the

[27] Kituyi, Mukhisa, 2017. "Power Dynamics and Capacity in the Negotiation of Economic Partnership Agreements Between the UK and African Countries." Royal African Society, January, http://www.royalafricansociety.org/sites/default/files/files/APPG%20report%202017-%20Future%20of%20Africa%20UK%20Relations%20Post%20Brexit%20v2.pdf, accessed on September 27, 2018.

[28] 2019. "Foreign Direct Investment to Africa Defies Global Slump, Rises 11%." UNCTAD, June 12, https://unctad.org/en/pages/newsdetails.aspx?OriginalVersionID=2109, accessed on March 4, 2020.

[29] Mendez-Parra, Max, Sherillyn Raga and Lily Sommer. 2020. "Africa and the United Kingdom: Challenges and Opportunities to Expand UK Investments." United Nations Economic Commission for Africa, January, https://www.odi.org/sites/odi.org.uk/files/resource-documents/africa_uk_investment.pdf, accessed on March 4, 2020.

UK aimed to become the largest G7 investor in Africa by 2022.[30] The UK is planning to generate up to £8 billion of vital public and private investment in Africa, with £3.5 billion provided by the UK's Development Finance Institution and a further £4 billion to be mobilized through private investment, particularly from the City of London.[31] A committee of investors in Africa was to be set up that would be entrusted with the task of making this objective a reality. In addition, the UK government was planning to convene an investment summit with African countries at which the investment targets to be achieved by 2022 would be settled.

Due to the delay caused by Brexit, the UK–Africa Investment Summit was postponed until January 20, 2020. Representatives from 21 African countries (including heads of government and state from 16 countries) attended this summit, with 27 deals agreed upon, totaling £6.5 billion. Some commentators categorized this summit as the largest-scale meeting in history between the UK and African businesses. The GDP of the participating 21 African countries accounts for 80% of that of Africa.[32] At this summit, UK PM Boris Johnson positioned Africa as the UK's "priority investment partner" and committed to greater openness toward Africa, as well as increasing its investment and foreign aid to Africa. Not long before the UK was going to leave the EU, this summit clearly signaled that the UK government intends to strengthen its trade and investment relations with Africa and seeks more opportunities and alternative partners for the future. This is primarily to avoid as much as possible any shocks to its economy, as well as to lay down foundations for a new partnership between the UK and Africa.[33]

[30] 2018. "PM Announces Ambition for UK to be Largest G7 Investor in Africa by 2022." HM Government, August 29, https://www.gov.uk/government/news/pm-announces-ambition-for-uk-to-be-largest-g7-investor-in-africa-by-2022, accessed on September 22, 2018.

[31] 2018, "UK Launches Ambition to Generate Billions More Investment in Africa to Trigger Transformational Growth." HM Government, August 28, https://www.gov.uk/government/news/uk-launches-ambition-to-generate-billions-more-investment-in-africa-to-trigger-transformational-growth, accessed on November 11, 2018.

[32] Qiang Wei, 2020. "The UK-Africa Investment Summit Focuses on Opportunities After Brexit." *People,* January 28. http://uk.people.com.cn/n1/2020/0128/c352308-31563925.html, accessed on March 7, 2020.

[33] 2020. "UK Government Statement on UK-Africa Investment Summit." UK Government, January 20, https://assets.publishing.service.gov.uk/government/uploads/system/uploads/attachment_data/file/859314/2020_01_20_AIS_-_UK_Government_Statement_-_Final_Version.pdf, accessed on March 5, 2020.

It is undeniable, however, that changes and adjustments will take place, especially concerning the prioritized areas, approaches to, and major investors in Africa.

To begin with, in recent years, a diversified tendency has already taken shape in the areas in which the UK invests most heavily, which is demonstrated by the transformation of the focus of the UK's investment from a traditional economy toward emerging sectors.[34] After the UK withdrew from the EU, the field of science and technology is expected to become the focal sector of the UK's investment in Africa. Theresa May declared the UK's intention to build "innovation partnerships" with South Africa, Kenya, and Nigeria during her 2018 visit to Africa,[35] partnerships that include an innovative team composed of UK experts across governmental and science, technology, and innovation sectors to provide a "one stop shop" for UK excellence and tailored support; a £32m DFID tech accelerator scheme covering digital technologies, health technologies, mobile technologies, and clean energy technologies; and a skills and entrepreneurship program to help African entrepreneurs demonstrate their ideas on the world stage. All of these were launched in 2019. At the 2020 UK–Africa Investment Summit, the UK government further stipulated that it will prioritize green energy, youth education and training, and national infrastructure.

Secondly, the UK will attach greater emphasis on investment, especially in the mining and infrastructure sectors. At the Africa Mineral Investment Conference that was convened in February 2018 (the largest-scale investment conference in the mining sector), the UK government proposed to establish a "Committee of Infrastructure Investment in Africa" in order to combine its investments in the mineral and infrastructure sectors and to contribute to the development of Africa using an integrated approach.[36] In practice, the UK Department of International Development and the

[34] An Chunying, 2011. "UK's Investment Cooperation with Africa." *Journal of International Economic Cooperation*, no. 8, p. 42.

[35] 2018. "Ambitious New Innovation Partnerships with African Countries." HM Government, August 29, https://www.gov.uk/government/news/ambitious-new-innovation-partnerships-with-african-countries, accessed on September 29, 2018.

[36] 2018. "UK Government Presents the Africa Infrastructure Board at the Mining Indaba." CNBC, February 9, https://www.cnbcafrica.com/apo/2018/02/09/uk-government-presents-the-africa-infrastructure-board-at-the-mining-indaba-an-initiative-to-offer-holistic-approach-to-infrastructure-development-in-africa/, accessed on October 11, 2018.

government of Angola co-funded a reconstruction project of an iron mine, which also included a refinery, an extension of an existing railway, and an expansion program for a port, and the establishment of a power plant with a capacity of 600MV, all of which highlight the comprehensive approach deployed by the UK in its investments in Africa.

Finally, the UK government has been making ongoing adjustments to the functions allocated to the government and private actors in terms of their investments in Africa, with the latter undertaking greater responsibility. In the 2017 Economic Development Strategy published by the UK Department of International Development, it pointed out that, as one of the world's largest capital markets and a global center of financial expertise, the UK has a central role to play in channeling private capital to developing economies.[37] This was echoed by International Development Secretary Penny Mordaunt in a speech delivered in April 2018, stressing the pivotal role that private capital and the city of London would play in investing in the Commonwealth countries.[38] Such a transformation of the UK government's existing stand originates from increasing domestic pressures on its policies for developmental and official aid, especially from the right-wing media and on the part of the Conservative members. This pressure has forced the UK government to rely principally on private actors in its investments in Africa and to emphasize the so-called "values" that such investment should generate, as well as contributions that investment might have to the formation of a business-friendly environment.

In view of the UK's strength in the financial sector, especially with a view toward the city of London as one of the world's financial centers, it is expected that the UK will substantially expand its investment in Africa in the future. However, we must also recognize that Africa is not listed as any of the priority investment destinations by the UK, and investment in Africa accounts for only a small part of that in the EU, US, and Asia. In addition, within Africa, the UK is investing most heavily in

[37] 2017. "Economic Development Strategy: Prosperity, Poverty and Meeting Global Challenges." DFID, January. https://assets.publishing.service.gov.uk/government/uploads/system/uploads/attachment_data/file/587374/DFID-Economic-Development-Strategy-2017.pdf, accessed on November 7, 2018.

[38] 2018. "International Development on UK aid—The Mission for Global Britain." HM Government, April 12, https://www.gov.uk/government/speeches/international-development-secretary-on-uk-aid-the-mission-for-global-britain, accessed on November 13, 2018.

the Sub-Saharan countries that make up its former colonies, particularly South Africa, and in North African countries such as Egypt and Morocco. Finally, the turmoil faced by a number of African countries may become a detriment to the UK's further investment in the continent.

Development Aid

The UK is the third largest donating country in the world. Its official development aid (ODA) reached $18.3 billion in 2017.[39] Development aid is not only one of the most significant approaches through which the UK aims to achieve its targets under the auspices of "Global Britain," but is also one of the most important instruments with which the UK is prepared to respond to global challenges such as poverty, illness, climate change, terrorism, and immigration issues. The UK government insists that it will maintain its commitment to spending 0.7% of national income on foreign aid, in spite of the growing opposition from within the Conservative Party and among certain elements of the media, and stick to the sustainable development goals put forwards by the UN.[40] Foreign aid to African countries represents 21% of the UK's total ODA (2016),[41] while half of its bilateral foreign aid (50.6%, 2018) went to Africa,[42] among which, Ethiopia, Nigeria, and Sierra Leone are the three top recipients.

When the UK was a member of the EU, a large part of its foreign aid was arranged through the EU, thus necessitating a shared goal between the UK and EU on policy directions. Brexit will, on one hand, affect adversely the UK's ability to wield its influence by being a member of

[39] 2018. "Aid Spending by DAC Donors in 2017." Development Initiatives, April, http://devinit.org/wp-content/uploads/2018/04/Aid-spending-by-DAC-donors-in-2017.pdf, accessed on November 13, 2018.

[40] 2018. "May to Resist Pressure to Cut Britain's Foreign Aid Commitment." *The Guardian*, February 12, https://www.theguardian.com/politics/2018/feb/12/theresa-may-committed-uk-foreign-aid-spending-pledge-oxfam, accessed on November 12, 2018.

[41] 2017. "Statistics on International Development 2017: Final 2016 UK ODA Spend Statistics." Department for International Development, November, https://assets.publishing.service.gov.uk/government/uploads/system/uploads/attachment_data/file/660062/SID-2017b.pdf, accessed on November 12, 2018.

[42] 2019. "Statistics on International Development: Final UK Aid Send 2018." Department of International Development, September, https://assets.publishing.service.gov.uk/government/uploads/system/uploads/attachment_data/file/857904/Statistics-on-International-Development-final-aid-spend-2018d.pdf, accessed on March 5, 2020.

a big bloc such as the EU, while on the other hand, unconstrained by the criteria set by the EU, it has provided the UK with unprecedented opportunities to freely define its priority targets and the approaches to foreign aid that are based on its own national interests. A variety of signs have emerged telling us that Brexit will lead to some changes in both the approaches to and priorities in the UK's foreign aid to Africa.

During her visit in Africa, Theresa May reiterated the UK's pledge to continue its aid to the world's least developed countries, including those in Africa. However, she asserted that "I am unashamed about the need to ensure that our aid programme works for the UK ......and support our national interests," and that it "is fully aligned with our wider national security concerns."[43] The present PM, Boris Johnson, has also spelled out his belief that the UK's foreign aid policy should focus on promoting overseas trade interests and serve the political and commercial interests of Britain. He even went so far as to suggest that the Department for International Development (DfID) should be brought back to the Foreign Office. [44] It can be seen from these statements that the UK's future foreign aid will center around its national interests, especially the UK's foreign trade. In fact, the UK put forward its concept of "aid for trade" as early as 2015 in a policy paper on "UK Aid: Tackling Global Challenges in the National Interest." The publication, "UK Official Development Aid: Value for Money Guidance," published in May 2018, stipulated that foreign aid must achieve "value for money."[45] One of the typical examples in this respect is the Trademark East Africa (TMEA) program funded by the UK Department of International Development, which, through foreign aid, is aimed at cutting down trade barriers and promoting business transactions with East Africa (particularly Kenya).

[43] 2018. "Theresa May's 'UNSHAMED' Promise: Foreign Aid in Africa MUST Work for UK After Brexit." *The Express*, August 28, https://www.express.co.uk/news/uk/1009488/Theresa-May-South-Africa-Brexit-trade-visit-latest-news-update-today, accessed on November 28, 2018.

[44] 2019. "Trade and Foreign Aid: Will Boris Johnson Bring an End to DFID?" *The Guardian*, July 24, https://www.theguardian.com/global-development/2019/jul/24/trade-foreign-aid-boris-johnson-dfid, accessed on March 6, 2020.

[45] 2018. "UK Official Development Aid: Value for Money Guidance." HM Treasury & Department for International Trade, May, https://assets.publishing.service.gov.uk/government/uploads/system/uploads/attachment_data/file/712367/ODA_value_for_money_guidance.pdf, accessed on November 13, 2018.

It is undoubtedly true that, aside from economic interests, spreading values such as equality and humanism is one of the main principles and practices that the UK adheres to in its offering of development aid. In particular, for many years the UK has been dedicated to realizing gender equality and improving women's status in Africa, especially through girls' education programs for poor families. At the 2018 G7 summit, Theresa May announced that the UK would enable 1.5 million marginalized girls to complete education from primary to high school through the implementation of the "Girls' Education Challenge" program over a period of eight years. However, trade still remains the primary focus of the UK government in carrying out activities in this respect. For instance, the UK government launched a "SheTrades Commonwealth" initiative in order to "support females to participate in the Commonwealth trade" jointly with the International Trade Centre, in Kenya, Ghana, and Nigeria. The UK government believes that trade can create jobs, stimulate investment, and help less developed countries get rid of their reliance on aid. It is exactly this fundamental idea that underlies the UK government's aid to Africa; that is, it is intended to help African countries build the capacity to resolve their own problems and find solutions through economic growth. In its Economic Development Strategy issued in 2017, the UK government made it clear that the ultimate aim of foreign aid is "supporting countries to mobilise their own domestic resources and reduce reliance on aid," and that the UK would focus on "trade as an engine for poverty reduction."[46]

Cooperation in Combating Terrorism and Security Affairs

To strengthen cooperation with African countries to combat both terrorism and illegal immigration is another concern for the UK after Brexit, since "it is firmly in our national interests – terrorism, illegal immigration and modern slavery is not confined within the border of any one state."[47]

[46] 2017. "Economic Development Strategy: Prosperity, Poverty and Meeting Global Challenges." Department for International Trade, January. https://assets.publishing.service.gov.uk/government/uploads/system/uploads/attachment_data/file/587374/DFID-Economic-Development-Strategy-2017.pdf, accessed on November 15, 2018.

[47] "Written Evidence—Foreign and Commonwealth Office (FPW0027)." UK Parliament, http://data.parliament.uk/writtenevidence/committeeevidence.svc/evidencedocument/international-relations-committee/foreign-policy-in-changed-world-conditions/written/79900.html, accessed on November 28, 2018.

Taking into account the severe difficulties with which the UK is confronted in combating terrorism, it published an official Strategy for Countering Terrorism in June 2018.[48] Terrorism is still identified as the single most serious threat to the UK's security, especially Islamic terrorist organizations represented by Daesh and Al Qaeda, which were the central concerns of the two Security and Defence Strategies issued in 2010 and 2015, respectively. The UK has also singled out Al-Shabaab in Somalia, an affiliate of Al Qaeda and Boko Haram and an affiliate of Daesh, as among the gravest terrorist threats in Africa. The UK government has carried out a wide range of operations in conjunction with various African countries to combat terrorist organizations, including both bilateral cooperation and military assistance through the African Union and the UN. These measures include such actions as providing assistance to Nigeria in tackling threats from Boko Haram, helping Somalia in disarming Al-Shabaab, dispatching 450 personnel to the UN peacekeeping force to Somalia and South Sudan, and providing emergency assistance to Somalia, South Sudan, and Nigeria. In August 2018, the UK signed the first-ever "Security and Defense Partnership Agreement" with Nigeria to help the latter in countering terrorist organizations such as Boko Haram and Daesh. In addition, the UK has helped some other African countries and the African Union to train military personnel and provide weapons.

The UK government places particular emphasis on helping those African countries exposed to a high degree of terrorist threats to enhance their own capacity to respond to conflicts and risks. In this regard, the UK hopes "to build partnerships with African countries, combat the common threats and support the African countries to play leading roles in resolving its own problems,"[49] which is manifested first and foremost in the UK's increased financial support to Africa. In the 2019/2020 financial year, the UK raised its financial aid to Africa to £31.5 million by drawing on its Counter Terrorism Programme Fund (CTPF). Additionally, the UK encourages the African countries to realize sustainable peace and stability

[48] 2018. "CONTEST: The United Kingdom's Strategy for Countering Terrorism." HM Government, June, https://assets.publishing.service.gov.uk/government/uploads/system/uploads/attachment_data/file/716907/140618_CCS207_CCS0218929798-1_CONTEST_3.0_WEB.pdf, accessed on November 23, 2018.

[49] "Written Evidence—Foreign and Commonwealth Office (FPW0027)." UK Parliament. http://data.parliament.uk/writtenevidence/committeeevidence.svc/evidencedocument/international-relations-committee/foreign-policy-in-changed-world-conditions/written/79900.html, accessed on November 28, 2018.

through "political progress," for which the UK set up a North Africa Joint Unit in charge of supervising the operation of the "North African Good Governance Fund" (NAGGF), which in 2018 totaled £40 million. Thirdly, the UK government is taking great pains to enhance the operating capability of the African states and the African Union, believing that they should be the primary actors in solving the crisis with which Africa is faced. For instance, in May 2017, the UK hosted an international conference on issues affecting Somalia, focusing on the building of Somalia's security structure, including a future program for its military force and approaches to combating terrorism, as well as protecting human rights.

In addition to the fight against terrorism, actions undertaken in response to illegal immigration are another area of focus for the UK's African policy. Apart from providing humanitarian aid and other forms of assistance to the refugees and their host countries, the UK government has endeavored to reduce the number of refugees through programs targeting the development of the local economy. Take the case of Ethiopia as an example. It is the second largest recipient of refugees in Africa and one of the most important routes to Europe, which has received a large number of refugees from South Sudan, Somalia, Sudan, and Eritrea, and several other African countries. The UK wishes to facilitate its economic growth and stability by helping it realize industrialization in order to cut down the number of immigrants and refugees.

4 Conclusions

The UK formally withdrew from the EU on January 31, 2020. In order to maintain its influence on the world stage, it subsequently coined the term "Global Britain" as its core diplomatic idea.

Within such a framework, the UK has re-oriented the direction and focus of its foreign policy and its relationships with other international players, among which Africa's position as a strategic partners has been upgraded. The UK has long-established historical relations with a number of African countries, while Africa as a whole possesses enormous population dividends, rapidly growing economic strength, an ever-expanding market, and rich resources, all of which signify precious opportunities for the UK. The African continent will thus become an indispensable partner to the UK across a variety of economic and political fields, and using these advantages, the UK will try to achieve its objective of maintaining and enhancing its influence and visibility in the world.

As for areas of concrete policy, any future UK policy toward Africa is certain to still prioritize trade, investment, development, and security, which conveys not only a sense of historical continuation, but also corresponds with the current development goals and national interests of the UK. The UK will continue to expand its trade with and investment in Africa, to provide aid to African countries, and to conduct cooperative endeavors with both individual African countries and the African Union in security affairs. However, changes will take place as to the approaches and channels in which the UK will build partnerships with Africa. In the case of investment, private actors will play an increasingly dominant role. In the area of foreign aid, trade interest will act as the fundamental starting point. And in the area of security cooperation, combating terrorism will be on top of the agenda. It is evident that national interests will the most pivotal factors for the UK's future policy toward Africa, in spite of the limited role of its so-called morals and values.

It is worth noting that the success of the UK's future African policy is dependent upon a series of complex factors both at home and abroad, especially in terms of the uncertainties surrounding its internal and external environments. These include the future economic and political situations of the UK after Brexit, the UK's ability to wield its hard and soft power in the global arena, and its future relations with Europe and the US. At the same time, the UK is faced with rivalries from the other international players, which will put great pressure on its African policies. In sum, Brexit not only provides opportunities for the UK to extend and deepen its relations with Africa, but also poses a number of challenges. It can be asserted that the UK will definitely expand and reinforce its existing objectives and directions in its African policy, while on the other hand, uncertainties and challenges still lay ahead in its renewed engagement with Africa.

CHAPTER 3

Inheritance and Breakthrough in the "French-African Special Relationship" During Macron's Presidency

Shuyi Peng

France maintains a very special relationship with Africa, and its Africa policy has a long historical continuity. At the same time, France's Africa policy has also exhibited some changes throughout different historical periods and under different administrations. Almost every French president since Charles de Gaulle has spoken of a break with traditional Africa policy; however, they all failed to establish any truly new solutions. Will there be any real change in France's Africa policy during Macron's presidency, the first leader to be born after the French colonies became independent?

This paper partly adopts the same author's paper, "From Charles de Gaulle to Macron: The Trajectory and Internal Logic of French Policy Changes in Africa," originally published in *West Asia and Africa*, no. 2, 2019.

S. Peng (✉)
Chinese Academy of Social Sciences, Beijing, China
e-mail: wxy03@ssap.cn

X. Li and C. An (eds.), *The Changing World and Africa*,
https://doi.org/10.1007/978-981-16-4983-7_3

1 The Historical Evolution of France's Africa Policy

The Establishment of the "France-Africa Special Relationship"

Prior to the start of World War II, France had 21 colonies in Africa, and maintained a comprehensive presence in and influence on the continent in everything from politics to the economy, from defense to language and culture. After the end of World War II, despite the independence of its colonies, France maintained a special relationship with African countries for the sake of protecting its vested interests in Africa and highlighting France's status as a great power—all of which was supported by African colonization. This peculiarity is highlighted by the establishment of a special institution in the presidential palace by Charles de Gaulle, the first president of France's Fifth Republic. Known historically as the "African Affairs Office," it was used to establish extensive personal contacts with African heads of state. France's relations with African countries are handled by this opaque network of personal relationships. The Foreign Affairs Ministry has no authority over French-African affairs. This network, first established by de Gaulle, has been in effect throughout the presidency of Georges Pompidou, Jacque Chirac, and others, emerging as a typical example of France's unconventional Africa policy. Under the network, there have been numerous exchanges of interest, and even occasionally scandals.

To maintain the "France–Africa special relationship," France has established a mature set of policy tools epitomized by the "three pillars." The first is development aid. Immediately after the independence of its colonies, France set up a Ministry of Cooperation to extend France's presence in Africa in the form of "technical cooperation," which means that when France withdrew the tricolor flag from the African continent, it cast a tightly woven net of cooperation in its place. The second is military assistance and cooperation. France had about 60,000 troops stationed throughout Africa before the independence of its colonies. After their independence, the situation remained largely unchanged. "From 1960 to 1994, France signed military agreements with 27 African countries, and at one point France had military bases in more than 20 countries. It is known as the African gendarme."[1] The third is currency linkage and control. Following the Bretton Woods system, France established the CFA (African Financial Community) franc zone for central and West Africa and adopted the franc (known as "CFA franc") in this region. While safeguarding African countries' financial order, this system has made these countries highly dependent on France in terms of their economies. At a time when the franc has retreated from the French mainland with the advent of the euro, the franc is still in use in several African countries.

In summary, after the independence of its African colonies, French-African relations did not transform into a real international relationship. Felix Houphouët-Boigny, former president of the Ivory Coast, coined the word "Françafrique"[2] to describe the "special relationship between France and Africa." The literal translation of the word is France–Africa, whereas some interpret it as France–Africa common prosperity. The issue of the "France–Africa special relationship" triggered a lot of questions on both sides. France has been criticized for not seeking to establish normal diplomatic relations with African countries after their independence; instead, it conspired with African leaders so as to continue controlling these countries, gaining economic interest through manipulation and underground transactions, transforming the relationship from a suzerain state and colony into the father–son type "consignor" and "consignee " type, which is also essentially colonialism. There is little like this relationship to be found elsewhere across the globe.

The Inheritance and Breakthrough of the "France–Africa Special Relationship"

The domestic and international situation in France evolved after the presidency of Charles de Gaulle; several French presidents have spoken of changing and normalizing the "France–Africa special relationship," but their promises have never quite been delivered. A typical example is that François Mitterrand, a left-wing Socialist Party leader who came

[1] Sun Degang, 2012. "Overseas Military Bases in French African Strategy: A Historical Investigation." *Journal of Tongji University* (Social Science edition), April, vol. 23, no. 2.

[2] Gounin, Yves, 2009. *La France en Afrique: Le Combat des Anciens et des Modernes*, éd.de Boeck, Bruxelles: 27.

to power for the first time in 1981 and changed the French government's ideology; however, the country's Africa policy was not altered correspondingly—Mitterrand's left-wing international and humanitarian approach was not significantly reflected in the Africa policy. Mitterrand's policy simply appeared to be an extension of Charles de Gaulle's, which can be most vividly expressed as "Africa lost its father Charles de Gaulle, but ushered in its uncle, François Mitterrand."[3]

The end of the Cold War shook the foundation of the "France–Africa special relationship" for the first time. The closing of the Cold War weakened the prominent role Africa had previously played in the geopolitical and ideological sphere for France; France, with its own sluggish economy and sizeable deficits, found it difficult to sustain its enormous spending in Africa. Furthermore, the Rwandan genocide cast doubt on the role of the big powers to act as "African gendarmes." These geopolitical changes caused then-President Jacques Chirac to transform France–Africa relations from what was then a "father–son style" to a "new partnership" after Mitterrand's ruling, and adopt measures to reduce military intervention, strengthen economic and trade exchanges, cooperate to a greater extent with African countries, and expand the scope of aid and France's influence in non-francophone regions. But in his second term, Chirac reassessed Africa's role and the world situation, deciding that Africa was still essential to France as a counterweight to the US. He then abandoned his previous policy and returned to the previous commitment to maintaining the special relationship between France and Africa.

Since then, with the birth and subsequent development of globalization, the increasing number of forces at play on the African continent has meant that France's influence on Africa and Africa's dependence on France have both weakened. With a view to adapting to the new reality that Africa can no longer be monopolized by France, and also avoiding Africa distancing itself further from French paternalism, Nicolas Sarkozy, who came to power in 2007, declared his intention to innovate the France–Africa relationship. Sarkozy promised that France would no longer intervene in African affairs and that their relationship would change from "father and son" to "partners." But with the outbreak of the Libyan crisis in 2011, France took the lead in launching an airstrike upon Libya. Then came the invasion of Cote d'Ivoire, where Laurent Gbagbo

[3] Marchesin, Philippe, 1995. «Mitterrand l'Africain», *Politique africaine*, no. 58, June: 5, quoted from Yves Gounin, op.cit., p. 39.

was replaced by a pro-Western Alassane Ouattara. Sarkozy eventually returned to "interventionism," just as his predecessors always had.

After gaining power in 2012, Francois Hollande, then-leader of the left-wing Socialist Party, also vowed to break with his predecessors' Africa policy and normalize relations with Africa. In the first stage of his administration, Hollande attached great importance to ideology, emphasizing such values as "democracy," "human rights," and "good governance," taking a nuanced view toward the countries considered not up to France's political standards. But since interfering in the crisis in Mali, Hollande has renounced his previous policy of non-interventionism in order to safeguard France's interests in Africa; no longer foregrounding issues of ideology, he once again prioritized "security" instead.

In short, from Mitterrand's government up through the present day, France's Africa policy has been full of contradictions: on the one hand, the country wants to break with the old, interventionist Africa policy; on the other hand, France is afraid of losing its interests and influence in Africa. Once national interests are involved, it will not hesitate to return to the old Africa policy initially laid by Charles de Gaulle, thus constantly swinging between "normalization" and "returning back to Africa." For example, from Sarkozy on, France has carried out institutional reform, abolished the African Affairs Office, gotten rid of the impact of the old personal relationship network on the France–Africa policy, and gradually brought the France–Africa relationship into normal diplomatic channels using the Ministry for Europe and Foreign Affairs. In order to counter the decline of its economic influence in Africa, France has practiced economic diplomacy, especially to strengthen economic and trade ties with Africa's English-speaking regions. On the whole, however, the "special relationship" between France and Africa has been maintained, and the normalization of relations is still only tentatively verbal—and widely questioned everywhere throughout France.

2 France's Orientation Toward Its African Policy Since Macron Took Office

In May 2017, at the age of 39, Emmanuel Macron replaced Hollande as the new French president. On November 29, the young president delivered a speech at the University of Ouagadougou in Burkina Faso during his first visit to Africa, expounding upon his African policy. His idea was quite a novelty—"there is no African policy." The implication is obvious;

that is, under Macron, African countries were to be treated like the rest of the world. With Africa not being regarded differently, there will no longer be a "France–Africa special relationship." As the first president born after 1960 (when most French colonies gained their independence), he said, "Like you, I am from a generation that has never known a colonized Africa." This means that he is no longer burdened with the special relationship between France and Africa in the colonial period. Instead, he is looking forward to the future, breaking with the strong "patriarchal" overtone of France–Africa relations that has been prominent since the Fifth Republic, changing the image of France in Africa, and opening a new chapter of the relations between France and Africa.

Symbolic Break

Macron has shown the courage of an innovator with his vision for the future. He has indeed taken steps to overwrite the previous France–Africa special "father–son" style relationship and normalize it. For instance, he apologized for France's colonial crimes in Africa. Additionally, under his direction, more than a dozen young and middle-aged people from all walks of life (more than half of whom were African expatriates) were selected from civil society to establish the Council for Africa (CPA) and were regularly consulted on the formulation of African policies, demonstrating that he is willing to listen to public opinion and stay away from shady practices. He returned 26 objects to Benin that had been looted in the war; he declassified some of the previously classified documents related to the assassination of Burkina Faso's former leader, Thomas Sankara, and returned them to the country; he held talks on cultural relations with Africa in Lagos, Nigeria's economic capital (July 2018); he announced that France would try its best to stay away from the France–Africa summit, a bilateral mechanism, and handle Franco-African relations within an EU–Africa framework; and he attended the EU–Africa Summit in Abidjan, the capital of Cote d'Ivoire, on November 29, 2017, hoping to bring France–Africa relations into the framework of the EU–Africa partnership. In particular, he took the initiative to repair the chilly relations between France and Rwanda that had been caused by massacres, and endorsed Louise Mushikiwabo, Rwanda's foreign affairs minister, to become the new secretary-general of the International Organisation of La Francophonie (OIF), amid doubts. In Rwanda, which was not a former French colony and where French (though one of the official languages) is spoken at a rate of only 5–10%, Macron's conciliatory gesture was unmistakable. Taken as a whole, these measures are largely symbolic, such as the fact that regular meetings between the CPA and the President are not guaranteed and play a rather limited role.

The most powerful reform was to rename the CFA franc—used by eight West African countries for decades—to the Eco, erasing French colonial overtones while easing financial controls over eight West African countries. These nations need not continue to deposit half their foreign currency in the French treasury, and France no longer send representatives to the local monetary authority, Communauté Financière Africaine (BCEAO), as a sign of a clean break from colonial times. However, the Eco remains pegged to the Euro, which has changed little from France's previous financial guarantees to these eight West African nations, therefore, this reform may be less "historic" than Macron claimed.

On top of all this, is Macron's Africa policy really a departure from the past? Macron addressed this question as part of a speech during the annual gathering of French ambassadors on August 29, pointing out that restoring security is the top foreign policy priority for France, a policy that combines the three pillars he described as "3D"—defense, diplomacy, and development.[4] Its Africa policy is also essentially an exercise of these principles and is focused on three main issues—security, development, and the economy.

Making Counter-Terrorism and Security a Top Policy Issue

Since the crisis in Mali, and especially with the deterioration of the security situation in the Sahel region of West Africa, "anti-terrorism and security" has become the primary concern of France's Africa policy, which Macron has inherited. Speaking at the annual meeting of France's ambassadors in August 2017, Macron said that countering Islamic terrorism was a priority for French diplomacy in order to ensure the security of French citizens. Even before his speech, he had made a series of moves to highlight the importance of security in Africa, such as visiting Mali and inspecting a

[4] 2017. «Climat, terrorisme, Venezuela: Macron dévoile ses projets pour la diplomatie.» *Le Parisien*, August 29. http://www.leparisien.fr/politique/climat-terrorisme-venezuela-macron-devoile-ses-projets-pour-la-diplomatie-29-08-2017-7221093.php, accessed on April 1, 2020.

French military base in Gao less than a week after taking office. Jean-Yves Le Drian, a former defense minister and commander of France's military operations in the Sahel, was appointed Minister for Europe and Foreign Affairs. Shortly after taking office, Macron also visited several North African countries on several occasions to discuss anti-terrorism cooperation with their heads of state. On Christmas Eve, Macron visited a French military base in Nigeria to spend the holiday with French and German soldiers involved in the fight against Islamic extremism in the Sahel. Furthermore, Macron pledged to provide financial and personnel support to the Group of Five Sahel States (G5 Sahel) Joint Force—newly established in early 2017—to combat trans-border terrorism. After the security situation began deteriorating in 2019, Macron attended a summit with G5 Sahel leaders in January 2020, aiming to integrate the area's anti-terrorism forces—Operation Barkhane, named after a crescent-shaped sand dune, the Group of Five Sahel States, and other countries' security forces—into the Alliance Sahel, so as to ensure a unified command structure and improve efficiency. Macron also announced the deployment of about 220 additional troops to the Sahel, and later declared an increase of 600 troops, increasing the number of French troops there from 4,500 to 5,100.

Solving Security and Migration Issues Through "Development"

"Development" was the second most important component of Macron's Africa policy, after "security." The purpose of this issue is to address the origin of the problem with refugees and illegal immigration in France, as well as its security problems in the Sahel region.

To this end, Macron has adopted two measures. One is to strengthen existing border control and restrict the entry of illegal immigrants while focusing on promoting local development in Africa, with the hope of curbing the generation of refugees and illegal immigrants. "Our relationship cannot be simply to ease pressing problems, but it is to help Africa succeed and, at the same time, to ask it for the responsibility to have a real policy of shared responsibility, to help succeed and at the same time to bring back to the border those who have come and who do not have a residence permit,"[5] Macron said during his election campaign in 2017. His concept of development covers education, health, gender equality, environment, sustainable urban development, culture, sports, small and medium-sized enterprises and other aspects, and targets young people. Africa is the world's youngest and fastest-growing continent, with 70% of the population under the age of thirty. Poor education, slow economic growth, low youth employment, and poor prospects are the main reasons why so many people are willing to take the risk of sneaking into Europe. Only by reversing the situation of poverty and backwardness in Africa can young people feel secure enough to stay and develop there.

At the same time, "development" is the fundamental direction for security governance in the Sahel region. At the 2017 meeting of France's ambassadors, Macron pointed out that "more needs to be done in terms of development, especially in the Sahel." The continuing unrest in the Sahel is directly related to the region's poverty, food crisis, underdeveloped education sector, and slow economic development. The Sahel region is one of the poorest regions in the world and suffers from low levels of governance. After national governance failed, leaving an institutional vacuum, terrorist organizations seized the opportunity to penetrate civil society. Some of these organizations and individuals have even taken on the role of state protector, and provide basic living supplies for the poor, attracting large number of followers, with some even seeking asylum with these terrorist organizations.[6] Dropouts due to poverty, school closures due to security problems, and unemployment due to sluggish economic growth have created a large number of unmotivated and discriminating teenagers who are vulnerable to extreme thinking. In particular, the extremist group the Islamic State has turned its attention to the Sahel region after suffering heavy losses in the Middle East, spreading its message among young people via the internet and other forums. Therefore, military means alone fundamentally cannot solve the security problems in the Sahel region. Only by taking comprehensive measures and improving the economic and social conditions can the soil breeding terrorism be fully eradicated.

[5] «Discours d'Emmanuel Macron1er avril 2017.» Website of La République En Marche. https://en-marche.fr/articles/discours/meeting-macron-marseille-discours, accessed on April 5, 2020.

[6] Liu Qingjian and Fang Jincheng, 2014. "An Analysis of the Spread of Terrorism in the Sahel Area of Africa." *Contemporary International Relations*, no. 11.

To achieve these goals, Macron has used a combination of policy tools primarily targeting young people. To this end, small and medium-sized enterprises are prioritized to promote employment. France has pledged to invest €1 billion to support small and medium-sized enterprises in Africa and to tackle the problem of young people's employment. Alain Joyandet, France's former Secretary of State for Cooperation, noted that the only way to reduce poverty in Africa is to build joint France–Africa cooperation, with the support of the French Development Agency, on the basis of business exchanges.[7] At the "Viva Technology" (VivaTech) summit held in May 2018 in Paris, designed to foster digital innovation, Macron announced the French development agency would allocate €65 million in funds to promote the development of African digital start-ups, including providing technical support, training, running programming schools, etc., in an effort to accelerate the till-now slow development of start-ups in Africa and to provide young people with greater entrepreneurial and employment opportunities.

In a very different arena, sports have been approached as a method for boosting education, enlightenment, and development. Given that 70% of the African population is under the age of 30, sports are seen by France as a major tool for sustainable development and a key investment area for the French Development Agency. "Africa's future should be designed with the imagination, expression and ability of young Africans, and sport is the easiest way to do it," officials said. During Macron's visit to Nigeria in July 2018, the French Development Agency (the official French development aid agency), signed a cooperative agreement with the American National Basketball Association (NBA), planning to invest €10 million in Nigeria, the Ivory Coast, Senegal, Morocco, and South Africa in order to build several basketball stadiums and other sports facilities, providing training for coaches and juvenile players, and integrate sports with education. While helping African teenagers improve their health and cultivate their cooperative consciousness and indomitable spirits, the plan would help them eliminate ignorance and embark on a path of sustainable development. It would also boost the sports industry, which accounts for 2% of PIB worldwide, and which, as the head of the French Development Agency pointed out, offers huge business opportunities.

[7] 2018. "Macron et l'Afrique (2/5): la realpolitik d'abord ?" RFI, August 5, http://www.rfi.fr/afrique/20180508-macron-afrique-25-realpolitik-abord-tchad-rdc, accessed on January 22, 2019.

Other measures include a talent visa, longer stays for eligible African students in France, support for girls from such places utilizing development aid (such as Niger), and extensive training for African teachers.

The goals of these policies are embodied in changes in to France's public development aid: Macron has pledged to increase development aid as a share of GDP from 0.43% in 2017 to 0.55% under his presidency by increasing investment in sustainable development projects such as education, culture, and sports, and further emphasizing the link between security and development.

Expanding the Market and Strengthening France's Non-economic Presence

Africa not only has the youngest population in the world, but also has an average annual population growth rate of about 5%. It has bright market prospects. Major countries in the world are rushing to tap the African market, and economic and trade activities in Africa are increasingly active. By contrast, France, long the region's leading foreign investor, is losing competitiveness and influence. French media point out that France's market share in Africa has fallen from 11% in 2003 to 5% in 2018. China's market share, by contrast, has risen from 3% in 2001 to 18% in 2018.[8] European countries are not to be outdone. Germany has overtaken France in trade with Africa. France is increasingly vulnerable. Therefore, the third major concern of Macron's Africa policy is to vigorously expand the African market and strengthen France's non-economic presence, which is primarily reflected in the expansion to non-Francophone countries.

Since taking office, Macron has visited Nigeria, Ghana, and other non-Francophone countries in East Africa, including Djibouti, Ethiopia, and Kenya, and plans to visit South Africa and Angola in 2020. With a large corporate delegation accompanying him on every trip, his incentive to open up markets is clear. The three East African countries, in particular, have long been outside France's traditional sphere of influence, with little domestic involvement by French leaders. French media believed that these three countries took this opportunity to get close to the US, Japan, China, and other countries. This is especially true of China, which

[8] Glaser, Antoine, 2018. «La Françafrique devient l'Africa France.» *La Croix*, July 4.

has invested heavily in these countries and has been expanding its influence. Macron wants to compete with China to "counter China's rising influence."[9]

Take Kenya as an example. As the economic powerhouse of East Africa, Kenya is highly valued by Macron. In March 2019, Macron visited Kenya with a delegation of entrepreneurs, including those from Danone, Électricité de France (EDF), Alstom, and Total. Macron explicitly proposed a new economic partnership with Kenya, making France a reliable and long-term economic partner for Kenya, and reached deals worth around €3 billion. Kenya's president also confirmed a major engineering deal with a French consortium aimed at improving Kenya's transport network. In Ethiopia, too, France has made a breakthrough: Ethiopia became France's third largest market in 2017, and French exports to the country have risen sharply, achieving record levels. A series of large French companies and groups have established regional bases in the country.[10]

Macron has also inherited the previous economic and foreign policies of Nicolas Sarkozy and has actively pursued them. In his August 2017 speech, Macron reiterated that "economic diplomacy is a priority" and called on ambassadors to redouble their efforts to help small and middle-sized French enterprises take root in foreign markets. He also explicitly wrote on Twitter: "Economic diplomacy is a priority. It starts with an effort to help our SMEs settle where you represent France." In coordination with the Ministry for Europe and Foreign Affairs, France has mobilized all its forces–financial institutions, social roles, bilateral organizations, multilateral organizations and so on–to pave the way for French companies' expansion in Africa. In Burundi, France joined forces with Germany and Belgium to limit competition from other countries. In Rwanda, where country-level relations have soured, France has used European and German power to indirectly safeguard its interests. As Antoine Glaser, an expert on France–Africa relations, pointed out, Macron wants to promote new economic diplomacy in Africa to revive flagging French companies.

[9] 2019. «Emmanuel Macron à la conquête de l'Afrique de l'Est.» *Le Parisien*, March 11.

[10] 2019. "Macron's Africa Visit Reveals Determination to Weaken China's Grip on the Continent." CNBC, March 16, https://www.cnbc.com/2019/03/16/macrons-africa-visit-reveals-determination-to-weaken-chinas-grip-on-the-continent.html, accessed on March 30, 2020.

3 Main Features of Macron's Africa Policy

Leveraging the Power of the EU

In terms of France's policies toward Africa, Macron is committed to leveraging the power of the European Union to solve issues on security, migration, and others through concerted efforts. In his first meeting with German Chancellor Angela Merkel after the election, he called on Germany and other European countries to join France in supporting counterterrorism in Africa. In a speech titled, "Sovereignty, Unity and Democratic Europe," delivered at Sorbonne University on September 26, 2017, he proposed to define a "common Mediterranean and African policy" as "one of the six key elements of European sovereignty," promoting the EU's participation in the fight against terrorism in Libya and the Sahel region and "Europeanizing" the fight against terrorism. Federica Mogherini, the High Representative of the European Union for Foreign Affairs and Security Policy and Vice President of the European Commission, said that "the stability and development of the Sahel region are crucial not only for Africa but also for Europe."[11] The president of the European Council also attended the January 2020 Summit between the Group of Five Sahel States and France on counterterrorism. As the report titled, "Macron, Diplomatic: A new French foreign policy?" of the Institute of International Relations and Strategies in France pointed out, Macron's diplomatic strategies pay more attention to national interests and have an obvious European orientation.[12]

But the EU is a complex community of interests; it is not easy for all members to agree on counterterrorism measures in Africa, and France's voice is often drowned out by many highly complex differences of opinion. At present, Germany, Italy, the Netherlands, and other countries are taking a more active hand in Africa, especially Germany. Germany and France respond more proactively on the EU level. German Chancellor Angela Merkel attended a summit held in Paris on December 13, 2017, organized around the fight against terrorism by France and the Group

[11] Li Dongyang and Xu Hui, 2018. "A Preliminary Analysis of the Anti-Terrorism Policy of the French Macron Government Towards the Sahel Region." *Africa Studies*, vol. 1, p. 61.

[12] 2018. Classicism That French Foreign Policy Tends to Be Flexible/Published on China Social Sciences Net, May 10: http://ex.cssn.cn/gj/gj_gjwtyj/gj_oz/201805/t20180510_4243861.shtml, accessed on January 10, 2019.

of Five Sahel states, where she pointed out that "Islamist terrorism is spreading. We cannot afford to wait. We must begin this fight as swiftly as possible." At that time, Germany was ready to provide about €1 billion between 2017 and 2021 to support the Sahel mission.[13] The EU as a whole has pledged €50 million in support of counterterrorism and arms support, and further measures may still need to be coordinated.

In the past decade, European integration has been plagued by the European sovereign debt crisis, the refugee crisis, the rise of populism, Brexit, etc., with serious Eurosceptic and anti-European sentiments. After taking office, Macron wants to work with Germany to restart the "Franco-German axis" as the "engine of European integration," revitalize Europe, and establish himself as the leader of Europe after the decline of Merkel's influence and Brexit. Terrorism in the Sahel region, refugees, and illegal immigrants pose a challenge not only to France, but also to the whole of Europe. With French and European interests at stake and France's desire to play a central role in European affairs, Macron is pushing for a European solution; counterterrorism, more than any other issue, is a relatively easy way to build a consensus and rally support.

Pursuing Pragmatism

Unlike his predecessor Francois Hollande, who made political demands on Africa at the beginning of his presidency, Macron seldom talks about politics in Africa; instead of emphasizing human rights, democracy, good governance, and other issues, he attaches more attention to practical economic interests. Antoine Glaser, a specialist on Franco-African relations, pointed out that "Emmanuel Macron is doing 'Realpolitik' in Africa,"[14] further noting "I think that the map of Africa for Emmanuel Macron is still where there are businesses, where there are elites who can promote French interests." He additionally claims, "we feel that with Emmanuel Macron, we have entered realpolitik. The realpolitik is: carry out your revolutions; we will not carry out these revolutions for you. So, there is not really a hostile reaction to the leaders in power [...] As long as these leaders are in power, France has interests to defend and these

[13] 2017. "Angela Merkel: 'Urgent' Need for Sahel Anti-Terror Force." DW, December 13, https://www.dw.com/en/angela-merkel-urgent-need-for-sahel-anti-terror-force/a-41773078, accessed on April 2, 2020.

[14] Glaser, Antoine, 2018. *La Françafrique devient l'Africafrance*, La Croix, July 4.

leaders have networks of influence in Paris, especially in major French groups."[15] It is under this guidance, he said, that Macron is committed to opening up Africa to non-Francophone regions, with the ambition of bringing Paris closer to the big regional players and providing new business opportunities for French entrepreneurs.

The Africa of today is very different from the Africa that existed during the presidency of Charles de Gaulle, who set the tone for France's African policy. Glaser noted that Macron "realizes that the [African] continent has been globalized and that he is interested in what Africa can offer to France, but does not consider France to be responsible for Africa; Françafrique has become Africa-France.".[16] This passage shows the change that has taken place in Africa and the corresponding change in the France–Africa relationship: previously, as much of Africa used to be colonized by France, France has been able to acquire markets and resources cheaply from Africa by virtue of the France–Africa special relationship; today, Africa has become part of the trend toward globalization, which has already posed serious challenges to France's near-monopoly, particularly in economic fields. Macron is clearly fully aware of this. Since France's own economy is in a long slump and in need of a boost, the country cannot afford to allow others to seize business opportunities in Africa, as it remains the most promising market globally. This is the main reason for Macron's pursuit of realpolitik and pragmatism.

Relying on Soft Power

Macron's policy practices in Africa employ both soft power and hard power diplomacy tools. While adhering to counterterrorism practices in the Sahel region, Macron relies on economic diplomacy to expand the market in non-Francophone Africa countries. Relying on "sports assistance," for example, Macron seeks to leverage Africa's development and improve local security. Overall, he has shown a tendency to rely more on soft power to improve France's image in Africa. The same logic is evident in Macron's pledge to support the development of Ethiopia's cultural heritage and partnerships in higher education. As an official from the French Development Agency pointed out, as an application of France's

[15] «Macron et l'Afrique (2/5): la realpolitik d'abord ?», op.cit.

[16] Glaser, Antoine, 2018. *La Françafrique devient l'Africafrance*, La Croix, July 4.

"soft power," sports are "about the change of France's image in Africa." The reform of the CFA franc is also a classic case of remaking France's image in Africa. In particular, in the future, the "Eco" system concerning the eight West African countries will fully learn from the institutional arrangements of the Eurozone, and stipulate that the fiscal deficit shall not exceed 3% of GDP, inflation shall not exceed 10%, and debt shall not exceed 70% of GDP, so as to promote the economic integration of the region. This export of European experience to Africa is also seen as an important part of Macron's "soft power" diplomatic strategy.

The most typical case is that, driven by Macron, France plans to carry out seasonal African cultural activities throughout the country (including native and overseas territories) from January to middle of December 2020. Representing 54 African countries, this project will fully demonstrate Africa's vitality in cultural, economic, technological, and culinary fields in the twenty-first century. This initiative aims to break down stereotypes about France–Africa relations, dispel doubts about France's presence in Africa, and improve France's image in Africa, with most of the funding being raised by France. Although the activities have been postponed due to the outbreak of COVID-19, there is no denying Macron's good intentions and diplomatic wisdom in this area.

Facing the Difficult Task of Combating Terrorism

Ever since Jacques Chirac was in office, France has wanted to reduce its non-military presence, turn direct intervention into support, help Africa build and strengthen its own defense forces, and reduce France's defense burden in Africa. Macron has inherited this national desire, pointing out during his presidential campaign that a long-term military presence in Africa is not in itself a goal for France and that such an arrangement will be phased out by an increased focus on medium- and long-term security operations. In December 2017, Florence Parly, French Minister of the Armed Forces, further stated that African security issues are, above all, African people's issues. At the summit attended by the heads of state of the five Sahelian countries in 2020, all countries once again committed to accelerating the independent counterterrorist capacity building of the five Sahelian countries. The problem is the deteriorating security situation in the Sahel region, especially the increase in terrorist attacks in the past year, where thousands of people were killed and millions displaced. In addition, in the latest terrorist attack in Niger in January 2020, nearly 100 soldiers were killed. Given this situation, France will have to continue to strengthen its military presence in the Sahel region, as it has just promised 600 additional troops.

France has been involved in the fight against terrorism in West Africa for more than seven years now, but terrorist activities have continued unabated. Especially in the past year, rampant terrorist activities have caused heavy casualties and aroused suspicion and rejection among the local people. Initially hailed as heroes, French troops are now regularly questioned whether the rise in terrorism is a consequence of their inaction, leading to further doubt about France's military presence. In the past, it is true that France applied a "double standard" when it came to counterterrorism–it chose whether or not to send troops based on its own interests. There are even doubts now about France's ability to fight terrorism at all, as it previously may not have fought at full capacity in order to keep West Africa dependent on it. Moreover, under the banner of anti-terrorism, France has also been criticized for stepping in some domestic political affairs in such countries as Chad and Mali. In February 2019, in particular, France launched an air campaign against Chad, which it said was anti-terrorist but in reality was mainly aimed at a political insurgency. All of this has led to a rise in anti-French sentiment and demonstrations in the Sahel region, particularly in Mali.

There are also different voices on the matter in France, as some people think that France is stuck in a quagmire of anti-terrorism. For instance, some officials pointed out that France's continued presence in the Sahel region is detrimental to France's image, and that France loses money as well as its reputation by remaining there. Dominique Moisi, a senior advisor of the Paris-based Institut Français des Relations Internationales, pointed out that French elites have already doubted the country's actions in the Sahel region, and now more and more people are thinking about whether France should reconsider its strategic priority by focusing on more pressing challenges such as the more powerful Asia, Iran, and Turkey.

Given France's traditional interests in the Sahel region and the growing threat of terrorism, doubts as to its competence will not lead to a French withdrawal, but Macron may find smarter ways to improve France's image and ways to "placate" local people.

4 Conclusion

The France–Africa relationship is one of the most special relationships between major countries. Due to the particular significance of Africa to France, the historical heritage between the two powers, i.e., the "France–Africa special relationship," also known as "Françafrique," is the biggest feature, the most interesting aspect, and the most controversial point of the relationship between France and Africa. As for the new generation of French presidents who have been in power since the twenty-first century, while both Sarkozy and Hollande have vowed to end this special relationship with its strong colonial overtones and establish a new partnership based on equality and transparency with Africa in the new context of globalization, this has failed to materialize. Macron, a young president with vision and drive, has said the same, but based on his policies toward Africa so far, he might go down the same path as his predecessors.

Considering the three major traditional policy tools sustaining the "France–Africa special relationship," the much-reviled French military presence in Africa hasn't changed. Despite France vowing to strengthen Africa's capacity for pursuing independent counterterrorist actions as a means of reducing its own burdens on the continent, because of the deteriorating security situation in the Sahel region and the challenges posed by refugees and illegal immigrants in Europe, France was not only unable to withdraw, but has actually had to strengthen its presence. In terms of the equally much-maligned problem of the CFA franc, although the birth of the "Eco" system would loosen France's financial controls over eight West African countries, analysts have pointed out that this is not an example of French "mercy;" instead, it comes more from the several years of extensive mobilization and vigorous promotion of pan-African social movements, African intellectuals, and ordinary citizens, and there is still a long way to go before Africa is totally out of France's financial control. Of course, we should not deny that Macron has complied with African public opinion. Therefore, on the whole, Macron's Africa policy still exhibits a strong sense of historical continuity, and it still takes defending France's interests in Africa as its starting point and orientation. Moreover, the side effects of historical heritages cannot be ignored. For example, on the issue of anti-terrorism, France seems to have fallen into the quagmire of counterterrorism. How best to overcome it poses a difficult problem. France also seems to take a more traditional approach as to how it views the role of other regional powers, especially China, and tends to treat China as an adversary, though some degree of cooperation does exist. At the same time, Macron has also made breakthroughs in his policy toward Africa. In particular, he hopes to leverage the power of the European Union, expand the market to a wider range of countries outside France's traditional sphere of influence, use soft power diplomatic tools more flexibly, improve France's image in Africa, and emphasize security and developmental aid to Africa. It may take quite a long time to expect any substantive changes in French policy toward Africa.

CHAPTER 4

The New Trends in Germany–Africa Relations

Jinyan Zhou

Compared with France and the UK, both of which have a long, colonial history, Germany has had relatively little influence over the African continent. Historically, Germany had colonies in Africa, comprising German East Africa (including Tanganyika, Rwanda, Burundi, Tanzania), German West Africa (including Cameroon and Togolese), and German South-West Africa (now Namibia). As a result of their defeat in the First World War, Germany lost control of its African colonial areas. Following a political reorganization after the historical crimes of the Second World War, Germany's foreign policy toward Africa has kept a low profile compared with that of the US, UK, and France. However, since 2014, Germany has witnessed dramatic changes in its foreign strategic objectives and

The original version of this chapter was revised: Affiliation of the chapter author has been updated. The correction to this chapter is available at https://doi.org/10.1007/978-981-16-4983-7_23

J. Zhou (✉)
Shanghai Academy of Global Governance and Area Studies, Shanghai International Studies University, Shanghai, China
e-mail: wxy03@ssap.cn

X. Li and C. An (eds.), *The Changing World and Africa*,
https://doi.org/10.1007/978-981-16-4983-7_4

become more active in global governance, with the increasing expectation for Germany to shoulder more international responsibility. At the same time, Africa's status in German foreign policy is becoming increasingly important. In January 2017, the German Ministry of Economic Cooperation and Development (BMZ) announced in a high-profile African policy paper, the "New Africa-Europe Relationship: a New Partnership for Development, Peace and a Better Future" (hereinafter referred to as the "African Marshall Plan"),[1] that Germany needs a paradigm shift in its policy toward Africa. This reveals Germany's decision to become more actively involved in Africa's development and security affairs. In addition, Germany took advantage of its presidency of the G20 in 2017, the African Summit in Berlin in 2019, and the rotating presidency of the Council of the European Union in 2020 to focus on African development. This paper will explore the new trends and internal mechanisms of Germany's development policy and security policy toward Africa.

1 Changes in Germany Development Policy Toward Africa

Since 2014, Africa, which had previously been marginalized in German foreign policy, has become increasingly important, with Chancellor Angela Merkel visiting Africa several times. In 2017, Germany declared a number of African initiatives during its G20 presidency. In 2019, a new Africa policy guideline was published to change the fragmentation of German policies toward Africa and form synergy among different German ministries.

A German Version of the Marshall Plan with Africa

Germany's Africa Marshall Plan, launched by the German Federal Ministry for Economic Cooperation and Development at the end of 2016, goes far beyond Germany's traditional development policy toward Africa. The Africa Marshall Plan is a comprehensive initiative covering economic, financial, trade, security, legal, environmental, and health

[1] "Afrika und Europa- Neue Partnerschaft fuer Entwicklung, Frieden und Zukunft", https://www.bmz.de/de/mediathek/publikationen/reihen/infobroschueren_flyer/infobroschueren/Materialie310_Afrika_Marshallplan.pdf, accessed on February 20, 2020.

policies.[2] The Marshall plan consists of three interrelated pillars: trade and employment, peace and security, and democracy and the rule of law. To promote trade and employment, Germany has decided to create new investment products for private investors (such as funds and bonds), to enhance the export potential of Africa's private sector, and to expand economic cooperation with Africa within the framework of the Mediterranean Union. To improve democracy and the rule of law in Africa, Germany is committed to promoting anti-corruption initiatives and human rights in African countries. In addition, the Marshall Plan also specifically involves German-African security cooperation to further strengthen existing levels of crisis prevention and intervention in Africa, an initiative that is implemented at two levels. First, on the German national level, to strengthen crisis prevention capacity building requires continuing to support Africa in improving its security architecture and highlighting peace consolidation, such as mediation and early warning systems; to systematically implement the German "Guidelines for civilian crisis intervention and peace-building" in Africa; to develop recommendations for a security partnership; and to launch initiatives to strengthen the skills of the security forces. Second, on the international level, (1) the EU–Africa Partnership for Peace and Security initiative will be launched at the 2017 EU–Africa Summit; (2) the EU African Peace Facility will be extended until 2020, mobilizing funds from sources other than the EU development fund and increasing the share of public capacity building; (3) after 2020, a dedicated EU funding mechanism will be established to expand military capabilities and finance peace operations; (4) an effective and reliable means of redistributing the burden between the African Union and the United Nations in terms of the financing of African-led peace operations will be facilitated; (5) diplomatic conflict resolution initiatives will be launched; (6) broader and more substantive approaches to post-conflict situations, including economic and employment, will be conducted to generate faster peace dividends; (7) the role of religious societies in promoting peace will be supported; (8) the "International partnerships for religion and sustainable development" will undergo

[2] "The Cornerstones of a Marshall Plan with Africa—A New Partnership for Development, Peace and a Better Future", Speech by German Federal Minister Müller at the German-African Business Summit, February 9, 2017 in Nairobi.

expansion; and (9) the export of arms to crisis areas will be halted and existing controls on the small arms trade will be improved.[3]

Germany believes that the African Marshall Plan is part of the EU–Africa partnership and therefore deserves support at the EU level. Drawing on the concept of a "development-security nexus," the European Union and Germany have common ground to fundamentally solve security problems—such as refugees—through economic development. In early 2017, the European Union announced that it would set up a 44 billion euro African Development Fund to help Africa develop its economy, eliminate risks, attract more foreign investment, and help Africa reduce the number of refugees and avoid the influx of more refugees into Europe.[4]

On the whole, the issue of the Africa Marshall Plan indicates that the future of German-African cooperation will focus on conflict prevention, driven by the African peace and security architecture (APSA), rely on multilateral organizations such as the European Union and the United Nations, and focus on soft means such as the economy, employment, diplomacy, and capacity building, in an attempt to eliminate unrest from the root causes and promote lasting peace and security in Africa through development.

Compact with Africa: Going Beyond Aid

The "Compact with Africa" is another Africa-related initiative launched by the German Ministry of Finance during Germany's rotating G20 presidency in 2017, known as the main pillar of the renewed partnership between G20 and the African continent. The goal is to attract more private investment in Africa, particularly in the area of infrastructure. Germany believes that African countries face the challenge of promoting economic diversification, thus there is an urgent need to accelerate industry and services in order to absorb a large, rapidly growing labor force. The lack of investment and the huge infrastructure gap in Africa

[3] http://www.bmz.de/en/countries_regions/marshall_plan_with_africa/chapter_03/01/index.html, accessed on February 20, 2020.

[4] "The European Union Will Launch a 44 Billion Euro Fund for African Development", published on the website of the Ministry of Commerce of China, (in Chinese), http://www.mofcom.gov.cn/article/i/jyjl/k/201702/20170202515035.shtml, accessed on February 2020.

are the main obstacles to economic transformation. Through a "compact" with specific countries, Germany aims to overcome bottlenecks in investment and improve infrastructure in Africa.

Compared with the ambitious African Marshall Plan, the concept of a Compact with Africa mainly aims to go beyond traditional development aid and encourage private institutions and companies to invest in Africa, with primary focus areas in industrialization, trade and investment, and infrastructure. The goal is to accelerate employment and inclusive economic growth. The Compact with Africa has the following characteristics: the Compact emphasizes the mobilization of private capital to complement existing public funding; fully respects the diversity of African countries and cooperates on the priority areas of development identified by African countries themselves; and focuses on African ownership and emphasizes policy coherence with existing African initiatives and cooperation mechanisms, such as the African Union Agenda 2063 and the Program Infrastructure Development for Africa (PIDA).[5]

The Compact with Africa attempts to bridge the shortcomings of traditional development assistance, which has historically focused on governance while neglecting the areas of production and investment. However, the concept behind the Compact is still deeply rooted in traditional "aid thinking," which insists on advancing investment in Africa through the improvement of good governance and, in particular, the business environment of African countries.

Besides the Compact with Africa, in 2019 the German Federal Government has initiated a "Reform Partnership" with Africa, announcing a 1 billion Euro Development Investment Fund for the period of 2019 to 2021. The Investment Fund is divided into two parts: the "AfricaGrowth" initiative and the "AfricaConnect" initiative. AfricaGrowth aims to improve access to venture capital for dynamic small and medium-sized enterprises and start-ups in Africa in order to create jobs and maintain sustainable economic growth. Through AfricaConnect, the goal of the German Federal Government is to promote expanded investment and business activities in the African market by providing advice and support

[5] Zhang Haibing, "The Transformation of German Policy Towards Africa from the Perspective of the Compact with Africa", *West Asia and Africa*, No. 2, 2019, pp. 68–84 (in Chinese).

to German companies. The African reform partner countries include Tunisia, Ghana, and Côte d'Ivoire, which signed in 2017, and Ethiopia, Senegal, and Morocco, which signed in 2019.

New Policy Guidelines for Africa

In 2019, the Federal Foreign Office of Germany issued new Policy Guidelines for Africa, which set five major objectives for the development of a comprehensive African policy.[6] First, promoting peace, security, and stability; second, sustainable development, growth, prosperity, and employment for all—which particularly means investing in prospects for Africa's youth and women; third, managing and shaping migration and tackling the cause of refugee movements; fourth, strengthening the rules-based global order together with Africa; fifth, deepening civil society partnerships.

Germany's new policy guidelines for Africa reaffirmed that it would strengthen trade and private sector investment and focus on job creation and economic participation in the future. The main initiatives underlying this commitment include: (1) support for economic integration in Africa and African Continental Free Trade Area (AfCFTA). The ongoing trade negotiations between the EU and Africa should be in conjunction with the AfCFTA, and the EU's trade preference for Africa (Everything But Arms) will continue. (2) Continue to support German reform initiatives in African countries, including the Compact with Africa, the Marshall Plan with Africa, and the Pro! Africa Initiative. (3) Support Africa's economic development through the increased promotion of foreign trade and investment. The policy will benefit German enterprises, and the German overseas Chamber of Commerce will help German enterprises understand the main obstacles to access for the African market and how to overcome it; the African business network (Africa Business Network) will mainly advise and support German small and medium-sized enterprises entering Africa. Other initiatives include support for sustainable environmental and social standards, sustainable agriculture and food processing industries, and support for health governance systems in Africa.

[6] "An Enhanced Partnership with Africa", https://www.auswaertiges-amt.de/blob/2203542/6274c1b95ddfe1126f9d466a8d9e10c5/190327-afrika-ll-volltext-data.pdf, accessed on February 20, 2020.

To conclude, Germany's policy toward Africa has been transformed to one characterized by more active involvement in Africa's development affairs. First, Germany's diplomatic rhetoric on Africa has gone from pure altruism to mutual benefit, no longer hiding its own national interest. In an interview with Die Zeit magazine before her visit to Africa in 2016, Chancellor Merkel said Germany should "express greater concern about the fate of Africa" and that "if we want to pursue Germany's own interests, then we must realistically say that the well-being of Africa has a bearing on Germany's own interests."[7] This reflects a fundamental shift in German policy toward Africa from a focus on humanitarian aid to political realities, even in the rhetoric. Germany's aid to Africa in the past was characterized by a more altruistic concern, but now it is also a basic political strategy that concerns Germany's own interests. Second, the change in rhetoric from "for Africa" to "with Africa" ("Loesungen" mit statt "fuer" Afrika) reflects the enhanced status of Africa in Germany's foreign policy. Germany intends to further strengthen its cooperation with Africa as an equal partner. Thirdly, since development is the key to tackling conflict from the root causes, the German government has reached a consensus that development and security should complement each other and development policies should be coordinated with security policies.

2 The Transformation of German Security Policy Toward Africa

During the Cold War, the Federal Republic of Germany kept a low profile on the international relations stage, and Africa was always on the outskirts of German foreign policy.[8] Back then Germany did not have a real security policy toward Africa. Since the end of the Cold War, Germany's African policy was largely influenced by the US, France, and later the

[7] Marc Brost & Martin Klingst, "Afrikanische Lektionen", October 27, 2016, http://www.zeit.de/2016/43/angela-merkel-fluechtlinge-afrika-fluechtlingspolitik-fakten-fluchtursachen, accessed on February 20, 2020.

[8] Ulf Engel, "Germany Between Value-Based Solidarity and Bureaucratic Interests", in Ulf Engel & Gorn Rye Olsen eds., *Africa and the North*. London and NY: Routledge, 2005, p. 83.

European Union.[9] Germany's response to African security affairs was passive. Germany still did not have a clear definition of its own security responsibility and interests in Africa even as late as the beginning of the twenty-first century, so German-African security cooperation mainly followed trends established by the United Nations, NATO, the European Union, or France. The reason lies in Germany's lack of African security strategy, the country's self-positioning as a "civilized force (Zivilmacht)," the restraint culture in strategy, as well as the lack of a sphere of influence in Africa.

Since the reunification of the Federal Republic of Germany and the Democratic Republic of Germany in 1990, Germany began to participate in United Nations peacekeeping operations. However, in the past 30 years, the German government has been reluctant to participate in US-led military operations. In 2003, the German Red and Green Alliance government, led by Gerhard Schroeder, refused to join the Iraq war led by the US and UK. In 2006, Germany broke through taboos in security cooperation in Africa.[10] Germany decided to lead the European Union military operation in the Democratic Republic of the Congo (EUFOR RD Congo) in order to supervise the Democratic Republic of the Congo (DRC) elections. From July 2006 to November 2006, a total of 780 German soldiers participated in the six-year operation, the first time since World War II that the German Federal Defense Forces had participated in security operations in Africa.[11] The financial crisis broke out in the US in 2008, followed by the European debt crisis among major European powers, while the German economic situation maintained a relatively good situation. Both the external international environment

[9] Ulf Engel, "Networked Security Between 'Restraint' and 'Responsibility'? Germany's Security Policy Towards Africa", *Journal of Military and Strategic Studies*, Volume 17, Issue 2, 2016, pp. 51.

[10] Denis Tull, "Deutsche Afrikapolitik. Ein Beitrag zu Einer Ueberfaelligen Debatte", *Friedrich-Ebert-Stiftung*, 2014, p. 1.

[11] It is worth noting that the EU military operations in the Democratic Republic of the Congo do not mark a change in Germany's principle of restraint in the use of military forces. Germany did not participate in the EU Chadian-Central African Republic military operation (EUFOR Tschad/RCA) from March 2008 to February 2009, because it believed that the operation was more for France's self-interest. As the operation drew to a close by the end of 2008, the French and Belgian governments tried to send troops to the Democratic Republic of the Congo through a new military action plan in the European Union, which were vetoed by Germany and other EU member states.

and Germany's own aspirations led to Germany's growing participation in international governance, including in the security field. The release of the German Policy Guidelines for Africa[12] marks a significant change in Germany's cooperation in African security affairs, undergoing a transformation from passive response to active action.

The Policy Orientation of Germany's Security Policy Toward Africa

Germany has repeatedly emphasized its equal footing with African partners in the new era and believes that Germany should adopt a comprehensive strategy toward Africa, focusing on a combination of development cooperation and peace and security, crisis prevention, multilateralism, and the prudent use of military means.

First, the German development and security policies, which have been divided and independent from each other in the past, have been gradually undergoing a process of integration. Concerning the relationship between development policy and security policy, we tend to see four models in use[13]: (1) the distancing type, for maintaining independence from goals of other goals; (2) the complementary type, for identifying a division of labor based on common goals, (3) the cooperative type, for ensuring close coordination of development, security, and foreign policies, such as the operation of the African conflict Prevention Mechanism of the UK and the African Peace Facility of the European Union; and (4) the subordination type, that is, when the development policy is subordinate to a short-term security agenda, serving the self-interest of donor countries. As far as Germany is concerned, for a long time the relationship between the German development policy and security policy belonged to the first model, being independent from each other. The major foreign policy departments mainly included the Ministry of Foreign Affairs, the Ministry of National Defense, and the Ministry of Development Cooperation, in which the Ministry of Foreign Affairs was responsible for

[12] "Policy Guidelines for Africa of the German Federal Government", http://www.auswaertiges-amt.de/cae/servlet/contentblob/755086/publicationFile/202381/Afrika_Leitlinien_download.pdf, accessed on February 20, 2020.

[13] Stephan Klingebiel, Tina Marie Blohm, & Rosa Eckle u. a., "Donor Contributions to Strengthening the African Peace and Security Architecture", Final Report of the DIE Country Working Group Ethiopia/South Africa, Deutsches Institut fuer Entwicklungspolitik, 2007, https://www.die-gdi.de/uploads/media/Studies_38.2008.pdf, p. 5, accessed on February 20, 2020.

diplomatic mediation, the Ministry of National Defense was responsible for traditional security cooperation, and the Ministry of Development Cooperation was mainly responsible for Germany's international responsibilities, showing a trend of distancing. In 2014, after the great diplomatic review, Germany reached a consensus that, in addition to the Ministry of Foreign Affairs, the Ministry of Defense, and the Ministry of Development Cooperation, other ministries should also assume the functions of German foreign affairs and shoulder Germany's new responsibilities in international affairs. Since security and sustainable development are closely related, the importance of development to security has been highly recognized by the German Government. As a result, Germany's new security policy has repeatedly stressed that the conflict should be resolved from the "root cause." Meanwhile, German academic circles are also aware of the importance of security policies to development. Dirk Messner, Director of the German Development Institute, believes that the coordination of development aid and security cooperation is indispensable and that any given development policy can be effective only if it "contributes to the security and stability of the country at the same time."[14] As a result, Germany's new security policy toward Africa focuses more on the level of coordination among different policies and ministries.

Second, German-African security cooperation adheres to the principle of prudent use of military power, with a focus on crisis prevention. The most direct and notable result of the 2014 German diplomatic review was the newly established Department for Crisis Prevention, Peace-building, and Humanitarian Assistance within the Ministry of Foreign Affairs, aimed at creating a body capable of responding to the crisis more effectively and earlier, coordinating foreign and security policies, development cooperation, the environment, education and training, trade and investment agreements, as well as tackling the issue of migrants and refugees. By highlighting the management of human resources and capacity building, efforts will be made to bring together all parties to manage and respond to crises. In general, crisis management in Western countries includes early crisis identification, crisis prevention, stabilization of the situation, implementation of humanitarian assistance, and post-conflict assistance. The focus of German-African security cooperation is to identify crises and

[14] Joerg Faust & Dirk Messner, *Entwicklungspolitik und Internationale Sicherheit. Die EU-Sicherheitsstrategie als Herausforderung für die Entwicklungspolitik*. Baden-Baden: Nomos, 2005, p. 141.

violent conflict situations in a timely manner and to develop appropriate responses according to different conflict situations.

It is worth noting that the shift in Germany's policy toward Africa to a more positive one has not fundamentally changed Germany's culture of restraint and prudence in its military deployment. In Mali, for example, in February 2014, Germany decided to increase the number of its soldiers participating in Malian military training from 180 to 250. In February 2015, the Bundestag again adopted a resolution to increase the number of German military trainers stationed in Mali to 350. However, compared with the 1,400 French and 6,000 African soldiers in Mali, Germany's military engagement continues to be characterized by restraint and a low profile. The peacekeepers of Germany designated to the EU military training Mission in Mali (EUTM Mali), the UN Multidimensional Integrated Stabilization Mission in Mali (Minusma), and the EU Sahel capacity training Mission (EUCAP Sahel Mali) are mainly unarmed soldiers, involved in civilian operations such as military training and humanitarian assistance, with only a small number of armed German personnel stationed in the United Nations' mission.[15]

Thirdly, German-African security cooperation emphasizes multilateralism, participating in multilateral security cooperation under the leadership of the EU, the UN, and regional organizations in Africa. One of the connotations of Germany's "civilized forces" is multilateral cooperation and institutional building.[16] It is expected to achieve the "civilization" of international relations through controlling the use of force, strengthening the rule of law, the peaceful settlement of conflicts, and respect for human rights. Therefore, Germany has repeatedly stressed that the resolution of conflicts in Africa should be achieved by enhancing Africa's own capacity and mechanism-building, supporting the role of the AU and African subregional organizations (such as the Economic Community of West African States, the East African Economic Community, the Southern African Development Community, etc.) in maintaining regional peace and security, and promoting African countries to promote African solutions for African problems. Another example of Germany's preference for multilateralism in German-African cooperation is the fact that

[15] http://www.auswaertiges-amt.de/EN/Aussenpolitik/Laender/Aktuelle_Artikel/Mali/150226_EUTM.htm, accessed on February 20, 2020.

[16] Adrian Hyde-Price, *Germany and European Order: Enlarging NATO and the EU*. Manchester: Manchester University Press, 2000, p. 121.

Germany, as the EU Council presidency and host of G8 Summit in 2007 and the host of G20 in 2017, promoted a focus on security and development in Africa and deepened discussions on relations with Africa through a multilateral framework.

Contents of German-African Security Cooperation

The primary policy goal of German-African security cooperation is to promote African regional integration and support the African Peace and Security Architecture (APSA) initiative. Under the guidance of this policy, German-African security cooperation focuses on non-traditional security, including peacekeeping, police training, conflict prevention, and so on.

First, Germany participates in peacekeeping operations in Africa. German peacekeeping operations in Africa are carried out under a multilateral framework provided by the EU or the UN. Germany is not unilaterally involved in African security affairs as a sovereign state, which is very different from France's unilateral actions in Africa. Germany's initial participation in UN and EU peacekeeping operations was modest, mainly to demonstrate its support for multilateral organizations, but has been increasing since 2014.

In recent years, Germany is aware of the importance of overseas military participation, setting the reform goal of the German Federal Defense Force so as to ensure that 10,000 armed forces are able to operate abroad, three times that of its current overseas presence (the total number of Federal Defense Forces currently stands at 177,000, of which 3,251 soldiers carry out peacekeeping tasks abroad).[17]Conflict resolution is a priority for the German Federal Defense Force.[18] Moreover, the deployment of Germany's participation in overseas peacekeeping has also been adjusted due to the changing global situation, with a focus on increasing participation in African peacekeeping.

Second, Germany helps Africa enhance its own security capacity building. Germany expects that Africa will be able to solve its security problems on its own. However, the African Union and African subregional organizations still have relatively limited capacities and, despite

[17] Federal Ministry of Defence of Germany, https://www.bundeswehr.de/portal/a/bwde/start/einsaetze/ueberblick, last visit date: February 20, 2020.

[18] "Peacekeeping Contributor Profile: Germany", http://www.ng.org/2014/04/03/contributor-profile-germany, accessed on February 20, 2020.

some progress, are unable to effectively guarantee their own peace and security in the short term. In view of this, Germany is committed to helping Africa establish an African Peace and Security Architecture and to cooperate with African regional organizations, individual African countries, non-governmental organizations, and other players in the field of security. The German efforts include: (1) peace mediation: supporting dialogue and mediation in the peace process and developing a path for future development; (2) promoting the rule of law: training of judges and lawyers and providing advisory services to the police; (3) democracy-building: promoting political participation, such as monitoring elections and education; and (4) security institution reform: helping to build stable and accountable security forces in Africa.

Third, Germany actively participates in the fight against transnational crime, health governance, refugee asylum, and other forms of non-traditional security governance. Take the joint fight against piracy in Somalia as an example: Germany has been involved in the fight against piracy on the Horn of Africa since 2008 and has led the EU fleet many times, which is also a rare exception to Germany's normal restraint in the use of military power. In recent years, the joint action of the international community against piracy has achieved remarkable results, to which Germany has also made significant contributions.

In the area of public health governance, Germany, when it chaired the G8 in 2007, the G7 in 2015, and the G20 in 2017, all listed the African continent as a priority for the global health governance agenda. Especially after the outbreak of the ebola virus in Africa, Germany's global health policy has undergone tremendous changes, becoming an active advocate of global health governance. Disease control has been taken as one of the country's foremost foreign security policies. According to the OECD Development Assistance Committee, nearly 20% of Germany's development assistance to sub-Saharan Africa went to social development areas such as education and health care in 2015.[19]

With regard to the provision of asylum to refugees, Germany has gradually become aware of the seriousness of the refugee problem. In 2016 alone, 890,000 refugees poured into Germany.[20] From 1986 to

[19] http://www.oecd.org/dac/stats/aid-at-a-glance.htm, February 20, 2020.

[20] "German Chancellor Merkel Begins Three-Day Africa Trip", http://www.dw.com/en/german-chancellor-merkel-begins-three-day-africa-trip/a-35999606, February 20, 2020.

1996, Algeria, Ghana, Nigeria, Togo, and the Democratic Republic of Congo were among the top ten countries of origin for asylum-seekers in Germany. By 2002, Algeria had become one of the top ten countries of origin of refugees in Germany; Nigeria ranked among the top ten countries of origin for German refugees in 2004, 2007, and 2009; Somalia became one of the main countries of origin of German refugees in 2013 and 2014; since 2013, Eritrea has become the largest source of African refugees in Germany. The number of refugees entering Germany in 2014 and 2015 reached 13,198 and 10,876, respectively.[21] Germany believes that achieving sustainable development in the countries from which refugees originated is an effective way to address refugee problems and terrorist threats. In 2015, Germany pledged 5 million euros to the African Union to support "Skills Initiative for Africa."[22] Germany and African immigration or transit countries are committed to three areas of work: strengthening private investment to achieve sustainable economic growth and job creation; developing infrastructure, particularly in the areas of transportation, communications, and energy; raising the level of vocational and technical education and training of local workers so as to better meet the needs of the job market.[23] As a whole, the German-African refugee partnership aims to provide employment opportunities, youth development opportunities, etc., for refugee-seeking countries to strengthen sustainable development of Africa in order to address the root causes of conflict and reduce the number of refugees, including those arriving in Germany. In addition, non-traditional security cooperative exercises between Germany and Africa include climate change and ecological security.

To sum up, cooperation between Germany and Africa in the field of security has gradually increased. For historical reasons, it has been difficult for Germany to assume a leadership position in the past, but with

[21] Bundesamt fuer Migration und Fluechtlinge, "Das Bundesamt in Zahlen 2015: Asyl, Migration und Integration, p. 19, https://www.bamf.de/SharedDocs/Anlagen/DE/Publikationen/Broschueren/bundesamt-in-zahlen-2015.pdf?__blob=publicationFile, accessed on February 20, 2020.

[22] "Germany Pledges 65 Mio EUR to the African Union", http://www.addis-abeba.diplo.de/contentblob/4679458/Daten/6128266/151210_PR_65_Mio_EUR.pdf, accessed on February 20, 2020.

[23] "Stepping up Engagement for the African Union", https://www.bundesregierung.de/Content/EN/Reiseberichte/2016/2016-10-11-merkel-in-aethiopien_en.html, accessed on February 20, 2020.

the growth of the younger generation and the rising expectations for Germany to take more responsibility, Germany will gradually take on a leadership role adapted to its national strengths, though this will be a process of gradual growth. With the recognition of external expectations, global responsibilities, and their own limited capacity, non-traditional security has become the starting point for German security governance leadership in Africa and the focus of future German-African security cooperation. Although Germany is involved in some military affairs in Africa, it tends to primarily participate in civilian peacekeeping operations in Africa, such as Germany's support for the capacity building of African troops and police and the provision of relevant professional and technical civilian support. It is also in line with the international image Germany wants to create: it does not want to be a champion of hard power, but a model of soft power.

3 Background of Germany's African Policy Transformation

Since the second decade of this century, Germany's development and security policy toward Africa has become more proactive, utilizing a variety of tools such as traditional aid, trade and investment, crisis prevention, mutualism, and so on. The motivation for Germany's deepening cooperation with Africa lies in its own desire to participate to a greater degree in international affairs and have a greater international voice, as well as the realistic demand for coping with its ongoing refugee crisis.

Security and Refugee Concerns Enhance the Importance of Africa Affairs

Germany's increased participation in African cooperation, in particular security cooperation, is mainly due to the changing security situation in Africa. Many conflicts in Africa are not confined to a country's border but are, instead, transboundary, thus posing transnational risks. To begin with, the spread of political unrest is worrying. Since the end of 2010, Egypt, Tunisia, Nigeria, Mali, and other countries have experienced different degrees of political instability due to the spillover effects of

North African upheaval and regional instability.[24] In addition, elections on the African continent often result in chaos; for example, Côte d'Ivoire, the Central African Republic, and the Gambia all have triggered political crises as a result of the general elections. In recent years, violent conflicts triggered by elections and heated debates as to the third term of leaders have been important factors contributing to instability on the African continent.[25] Second, violent and terrorist incidents occur frequently. In recent years, as a result of regional instability, terrorist and extremist organizations (e.g. al-Qaida in the Maghreb, Ansar Dine, Ansar Bait Al-Maqdis in Jerusalem, MUJAO in West Africa, Al-Shabaab in Somalia, and Boko Haram in Nigeria) pose serious security threats to Africa and other regions. At present, violent conflicts in Africa have not declined and the nature of violent conflicts in Africa has changed, with a correspondingly increased number of transnational conflict networks, such as jihadists in Mali and Somalia. Take the year of 2015 as an example: one-fourth of the world's political conflicts and nearly half of the violent conflicts occurred in sub-Saharan Africa during that year.[26] Third, public health incidents threaten the African continent and the world at large. In 2014, Ebola was spreading in countries such as Liberia and Sierra Leone, not only causing casualties in Africa, but also spreading rapidly to Europe and the US. In 2020, the rapid and extensive spread of Coronavirus (COVID-19) in most countries in the world resulted in considerable mortality in Europe and the US and affected African countries and the developing world. Public health crises such as these not only have significant security impacts on African countries, but also disrupt trade, tourism, and economic development between Africa and Germany. African security is in the interest of all countries in the world, including Germany.

[24] African Union, "Report of the Chairperson of the Commission on Current Challenges to Peace and Security on the Continent and the AU's Efforts: Enhancing Africa's Leadership, Promoting African Solutions", extraordinary session of the Assembly of the Union on the state of peace and security in Africa, Addis Ababa, 25–26 May 2011, Ext/Assembly/AU/2.

[25] African Union, "Report of the Chairperson of the Commission on the Prevention of Unconstitutional Changes of Government and Strengthening the Capacities of the African Union to Manage Such Situations", 16th Ordinary Session of the Executive Council held in Addis Ababa, Ethiopia, on 25–29 January 2010, EX.CL/566 (XVI).

[26] Heidelberg Conflict Barometer 2015, p. 61, https://www.hiik.de/en/konfliktbarometer/pdf/ConflictBarometer_2015.pdf, accessed on February 20, 2020.

In particular, the refugee problem has made Germany hyperaware of its own safety. The influx of African refugees and immigrants poses a specific and direct threat to Germany's social security. Violent incidents, including the harassment incidents that occurred during the Christmas in Cologne and Berlin Christmas market, resulted in many deaths and injuries in Germany. Germany is no longer an isolated island of relative peace and security. Like the US, Spain, the UK, Belgium, and France, Germany has become one of many countries that have suffered terrorist attacks. Especially in 2016, with the arrival of large number of refugees, Germans were directly feeling the pains of insecurity.

As result, African security affairs are no longer far away from Germany. German political and academic circles are calling for Germany to reform its policies toward Africa, hoping that Germany's security policy will change to a positive and powerful one. For example, Robert Kappel, a scholar at the German Centre for Global and Regional Studies (GIGA), called for Germany to adopt a proactive African policy.[27] Hans-Peter-Bartels, Parliamentary Commissioner for the Armed Forces, clearly stated in 2014 that "we should not be bystanders in Africa."[28] In 2014, scholars from the German Institute for International and Security Affairs (Stiftung Wissenschaft und Politik, SWP) indicated that the German government should strengthen security cooperation with Africa for several reasons. First, the stability of Africa concerns Europe. Conflicts in countries such as Somalia, Mali, the Central African Republic, and the Sudan can cause humanitarian disasters, leading to problems such as refugees, drug trafficking, and human trafficking, all of which may spread to Europe. Second, Europe has an obligation to help Africa build its security architecture by supporting African countries and regional organizations. Third, Germany's security thinking needs to adjust to reality. Germany prefers to provide military training, but the case of Mali shows that training cannot take place without large-scale military intervention. The German security policy in Africa cannot only rely on France. Fourthly, the UN should

[27] For example, Robert Kappel, "Auf dem Nullpunkt, Deutschland Braucht Eine Aktive Afrikapolitik. Fuenf Vorschlaege für Eine Koordinierte Strategie", http://www.ipg-journal.de/rubriken/aussen-und-sicherheitspolitik/artikel/auf-dem-nullpunkt-1070/, February 23, 2017.

[28] Von Thorsten Jungholt, "Wir Koennen in Afrika Nicht Nur Zuschauer Sein", January 21, 2014, https://www.welt.de/politik/deutschland/article124054944/Wir-koennen-in-Afrika-nicht-nur-Zuschauer-sein.html, accessed on February 20, 2020.

be strengthened. Germany has always emphasized the United Nations as the leading force for world peace. Half of the United Nations' peacekeeping operations are in Africa. But the West's material and human support for the United Nations is inadequate. Fifth, it is conducive to developing further economic cooperation between Germany and Africa. These five points basically explain the motivation of the transformation of German-African security cooperation from a low key to a positive one.[29]

International Expectations Push Germany to Take More Responsibility

By 2014, Germany's self-positioning in the world was as a middle power and European country. The Euro crisis and Brexit have changed the balance of power that the EU used to hold alongside Britain, France, and Germany. Germany has become the most powerful country in Europe, politically and economically. Given Germany's place in the international community, outside observers have criticized Germany's dilemma as "Germany's role in the world is ambiguous."[30]

As one of the most important global economies after the US and China, Germany lacks a clear diplomatic strategy, especially when it comes to military affairs. "Germany is too strong compared to other players, but not enough as a leader."[31] "Ambiguity is both the main feature and the main problem of German foreign policy. In almost every area of foreign policy, Germany has failed to project a fully reliable and predictable image. This is in opposite of Germany's self-perception, a particularly good partner."[32] In other words, outsiders think that Germany's participation in international affairs is not consistent with its status as a major power, expecting Germany to play a greater role.

[29] Claudia Major & Christian Moelling, "Fuenf Gute Gruende fuer Ein Deutsches Engagement in Afrika", https://www.euractiv.de/section/entwicklungspolitik/news/funf-gute-grunde-fur-ein-deutsches-engagement-in-afrika/, accessed on February 20, 2020.

[30] "Warum Deutschland in der Welt Aktiver Werden Muss", June 17, 2015, http://www.wiwo.de/politik/deutschland/aussenpolitik-warum-deutschland-in-der-welt-aktiver-werden-muss/11927776.html, accessed on February 20, 2020.

[31] Ibid.

[32] Von Jan Techau, "Zu Europa und Westbindung bekennnen!" http://www.review2014.de/de/aussensicht/show/article/europa-und-westbindung-absolutsetzen.html, accessed on February 20, 2020.

In recent years, with the ongoing decline of the West characterized by Brexit and "America first," Germany has become the hope of the West to maintain the existing international order. International public opinion and expectations boost Germany's security policy toward positive change.

Domestic Public Opinion Forced Germany's Diplomatic Transformation

In 2014, Frank-Walter Steinmeier (the current German president) took over as the new Federal Foreign Minister and began to discuss with then-Defense Minister von der Leyen (current president of the European Commission) and the security commissioners about how Germany should assume more international responsibilities. The debate on German foreign and security policy began with the launching of "Review 2014—Further Thinking about Foreign Policy."[33] A fundamental review of Germany's own foreign strategy was carried out, with national discussions on Germany's position in the world and the transformation of Germany's foreign policy. The reason why the whole of German society is mobilized to participate in the discussion of diplomatic review is to win public support for the transformation of German diplomacy in a more active direction and to explore the positioning of German foreign policy and Germany's responsibility in a more inclusive manner, so as to achieve a self-understanding of German foreign policy and create more strategic space for German foreign policy. The German people's understanding of Germany's position and responsibility in the world will affect Germany's participation in security governance in Africa.

The core question for the Review is whether Germany should transform from a follower to a leader (vom Trittbrettfahrer zur Fuehrungsmacht),[34] with the purpose of seeking a balance between restraint and commitment in German foreign policy (Balance zwischen Zurueckhaltung und Engagement). The Review focuses on two big questions: "If German foreign policy is wrong, where is it wrong?" and "What foreign policy must change?" Thousands of German citizens and staff from

[33] http://www.aussenpolitik-weiter-denken.de/de/themen.html, accessed on February 20, 2020.

[34] Rolf Muetzenich, "Deutschland - Vom Trittbrettfahrer zur Führungsmacht Wider Willen?" *Zeitschrift fuer Außen- und Sicherheitspolitik*, January 2015, Volume 8, Supplement 1, pp. 273–287.

various German ministries, as well as more than 50 leading foreign policy experts and scholars from around the world, have been involved in the debate and discussion.[35] As a result of the consultation, the outside world and parts of the German public expect Germany to become a leading country in Europe with the willingness to assume international responsibility for the maintenance of a peaceful and free international order. Germany, while well aware of the gap between expectations and political reality, despite knowing that "other countries have high — and sometimes exorbitant — expectations of Germany," has decided to shoulder its responsibilities to Europe and the world.

As a result of this great diplomatic reflection, Germany is in the process of repositioning its role in the world and has decided to take advantage of Germany's economic, civilian, and multilateral strengths to play a greater role in world affairs and to build a global governance system with partner countries. As a result, 2014 became an important turning point in German foreign policy and cooperation with Africa, and Germany began to redefine its role in the world and its responsibilities in Africa, beginning to reorient its policy toward Africa. Prior to this, there was a lack of strategic consensus between the German Ministry of Foreign Affairs, the Ministry of Defense, and the Ministry of Development Cooperation; Germany rarely mentioned its own interests and positions in Africa; Germany did not have a clear position on its own interests and positions in Africa, often simply responding to EU and French initiatives and prone to falling into a passive, reactive German model of security cooperation; and avoiding as much as possible the use of military forces in security cooperation with Africa.

Furthermore, the desire of a younger German generation to participate in international affairs to a greater extent than previous generations and have a greater international voice has also affected, to some extent, the shift of German foreign policy to a more positive direction. Therefore, with the rise of Germany's national strength in the international arena, the country's previously conservative method of participation in foreign affairs no longer met the needs of the country's current development. This serves as the most fundamental reason for Germany's increased cooperation with Africa.

[35] www.review2014.de, accessed on February 20, 2020.

4 Developmental Trends of German-African Security Cooperation

With the transformation of Germany's foreign strategy and the increasing importance of African affairs to Germany's foreign policy, as well as a common ground on "Germany should assume more international responsibilities," Germany will continue to strengthen security and development cooperation with Africa, as demonstrated in the G20 Compact with Africa Investment Summit held in Berlin in November 2019.

More Focus on African Security Governance

Germany's economic and political interests in Africa, historical responsibilities, and the reshaping of France-African relations have all affected the transformation of German security policy toward Africa. Germany recognizes its moral responsibility for security matters in Africa. Merkel made it clear when she claimed: "Some of the difficulties in Africa today are the result of colonialism. Look at Africa's national borders, which are not divided by ethnicity, but by resources."[36] The economic cooperation between China and other emerging countries in Africa showed Germany Africa's huge potential as a market for German exports and investment destinations. As of 2013, more than 800 of Germany's 400,000 companies have invested in Africa, with a turnover of 27.5 billion euros.[37] But Germany's FDI flows to just 13 of the 54 countries in Africa, and its trade with Africa accounts for only 2% of Germany's foreign trade.[38] Germany's economic participation in Africa has witnessed stable growth, with significant room for expansion in the future. The maintenance of African peace and security not only helps protect Germany's existing economic interests in Africa, but will also improve the business environment in Africa in order to attract more German private enterprises to invest in Africa and share the dividends of African growth.

[36] "Die Wurzel Aller Flüchtlingsprobleme", October 9, 2016, http://www.tagesschau.de/ausland/merkel-afrika-103.html, accessed on February 20, 2020.

[37] Jahresbericht, 2015/2016, Afrikaverein der deutschen Wirtschaft, p. 13.

[38] "The Cornerstones of a Marshall Plan with Africa—A New Partnership for Development, Peace and a Better Future", Speech by German Federal Minister Müller at the German-African Business Summit, February 9, 2017 in Nairobi.

Therefore, maintaining the security and stability of Africa has become the focus of Germany's cooperative trajectory with Africa now and in the future. Germany casts immigration as a "peace project"[39] in an attempt to turn the crisis into opportunity. Furthermore, Germany can use its own advantages to play a unique role in African security affairs. In the past, Germany's security initiatives in Africa relied mainly on the EU and NATO, particularly France. But Germany has realized that it should no longer be a mere follower, but should clearly define its own interests in Africa. Although Germany has less influence in Africa than France, Germany has accumulated certain resources and networks in Africa through its embassies, political foundations, and development agencies such as GIZ. Thus, Germany has the capacity to play a more active and autonomous role in security governance in Africa in the future.[40]

Promoting German-African Security Cooperation Through a Security-Development Nexus

In the German view, only sustainable development can achieve true sustainable security. "Without economic development, it will be difficult for Africa to achieve stability."[41] Thus the main approach for Germany to help achieve African peace and security is to promote peace through development. Regarding the refugee problem, Chancellor Merkel realized that the fundamental solution for the refugees crisis lies in the origin countries of refugees; for refugees in Africa, it is necessary to improve the living conditions of the African people so that they can live and work in Africa in peace and no longer form flows of refugees to Germany or other countries. It is on the basis of a deep understanding of the relationship between development and security that Germany launched the African Marshall Plan at the end of 2016 in an attempt to eliminate unrest from its root causes and to achieve lasting peace and security through development.

[39] Discussion with Ms. Maria Adebahr from German Foreign Ministry, Berlin, October, 2016.

[40] Denis Tull, "Deutsche Afrikapolitik. Ein Beitrag zu Einer Ueberfaelligen Debatte", Friedrich-Ebert-Stiftung, 2014, p. 4.

[41] Jahresbericht 2015/2016, Afrikaverein der deutschen Wirtschaft, p. 18.

Germany–Africa Still Faces Difficulties

With the UK leaving the European Union and US President Trump pursuing his "America first" policy, outside observers worry about whether Germany will remain the only big country in the world that still adheres to EU values. In the meantime, there is a growing expectation for Germany to take up the mantle of global responsibility. The German government is thus under great pressure. Germany is facing unprecedented dilemmas insofar as the question of how to meet the growing global expectations of Germany and assume the leading role of global governance, which is both an opportunity and a challenge for the country.

First, there is a gap between German responsibility and actual capacity. For example, the Marshall Plan for Africa, announced by the German Ministry of Economic Cooperation and Development, is an ambitious one. But the German government only pledged 1 billion euros in funding for the program in 2019. The biggest challenge to German-African new policies is how to achieve such ambitious plans with limited resources.[42]

Second, better coordination between foreign policy, development policy, and security policy has become a critical task for Germany. Since 2017, German ministries have issued different and distinct African policy initiatives, revealing the fragmentation of German policy toward Africa. In terms of Somalia anti-pirate action, for example, the German ministries involved include the Federal Ministry of the Interior, the Ministry of Defense, the Federal Ministry of Transport, Construction, and Urban Planning, the Federal Ministry of the Economy and Technology, the Ministry of Justice, the Ministry of Finance, and the Ministry of Foreign Affairs. Security cooperation can be achieved only through effective coordination.[43] There is a worry that the security cooperation as a priority for German-African cooperation will reduce the resources for developmental assistance to Africa. In a meeting with Malian President Ibrahim Boubacar Keita, Chancellor Merkel stated that Germany wants to increase its military support for Mali, but "not at the expense of (reduced) development

[42] Discussion with Dr. Christine Hackenesch from DIE during T20 conference in Johannesburg, South Africa.

[43] Hans-Georg Ehrhart, Heinz Dieter Jopp, Roland Kaestner & Kerstin Petretto, "Deutschland und die Herausforderung 'Vernetzter Sicherheit' bei der Pirateriebekämpfung: Governancestrukturen und-akteure", p. 13, http://www.maritimesecurity.eu/fileadmin/content/news_events/workingpaper/PiraT_Arbeitspapier_Nr19_2012.pdf, accessed on February 20, 2020.

assistance."[44] But it remains to be seen if securitization will actually harm development policy.

In summary, the effectiveness of German-African cooperation in the future is mainly based on the following factors: first, it depends on whether Germany's policy toward Africa can be linked with Africa's own development and security strategy and cooperation mechanism. The key is whether or not to give full play to the autonomy of African countries. Second, Germany, as a leading country of the EU, continues to give priority to multilateralism in its foreign development and security cooperation, and its development and security cooperation with Africa is bound to take place within the framework of the European Union. Therefore, Germany needs to strive to legitimize the German program into a European program and integrate it with the existing European-African cooperative programs. Third, it depends on policy coordination among German ministries. The mechanism for German cooperation with Africa is intertwined with German internal affairs, foreign affairs, national defense, and economic and development policies, which require all ministries and departments to cooperate with each other.

5 Concluding Remarks

Germany's more active transformation in foreign policy and the increasing importance of African affairs to Germany indicates that the future of German-African cooperation will continue to maintain the comprehensive use of trade agreements, development assistance, direct investment, humanitarian assistance, institution-building, political dialogue, and other policy tools to achieve development and security through political, diplomatic, and developmental means.

Germany's Africa strategy is not isolated, but requires the close coordination of Germany's own foreign, security, and development policies, as well as alignment with the strategies of multilateral organizations such as the European Union, the G20, and other strategic partners. Taking Germany's African Marshall Plan as an example, the key to its peace and security pillar depends on three aspects: first, whether or not German-African cooperation is based on equal consultation with Africa, whether

[44] "Sicherheit und Entwicklung Gehören Zusammen", https://www.welt.de/newsticker/dpa_nt/infoline_nt/brennpunkte_nt/article158653517/Sicherheit-und-Entwicklung-gehoeren-zusammen.html, accessed on February 20, 2020.

this cooperation can align with the AU Agenda 2063 designed by the African continent itself, and whether it can win the support of African governments and people. Second, it depends on whether the various decision-making and executive bodies of German-African cooperation, for example, the German Ministry of Economic Development and Cooperation, can maintain policy coordination with the Federal Ministry of Foreign Affairs, the Ministry of Defense, and the Ministry of Finance to integrate development, foreign, and security factors, as well as work together to achieve peace and development in Africa. Third, German-African security cooperative measures should be linked with the existing security cooperation between the European Union and Africa, in order to make further German-African cooperation more legitimate and inclusive and reduces resistance from within the EU to Germany's leadership role. It remains to be seen how Germany will affect Africa's development during its presidency of the Council of the European Union in 2020.

Moreover, the rise of the Alternative für Deutschland party (AfD), the surge of populism that is ongoing in German society, and the uncertainty of a post-Merkel era, combined with the rustling future of Europe have brought unpredictability to Germany. German-African cooperation should not only aim to pursue Germany's own national interests, but also support Germany's efforts to bear global responsibility. The successful implementation of Germany's African policy depends on whether or not Germany is able to deal with its own domestic crisis, continue to lead the integration of the European Union, and preserve the EU values of global governance in an uncertain era.

CHAPTER 5

An Analysis of Italy's Africa Policy Adjustment Since 2013

Kaixuan Li

Despite Italy being a latecomer capitalist country in Europe, as soon as unification was achieved in the nineteenth century, it colonized Africa by emulating other traditional European powers, such as Britain and France. After years of brutal and bloody wars, Italy established colonial rule in the Horn of Africa and Libya. After World War II, Italian colonies became independent. At the same time, due to the beginning of the Cold War, as part of the Western camp, Italy shifted the focus of its foreign policy to the process of European integration, establishing closer transatlantic relations and seeking greater influence in the Mediterranean. In addition, compared with other major European countries such as Britain, France, and Germany, Italy's lagging and unbalanced domestic economic development and less stable central government significantly affected its strategic considerations toward Africa. Therefore, although Italy enjoys the geographical advantage of being adjacent to Africa, it has not yet played a prominent role in European-African relations until recently.

K. Li (✉)
Chinese Academy of Social Sciences, Beijing, China
e-mail: wxy03@ssap.cn

X. Li and C. An (eds.), *The Changing World and Africa*,
https://doi.org/10.1007/978-981-16-4983-7_5

After the Cold War, Italy's interest in sub-Saharan Africa gradually increased. But it wasn't until 2013, when the center-left political alliance came to power, that Italy began to substantially adjust its Africa policy and to treat it as a foreign affairs priority in order to alleviate immigration/refugee pressure, expand overseas markets, strengthen the diversification of energy supply, and enhance international influence, etc. In addition, a series of background factors cannot be ignored: the rapid economic development of Africa in recent years, the achievement of the peace process, the deepening of the Italian economic and social crisis, and the major adjustments made by the world powers and the EU's policies toward Africa, etc. Based on an analysis of the main characteristics of Italy's Africa policy before 2013, this article attempts to interpret its main motivations and new trends in adjustments since 2013, as well as to analyze the short-term effects of the adjustments and constraints in achieving its medium and long-term goals. The prospects for Sino-Italian cooperation in Africa are briefly discussed.

1 Characteristics of Italy's Africa Policy Before 2013

After World War II, Italy's influence in Africa gradually shrank, eventually becoming almost completely negligible. In 1947, Italy signed the Paris Peace Treaty under pressure from the United Nations. Since then, Italy has gradually given up control of its African colonies. In November 1949, the UN General Assembly declared the independence of colonial Libya, which became truly independent in 1952. Italy signed the Addis Ababa Treaty with Ethiopia in 1956, promising to pay US$16.3 million in compensation for their atrocities during colonial rule.[1] Since then, Africa's position in Italian foreign affairs has declined significantly. At the same time, Italy's diplomatic focus has shifted to transatlantic relations. In April 1949, Italy signed the North Atlantic Treaty and joined NATO, as well as participated in the European integration process. Italy maintained its transatlantic partnership and influence in the Mediterranean by strengthening its influence on the European integration process

[1] Gabriele Abbondanza, *Italy as a Regional Power: The African Context from National Unification to the Present Day*, Rome: Aracne Editrice, 2016, p. 132.

and its close links with Latin American communities of Italian descent.[2] After the Cold War, the world's political and economic landscape entered a period of multipolar transformation and reconstruction. During this period, especially after a series of severe domestic political and economic crises, Italy underwent an adjustment and transition of its foreign policy, gradually seeking to expand its influence beyond European integration, transatlantic priority relationships, and beyond the Mediterranean region altogether. Sub-Saharan Africa has been reconsidered. From the 1990s to 2013, Italy's Africa policy has mainly exhibited the following three characteristics.

Fragmentation and Marginalization of the Italian Cooperation Mechanism with Africa

Firstly, in the sense of having a bilateral framework, Italy's Africa policy system has been fragmented and marginal. Italian governments collapse frequently, so the country's African policies have frequently lacked continuity. They neither developed an institutionalized Italian-African mechanism for cooperation nor effectively promoted closer Italian-African relations. From 1990 to 1992, due to corruption scandals, the era of Italy's First Republic came to an end with the collapse of its mainstream parties. During that period, Italy struggled to cope with internal difficulties and had no time to formulate a new foreign policy for responding to the changes in the post-Cold War era and its own development needs. In sub-Saharan Africa, Italy has only maintained good relations with South Africa and Mozambique. Although Italy strives to maintain its traditional influence on the Horn of Africa, in the relationship between Ethiopia and Eritrea, it can only play the role of a "neutral observer," rather than a peaceful leader.

In the Dini Technical Government that lasted from 1995 to 1996, the Italian Ministry of Foreign Affairs' focus on Africa remained on Mozambique and the Horn of Africa. The central-right government of Berlusconi, which maintains a certain skepticism toward Europe, has been in power from 1994 to 1995, 2001 to 2006, and again from 2008 to 2011, and has gradually put Africa on its agenda for Italian foreign affairs.

[2] Bernardo Venturi, "Africa and Italy's Relations After the Cold War", in Dawn Nagar and Charles Mutasa, eds. *Africa and the World: Bilateral and Multilateral International Diplomacy*, London: Palgrave Macmillan, 2017, pp. 170–172.

However, this change was due primarily to the demands for immigration control and energy security. After entering the twenty-first century, the center-right government committed to developing cooperative relations with North African countries, especially Libya, and to promoting the latter's border control in order to prevent illegal immigrants from pouring into Italy and to encourage the introduction of Italian investment. During this period, Italy carried out mineral resource extraction and laid energy pipelines in Nigeria, Angola, and other African countries, as well as participated in the construction of infrastructure to pursue economic benefits.

The center-left government, characterized by multilateralism and Europeanism and having been in power at the end of 1990s and from 2006 to 2008, attached greater importance to promoting cooperation between the European Union and the African Union. The center-left government has also tried to promote Italy's return to Africa, especially in sub-Saharan Africa, and to rebuild traditional Italian-African political priorities.

In short, until 2013, Italy was more inclined to place North African countries such as Algeria, Tunisia, Libya, and Egypt in the pan-Mediterranean strategic circle. Its cooperative projects have mainly focused on deterring illegal immigration and strengthening cooperation in logistics and energy. In sub-Saharan Africa, Italy has basically maintained its sporadic and marginal influence, and has kept its interests concentrated in the energy field.

Secondly, the developmental cooperation system between Italy and Africa is disorganized, and non-governmental entities have played a more important role than has a stable government. From 1987 to 1994, Italy assisted 117 countries in total. But there were neither clearly prioritized countries nor clear issues.[3] Italian left-wing organizations and Catholic organizations have played an important role in development cooperation with Africa. Due to their ideological position, left-wing organizations, such as the Italian Communist Party, have promoted Italy's development assistance to countries governed by Marxist parties, such as Angola, Tanzania, and Mozambique. In addition, the Italian Catholic group Sant'

[3] Bernardo Venturi, "Africa and Itlay's Relations After the Cold War", in Dawn Nagar and Charles Mutasa, eds. *Africa and the World: Bilateral and Multilateral International Diplomacy*, London: Palgrave Macmillan, 2017, p. 179.

Egidio Community has played an active role in promoting the peace process in Mozambique.

Due to the internal political and economic crisis of the early 1990s, the reform of the Italian development cooperation system has not been put on the agenda for over a decade. Although the financial resources needed for development cooperation fell to the lowest level in history in 1996, Mozambique and Ethiopia have always been priority countries. The priority issues were not clear in this period, however. For example, additional development assistance funds have been temporarily allocated to help the Congo and Uganda respond to emergency crises. In the twenty-first century, as Italy's central government has been more stable, its enthusiasm for development cooperation with Africa has increased. However, until the 2013 reforms, Italy has not established a special development cooperation mechanism, and multi-sector interwoven development cooperation with Africa remained scattered and inefficient.

Thirdly, Italy has relied on a multilateral framework and has participated in the African agenda with multilateral institutions such as the United Nations and the European Union. After World War II, Italy has shown a strong preference for multilateralism in international affairs, both for the reality and positioning of its own medium power. Despite Africa's relatively marginal position in its foreign policy, Italy still considers itself to be Africa's "main supporter of the peace process," a process it supports by exerting some influence on the United Nations' agenda in Africa and by participating in a few peacekeeping operations, such as those that took place in Somalia in the early 1990s. In the 1990s, European integration not only made remarkable achievements in the fields of economy, justice, and internal affairs, but also made a certain amount of progress in the construction of a European common diplomacy and defense mechanism. Therefore, Italy has gradually come to regard the EU as the most important stage for enhancing the power of discourse in international affairs. To that end, Italy began to carrying out its Africa policy even more under the EU's framework and principles, and continuously improving the development of cooperative resources allocated to Africa through the EU.[4] Especially after entering the twenty-first century, Italy's standards for defining development assistance priority targets are consistent with the program clearly defined in the Cotonou Agreement that was signed

[4] Maurizio Carbone, "Holding Europe Back: Italy and EU Development Policy", *Journal of Southern Europe and the Balkans*, Vol. 9, Iss. 2, 2007.

by the European Union with the Group of African Caribbean and Pacific Region Countries in 2000. In addition, Italy has also participated in the initiatives of other multilateral organizations on Africa, such as the initiative to reduce the debt burden of poor countries made by International Monetary Fund and the World Bank. From 2001 to 2012, Italy reduced the debts of sub-Saharan African countries by 3.6 billion euros.[5]

The Economic Demands on Africa Are Mainly Concentrated on the Energy Field

After the Cold War, due to information asymmetry and long-standing prejudices, Italian SMEs have invested very little in Africa, and bilateral trade dependence has been very low. After entering the twenty-first century, although the Italian government gradually attached greater importance to opportunities for large enterprises seeking to invest in Africa, it failed to outline a long-term strategy for its trade with Africa, nor did it form an institutional platform for deepening existing levels of cooperation between Italian and African companies. In general, Italian-African trade has not been a linear upward trend, but an undulating wave. Italy has nonetheless maintained relatively close trade ties with countries in sub-Saharan Africa, such as South Africa, Nigeria, and Angola.

Italy's investment and trade demands in Africa are mainly concentrated in traditional energy sources such as oil and natural gas. Italy imports large quantities of oil, natural gas, and other raw materials from countries such as Angola, Mozambique, and Nigeria. In the energy trade area, the energy giant Eni Group has played a pivotal role. The Italian government is a major shareholder in ENI, and ENI, too, has long had a revolving door between its business and government. ENI has been an extension of the state apparatus fulfilling Italian national interests. As former Italian Prime Minister Matteo Renzi once declared in an interview with national television: "ENI is a fundamental component of our energy policy, our foreign policy, our intelligence policy. By intelligence, I mean our secret services."[6] It is worth noting that with the expansion of the Eni Group's

[5] Ministero degli Affari Esteri e Cooperazione Internazionale, "Iniziativa HIPC Rafforzata", https://www.esteri.it/mae/en/politica_estera/economia/cooperaz_econom/debito_estero/sace/iniziativa_hipc.html.

[6] Fasil Amdetsion, "Strengthening Business Ties Between Italy and Africa", https://www.fairobserver.com/region/europe/italy-business-presence-in-africa-11129/amp/.

business, other related infrastructural enterprises have also begun to enter Mozambique, Nigeria, Angola, and other countries in the early twenty-first century.

2 The Main Motivation for Italy's Adjustment of Its Africa Policy Since 2013

Since Italy's central-left government came to power in 2013, the reasons for the "sudden" adjustment of its Africa policy have been diverse, mainly due to the following interest demands.

Relieve Domestic Immigration/refugee Pressure

Due to the closeness of the geopolitical structure of Italy and Africa, Italy has become the "frontline country" for the European refugee crisis. According to data released by the UN Refugee Agency, more than 620,000 immigrants/refugees poured into Italy from across the sea in 2014–2017. They mostly came from Nigeria, Eritrea, Senegal, Guinea, Mali, Sudan, Somalia, Gambia, and other sub-Saharan African countries.[7] The growing refugee crisis has put tremendous pressure on Italy's fiscal expenditures and social governance system. In 2017, Italy's total expenditures on refugee affairs reached 4.4 billion euros, accounting for about 0.5% of total social expenditures.[8] At the same time, the surge of illegal immigrants and refugees in the short term has also impacted Italy's security.[9] As a result, immigrants/refugees are increasingly regarded as "invaders," and a degree of populism mixing with European skepticism and xenophobia has taken off. Therefore, the Italian center-left government has initiated dialogues with Africa on border control measures that can effectively curb immigration in the short term. At the same time, these dialogues and measures have also involved increasing financial resources in development cooperation with Africa, promoting the employment of

[7] https://data2.unhcr.org/en/situations/mediterranean/location/5205.

[8] "Immigrazione, quanto costa davvero all'Italia?", https://quifinanza.it/finanza/immigrazione-quanto-costa-davvero-allitalia/211824/.

[9] Francesco Palazzo, "Immigrazione e Criminalità, Una lettura di dati statistici", https://www.penalecontemporaneo.it/upload/PALAZZO_2016c.pdf.

African youth, to solve the root cause of immigration/refugee issues and to ease the domestic political and social crisis in Italy.

Expand the African Market for "Made in Italy"

Given its high levels of dependence on external markets and weak recovery in recent years, it is urgent for the Italian economy to further expand into the African market. After the outbreak of the international financial crisis in 2008, the Italian economy declined significantly. Negative growth occurred in 2009, 2012, and 2013, and the Italian economy has not yet recovered to the level before the crisis. In addition to its insufficient domestic demands, varying degrees of recession or low-speed growth of its traditional trading partners may also affect the recovery of Italy. Sub-Saharan Africa, however, has maintained rapid development after the outbreak of the international financial crisis and is becoming the new center of world economic growth. According to data published by the International Monetary Fund, the annual GDP growth rate of this region basically remained above 3% (except for 2016, when it was 2.1%) since 2008, second only to Asia.[10] In addition, with the continuous advancement of the AfCFTA, the growth and potential demand that the unified African market promises to unleash would be significant to "Made in Italy."

Enhance the Diversification of the Energy Supply

Italy's domestic mineral resources are scarce and the level of its energy self-sufficiency rate is very low. In Italy, 93% of the crude oil and natural gas needs are met through imports.[11] Iraq in the Middle East, Libya and Algeria in North Africa, and Russia have always been Italy's main energy suppliers. For the energy security challenges encountered since the outbreak of the "Arab Spring" and "Ukraine Crisis," it has become more urgent than ever for Italy to increase its oil and natural gas import channels and to strengthen the diversification of its energy supply. Africa's rich

[10] "Real GDP Growth -Annual Percent Change", https://www.imf.org/external/datamapper/NGDP_RPCH@WEO/OEMDC/ADVEC/WEOWORLD/AFQ/SSQ.

[11] "Italy", https://www.eia.gov/international/analysis/country/ITA.

energy and mineral resources reserves have been receiving much attention. Oil and natural gas reserves account for 12% and 8%, respectively, of the world's total reserves.[12] Meanwhile, the political and security situation in sub-Saharan Africa, which is rich in mineral resources, has improved in recent years. This has prompted Italy to upgrade the region's position in its energy security strategy.

Reliance on Africa to Increase International Influence

Italy is trying to enhance its influence in multilateral affairs by winning the strategic support of Africa. As a medium power on the global stage, Italy has always maintained a prominent multilateralist orientation in international affairs. Due to the political and economic stabilization of most African countries and the important breakthroughs made in the process of regional integration, Italy has started to reconsider Africa's strategic value in multilateral affairs. Considering the UN's "one country, one vote" principle, it is important for Italy to win the support of African countries when voting for a non-permanent member. At the same time, the sustainable development of regional integration organizations such as the African Union, the East African Community, and the Southern African Development Community have also become strategic partners that cannot be ignored in multilateral affairs. In addition, with the fierce competition of world powers in Africa, African affairs are gradually becoming the focus of the global multilateral agenda. As then-European Commission President Jonker declared at the Africa-Europe High-level Forum in 2018: " Africa's challenge is Europe's, so are the hope and the future."[13]

In light of the above motivations, the Italian center-left political alliance started to adjust Italy's fragmented Africa policy after taking office in 2013. The center-left government has not only set about developing Italian-African relations under the existing bilateral framework while seeking to maximize national interests, but has also striven to strengthen its influence on African issues within the EU and other multilateral frameworks. This position is best expressed by Italy at the

[12] "Our Work in Africa", https://www.unenvironment.org/regions/africa/our-work-africa.

[13] "Juncker, Ue e Africa Hanno Destino Comune", http://www.ansa.it/sito/notizie/mondo/europa/2018/12/18/junckerue-e-africa-hanno-destino-comune_377b2088-f6a8-4b18-9656-b7b0e4df6be0.html.

Italian-African Ministerial Conferences: "The relationship between Italy and Africa is strategic and priority" and "Italy is the bridge between Europe and Africa."[14]

3 New Trends in Italy's Africa Policy Since 2013

Since 2013, Italy has adjusted its Africa policy by focusing on the establishment of summit diplomacy, bilateral political and commercial dialogue mechanisms, and by using developmental cooperation as a starting point to increase its influence on Africa. It has also highlighted immigration issues and emphasized economic benefits. In this round of adjustment, Italy continues its tradition of relying on the existing multilateral framework and has begun to pursue greater levels of influence in Africa.

Improve the Cooperation Mechanism with Africa

First of all, Italy has emphasized summit diplomacy and the construction of a bilateral dialogue mechanism. At the end of 2013, the Italian Ministry of Foreign Affairs and International Cooperation launched an "Italian Initiative," which included three ministerial meetings on the theme of agriculture, energy, culture, and ethnography held in 2014, to promote Italy to pay more attention to Africa and the development of bilateral relations between Italy and Africa.[15] On this basis, Italy has initiated summit diplomacy with multiple African countries. In July 2014, Prime Minister Matteo Renzi visited Mozambique, Congo (Brazzaville), and Angola. This was the first official visit of the head of the Italian government to the sub-Saharan Africa region since World War II. It is regarded as "Italy's return to Africa."[16] From 2015 to 2016, Renzi visited Ethiopia, Nigeria, Ghana, and Senegal in succession. With the aforementioned diplomatic

[14] Marco Galluzzo, "Italia-Africa, la Cooperazione Cresce 'Ma tutta l'Europa Deve Fare di Più'", https://www.corriere.it/esteri/18_ottobre_26/italia-africa-cooperazione-crescema-tutta-l-europa-deve-fare-piu-1c36909a-d923-11e8-bb5a-fd7ad32a316b.shtml; Andrea Gagliardi, "Mattarella: Italia Ponte con Africa, Gentiloni non Esclude Emergenze", https://st.ilsole24ore.com/art/notizie/2016-05-18/mattarella-italia-ponte-africa-primo-dovere-e-salvare-vite-umane-094805.shtml?uuid=ADBi3RK.

[15] "Iniziativa Italia-Africa", http://www.esteri.it/mae/tiny/1710.

[16] Italian scholars tend to exclude Prodi's trip to Addis Ababa in 2006 as the first official visit, because it is not a special visit to Africa but participation in AU activities, see

efforts, Italy successfully established a bilateral high-level dialogue platform with African countries in 2016, namely the "Italy-Africa Ministerial Conference," which strongly promoted the development of the relationship between Italy and Africa. The conference has been successfully held two times so far, focusing on topics such as peace and security, economic development, immigration, and the sustainability of the social environment; "the aim of the Conference is to find shared solutions to the biggest challenges in the area of peace, freedoms, democracy and security, and to single out common paths towards growth, especially by involving qualified Italian players from the world of industry and business, academia and non-governmental organizations."[17] The first bilateral ministerial conference in 2016 also focused on other important issues, such as "ownership" under African economic and sustainable development, African security, and refugees. In 2018, the second conference expanded even further, involving 54 African countries, mostly represented at the ministerial level, and 13 international organizations, including the African Union, for a total of 350 delegates. The second conference focused on issues such as the peace agreement between Ethiopia and Eritrea, as well as the promotion of youth employment and the empowerment of women through development cooperation.

Italy attaches increasingly greater importance to the dialogue and cooperation between regional African integration organizations. For example, through summit diplomacy, Italy has strengthened dialogue with the AU to win the latter's support in the areas of immigration, security, and development cooperation, and has consolidated cooperation with the East African Community by sending special envoys and providing support in areas such as agricultural processing and renewable energy.

Secondly, Italy has reformed the cooperative development system with Africa. In this round of development cooperation mechanism reform, Italy has completed its development cooperation mechanism with Africa to make it more efficient while increasing its financial investment in development cooperation with Africa. Italy renewed the development cooperation system with LAW 11 August 2014, n. 125, i.e. a general

Ettore Greco e Nattalino Ronzitti, "Rapporto sulla Politica Estera Italiana: Il Governo Renzi", https://www.iai.it/sites/default/files/iaiq_17.pdf, p. 108.

17 "Farnesina Hosted the Second 'Italy-Africa Ministerial Conference'", https://www.esteri.it/mae/en/sala_stampa/archivionotizie/comunicati/2018/10/italia-africa-alla-farnesina-si-apre-il-25-ottobre-la-conferenza-ministeriale.html.

framework on international development cooperation, to make it a newer and more modern foreign policy tool for Italy. In addition to encouraging and ensuring the participation of multiple subjects such as parliament, civil society organizations, non-profit organizations, socially responsible enterprises, immigration organizations, trade union organizations, business owners' associations, local governments, and colleges and universities, the major changes in the new system are mainly reflected in three aspects: (a) the establishment of inter-ministerial coordination committees and national committees for development cooperation to ensure the consistency, efficiency, coherence, and political priority of development cooperation; (b) assigning a vice minister of the Ministry of Foreign Affairs specializing in development cooperation.; and (c) establishing a financial security mechanism for the Development Cooperation Agency. Through this round of reforms, Italy has focused on increasing the proportion of financial investment in Africa's development cooperation. According to the data from the OECD, Italy's net official development assistance investment has increased significantly, from nearly 0.14% of its GNI in 2012 to 0.30% in 2017, and the portion flowing to Africa has doubled.[18] The International Development Cooperation Agency, which began operations in 2016, enjoys a high degree of autonomy and flexibility and plays a major role. It has 20 representative offices around the world, nine of which are located in Egypt, Tunisia in North Africa, and Burkina Faso, Ethiopia, Kenya, Mozambique, Senegal, Somalia, and Sudan in sub-Saharan Africa. In 2018, the International Development Cooperation Agency's disposable financial funds were 587 million euros, 53% of which was invested in Africa.[19]

Thirdly, Italy is pursuing greater influence under the multilateral framework. In the Three-Year Programming and Address Document (2016–2018), the Ministry of Foreign Affairs and International Cooperation of Italy proposed improving Italy's ability to lead development cooperation policy discussions and policy shaping in international organizations, and to increase the input of relevant responsible Italian actors,

[18] "Net ODA", https://data.oecd.org/oda/net-oda.htm#indicator-chart; "Development Aid at a Glance Statistics by Region, 2. Africa 2019 Edition", https://www.oecd.org/dac/financing-sustainable-development/development-finance-data/Africa-Development-Aid-at-a-Glance-2019.pdf.

[19] "Piano della Performance 2019–2021", https://www.aics.gov.it/wp-content/uploads/2019/02/AICS-PP-2019.pdf, p. 8.

thus ensuring the cohesion of multilateral development cooperation initiatives and national priorities.[20] Hence, Italy has foregrounded development cooperation as an entry point to increase its influence on the African agenda under existing multilateral frameworks, such as the United Nations and the European Union. For example, in January 2019, the Italian government, the United Nations Environment Programme, and the United Nations Food and Agriculture Organization jointly established the African Climate and Sustainable Development Center to focus on climate change issues and to give priority to promoting sustainable development in the Sahel. Italy has also actively sought to enhance its voice in the EU's policy on Africa. For example, after the Valletta Refugee Summit in 2015, the EU set up an "African Emergency Trust Fund" to specifically address immigration issues, and Italy became the second largest contributor, jointly launching the first immigration-development project with the European Commission. In terms of the concept of development cooperation with Africa, Italy has advocated going beyond the traditional "donor–recipient" model in order to establish equal partnerships and respect African autonomy. Italy has also advocated for becoming the leader of the "new model" of development cooperation within the EU.[21]

More Pragmatic Issues on African Cooperation

First and foremost, Italy highlights immigration issues. The issue of immigration has always occupied a central position in the adjustment of Italy's Africa policy. Especially after the outbreak of the refugee crisis, the Italian Ministry of Foreign Affairs has made it no secret that immigration is the core element of the relationship between Italy and Africa.[22] In 2016, Italy

[20] "Cooperazione Internazionale per lo Sviluppo, Documento Triennale di Programmazione e di Indirizzo 2016–2018", https://www.esteri.it/mae/resource/doc/2017/03/doc_triennale_2016-2018_-_finale_approvato.pdf, p. 55.

[21] "Intervento del Presidente Conte alla XVIII Edizione del Forum Internazionale dell'Agricoltura e dell'Alimentazione", http://www.governo.it/it/articolo/intervento-del-presidente-conte-alla-xviii-edizione-del-forum-internazionale-dell.

[22] "Migrations and International Relations: The Italy-Africa Agenda" in a conference at the Italian Foreign Ministry, https://www.esteri.it/mae/en/sala_stampa/archivionotizie/approfondimenti/2016/01/africa-alla-farnesina-il-convegno.html.

set up embassies in Guinea and Niger to deal with large-scale immigration in West Africa. In December 2016, Italy launched the Africa Fund,[23] providing financial support through special interventions of the Development Cooperation Agency on African development cooperation and immigration issues. At the beginning of 2017, Italy and Libya reached a memorandum of understanding on cooperation in the field of development, aimed at combating illegal immigration, trafficking in human beings, smuggling, and on strengthening the security of the borders between the State of Libya and the Italian Republic, clarifying Libya's border control responsibilities and Italy's commitment to provide Libya with more financial and technical support through development cooperation.[24] In 2017, Italy also created the African Information Exchange Center, as an exchange platform for immigration and development issues.

In addition to border control and other emergency response measures, Italy also advocates a more strategic vision of immigration in the era of globalization. Italy insists that the logic of development and good governance should be used to achieve economic and employment growth, thereby fundamentally solving the financial pressure and social governance dilemmas brought about by the immigrant/refugee problem. Therefore, the countries being prioritized for Italy's development cooperation with Africa are mainly those considered to be the origin of immigrants, as well as and transitional ones.[25] At the same time, priority issues have shifted from the original issues of inclusive growth, poverty reduction, sustainable agricultural development, food security, education, private sector development, natural resources, and environmental protection to ten other

[23] The Italian Agency for Development Cooperation, public agencies, the European Union, the Office of the United Nations High Commissioner for Refugees and other international organizations are contributors of the fund, see "Documento Triennale di Programmazione e di Indirizzo 2017–2019", https://www.esteri.it/mae/resource/doc/2018/02/doc_triennale_2017-2019_27.07.2017.pdf, p.14.

[24] Migranti: accordo Italia-Libia, il testo del memorandum, https://www.repubblica.it/esteri/2017/02/02/news/migranti_accordo_italia-libia_ecco_cosa_contiene_in_memorandum-157464439/.

[25] A total of 22 priority countries for development cooperation announced by the Italian Agency for International Development Cooperation in 2017–2019, including 11 African countries, which are Egypt and Tunisia in North Africa, Burkina Faso, Niger, Ethiopia, Kenya in Sub-Saharan Africa, Mozambique, Senegal, Somalia, Sudan and South Sudan.

major topics, including humanitarian assistance, migration and development, agriculture and food security, energy, health, education, global civic education, culture, development and industrial innovation, and juvenile justice.[26] As can be seen from the changes in the agenda, Italy aims to start with development and education to deal with the influx of immigrants and the subsequent issues of social integration, employment, and security in a more fundamental way. As Italian Prime Minister Conte's appeal at the second Italy–Africa Ministerial Conference in 2018 stated, "when we talk about their and our security, Italy and the EU must invest in the continent, get out of the emergency logic and make sure that immigration returns to be an opportunity and not a forced choice for the locals."[27] Italy hopes that by achieving sustainable development, it will help young people to have an alternative of development by taking root locally rather than immigrating abroad.[28]

Secondly, Italy emphasizes economic benefits. Unlike the previous lack of a clear and long-term economic and trade strategy with Africa, in recent years Italy has now established a multilevel platform for bilateral economic dialogue to clarify the demands for economic interests and promote the development of domestic enterprises in Africa. Italy has taken advantage of summit diplomacy to promote the expansion of large enterprises in the African market. From 2014 to 2016, almost all Renzi's visits to Africa were accompanied by a delegation composed of energy giant ENI Group, infrastructure leader Trevi Group, construction leader Salini Impregilo, and other agricultural and commercial enterprises. High-level government interactions have strongly promoted the growth of Italian businesses in energy, infrastructure, and other fields in Africa.

Italy has built a bilateral business dialogue platform to promote the ability of SMEs to enter the African market. Due to the promotion of the Italy–Africa Ministerial Conference, the Italy–Africa Business Week

[26] "COOPERAZIONE INTERNAZIONALE PER LO SVILUPPO, Documento Triennale di Programmazione e di Indirizzo 2017–2019", https://www.esteri.it/mae/resource/doc/2018/02/doc_triennale_2017-2019_27.07.2017.pdf.

[27] Marco Galluzzo, "Italia-Africa, la cooperazione cresce 'Ma tutta l'Europa deve fare di più'", https://www.corriere.it/esteri/18_ottobre_26/italia-africa-cooperazione-crescema-tutta-l-europa-deve-fare-piu-1c36909a-d923-11e8-bb5a-fd7ad32a316b.shtml.

[28] "Nasce a Roma il Centro per il Clima e lo Sviluppo Sosotenibile dell'Africa", https://www.aics.gov.it/oltremare/rubriche/eventi/nasce-a-roma-il-centro-per-il-clima-e-lo-sviluppo-sostenibile-dellafrica/.

started in 2017 and has gradually developed into an important platform for communication and dialogue between political decision-makers and businesspeople. The Italy–Africa Business Week provides SMEs with advantages in the fields of textiles and fashion, infrastructure, renewable energy, agriculture, new technologies, and bio-medicine opportunities in order to enter the African market. In addition, the South Africa–Italy Summit, which has been held since 2014, is regarded as a leading platform for strengthening economic ties between the two countries, as well as creating strategic business partnerships for companies in both continents.

Italy has emphasized the protection of the economic interests of business entities participating in development cooperation with Africa. In 2017, Italy clearly proposed in its international development cooperation plan that, on the basis of transparency, competition, and a sense of social responsibility, it should promote those profit-oriented Italian small and medium-sized enterprises that are actively participating in the priority issues of development cooperation with target countries.[29]

4 Effects of the Italy–Africa Policy Adjustment After 2013

At present, the demands of Italy's Africa policy on controlling immigration, improving economic interests, and enhancing multilateral influence have not yet failed. Italy has even played a leading role in the formulation of the EU's policy on migration in Africa. However, in the coming period, due to the relatively marginal impact Italy has on African political culture, the limited availability of financial resources, and the rightward tilt of domestic political balances, it is hard for Italy to play a more prominent role in the multilateral agenda for Africa.

Firstly, the pressure exerted by immigrants/refugees on Italy has been relieved to a certain extent. This is mainly due to the "Memorandum of Understanding" reached between Italy and North African countries, especially with Libya. According to data released by the UN Refugee Agency, from 2014 to 2017, more than 620,000 immigrants/refugees poured into Italy by sea. Most of them came from Nigeria, Eritrea, Senegal, Guinea, Mali, Sudan, Somalia, Gambia, and other West African

[29] "Cooperazione Internazionale per lo Sviluppo, Documento Triennale di Programmazione e di Indirizzo 2017–2019", https://www.esteri.it/mae/resource/doc/2018/02/doc_triennale_2017-2019_27.07.2017.pdf, p. 9.

and sub-Saharan African countries. Under the combined efforts of border control and development cooperation, the number of immigrants/refugees entering Italy by sea dropped to 119,000 in 2017, indicating a decline of 34.2% from 2016. In 2018 and 2019, the number dropped to more than 23,000 and 11,000, respectively.[30] At the same time, the top African countries of origin of immigrants have also changed. Currently, there are mainly Tunisia, Eritrea, Algeria, Côte d'Ivoire, and Sudan. The reduction in the influx of immigrants/refugees eased the pressure on Italy's financial expenditure and social governance and is predicted to continue doing so for the foreseeable future.

Secondly, the bilateral economic and trade relationship continues to heat up. With the continuous growth of direct investment and exports to Africa since 2013, Italy has quietly become one of the major countries competing in the African market. According to the World Investment Report 2019 published by the United Nations, as of 2017, Italy ranked sixth in the world and fourth in Europe, with US$28 billion in direct investment in Africa.[31] Italy has not only greatly increased its investment in energy development in Nigeria, Angola, Mozambique, and Egypt through the Eni Group, it has also increased investment in transportation, infrastructure, agriculture, food, etc. in the above countries and in the Horn of Africa. As far as trade is concerned, according to data released by the Italian Ministry of Foreign Affairs and International Cooperation, Italy's exports to sub-Saharan Africa increased by 7.02% in 2018, and basically maintained those same export levels through 2019.[32] Among these, the export of chemicals, food, clothing, and stone glass products have achieved steady growth since 2014. The increase in direct investment in Africa and the growth of exports to Africa by advantageous industries have injected a certain amount of growth momentum into the Italian economy, which is highly dependent on external markets.

Thirdly, Italy's influence on the African agenda has increased under a multilateral framework. As Italy's Africa policy has further refined the

[30] "Calano gli Sbarchi, Cambiano le Provenienze. Andamento degli Arrivi via Mare in Italia", https://www.ismu.org/dati-sugli-sbarchi-gennaio-2020/.

[31] "World Investment Report 2019", https://unctad.org/en/PublicationsLibrary/wir2019_en.pdf, p. 34.

[32] Osservatorio Economico—MAECI, "Aree Geografiche di Destinazione dell'Export Italiano", https://www.esteri.it/mae/resource/doc/2020/04/tabella4a_aree_geografiche_destinazione_dellexport_italiano.pdf.

priority countries and priorities of development cooperation and increased financial input, Italy has won the support of more and more African countries, and has enhanced its voice in the multilateral framework on Africa's sustainable development agenda as a result. Support from African countries has become an important guarantee for Italy to win a seat among the non-permanent members of the UN Security Council from 2017 to 2018 and exert greater influence. At the same time, within the EU, Italy's ability to shape policies on African migration has also been strengthened. In 2016, based on the basic principles of the *Migration Compact* submitted by Italy, the European Union launched the *Migration Partnership Framework* and the EEIP, both of which support Africa's development. In February 2017, with the support of the European Union, Italy took the lead in reaching a Memorandum of Understanding with Libya, the main purpose of which is to control illegal immigration, which has played a model role in the signing of bilateral agreements between other European and African countries to a certain extent. As the second contributor of EUTF, Italy's priority issues as migration and security might produce an impact that cannot be underestimated.

However, Italy is still far from being the powerful bridge between Europe and Africa and the leader of the "new model" of development cooperation with Africa that it would like to be. It still faces many constraints. Firstly, Italy has not yet formed the political advantages necessary to play the above role. As of March 2020, Italy has only 22 embassies in sub-Saharan Africa, while France, Germany, and the UK have 42, 39, and 33 embassies, respectively. Italy is unable to play the role of "coordination or peace-building" in the peace process in sub-Saharan Africa, for the insufficient political resources.[33] Secondly, Italy's cultural influence on Africa is also relatively marginal. Italy has only seven cultural offices in sub-Saharan Africa, and that cannot compare with Britain and France. Despite having added cultural, artistic, and other soft power factors to its cooperative development with Africa since 2019, currently, Italy's soft power in Africa is still relatively limited. Thirdly, Italy's political and economic situation is not conducive for it to play a more active role in the EU's Africa policy. After the 2018 parliamentary elections,

[33] Bernardo Venturi, "Un Rinnovato Impegno per l'Africa" in Ettore Greco, *L'Italia al Bivio.Rapporto sulla Politica Estera Italiana Edizione 2018*, https://www.iai.it/sites/default/files/iaiq_19.pdf, p. 155.

the decline of the center-left and the rise of populist parties have exacerbated the differences in Italy's policy toward Africa. Under the combined influence of political conservatism and pessimistic economic expectations, the tension between the "security" orientation, i.e. self-security prior to Africa's development, and the concept of "new model" of development cooperation with Africa will become more prominent.

5 Conclusion

Due to profound changes that it has undergone in the political and economic landscape both at home and abroad, as well as the rapid development of Africa, Italy has formed a new understanding of Africa's strategic value. Africa's rich resources and markets are an important foundation for Italy's economic recovery. At the same time, the series of risks facing Africa are also closely related to Italy's own security and development. Therefore, migration, security, energy, the economy, trade, investment, etc., have all become important issues in Italy's adjustments to its Africa policy. From an institutional perspective, there are two major developments in the relationship between Italy and Africa: the construction of a bilateral political and business dialogue platform and the improvement of the development cooperation system. For Italy, Africa is indeed an important strategic basis for enhancing its international and regional influence. But Italy must also make more adaptive adjustments and invest more political, economic, cultural, and social resources before it can become a "bridge" for European-African relations, much less serve as a leader of the new EU–Africa development cooperation model.

Since Italy adjusted its Africa policy—especially after signing the Memorandum of Understanding on the "Belt and Road" Initiative with China in March 2019—Italy has been closely following China's development cooperation and investment trends in Africa. In practice, Italy has proposed that both sides should strengthen their cooperative efforts in the fields of agriculture, food safety, agricultural machinery, and high-level infrastructure under the framework of the "Belt and Road" initiative from the perspective of further complementarity between Chinese and Italian companies and the needs of Africa. Chinese and Italian companies have already begun to explore ways to cooperate in these areas. In 2019, the China Electric Power Construction Group signed a memorandum of understanding with Bonifica, a subsidiary of Italy's IRI Group, which has technological advantages in infrastructure, water conservancy, energy and environmental rectification, and other business areas. Both sides are jointly building Africa's largest infrastructural project, the Lake Chad Water Transfer Project. In addition, the two sides will also carry out road construction, energy development, and agricultural product processing projects to promote local economic development. In this regard, the former Deputy Minister of Italian Economic Development, Michele Geraci, even advocated that it should be regarded as a model of justice and benefit in Sino-Italian cooperation.

This is because in Italy's view, although China's economic interests in Africa are mainly focusing on mineral extraction and the expansion of the "Made in China" market, China does not attach any conditions to infrastructure and urbanization development cooperation methods, which can be more effective in terms of promoting economic and social development in Africa, as well as promoting youth employment and alleviating poverty. Therefore, for the continued existence of Sino-Italian cooperation in Africa, China should not ignore Italy's concerns on migration issues and economic interests. Italy not only hopes to quickly expand the African market for its own purposes by leveraging China's political and economic network, but also has the expectation of improved efficiency and quality of development cooperation with Africa in order to fundamentally solve the migration/refugee problem.[34] China and Italy may explore more opportunities for cooperation in Africa, with the continuous improvement of Africa's status in Italy's foreign political and economic affairs and the heating up of Sino-Italian relations.

[34] Ilaria Maria Sala, *Italy's New Government Says It'll Fight Immigration by Helping China Invest in Africa*, https://qz.com/1366261/italy-aims-to-slow-immigration-by-helping-china-invest-in-africa/; Ambasciatore Ettore Francesco Sequi (La Stampa), *Lavorare con la Cina in Africa per ridurre i flussi di profughi*. https://www.esteri.it/mae/it/sala_stampa/interviste/2019/07/lavorare-con-la-cina-in-africa-per-ridurre-i-flussi-di-profughi-ambasciatore-ettore-francesco-sequi-la-stampa.html.

CHAPTER 6

Relations Between Portugal and Portuguese-Speaking African Countries: From the Perspectives of Economy and Culture

Kuai Peng Ip and Lixia Shen

Portugal was the first European colonial power on the African continent, with the earliest expansion overseas. At the end of the fifteenth century, the Portuguese invaded Equatorial Guinea on Central Africa's west coast, making them the oldest colonizing power. Since the sixteenth century, Portugal gradually became an important maritime power and established a huge colonial empire that encompassed Guinea-Bissau, Cape Verde, Angola, São Tomé, and Príncipe, as well as Mozambique. While these

This paper is translated by Tang Ge, Jia Zhixiu and Zhang Xu.

K. P. Ip (✉) · L. Shen
City University of Macau, Macau, China
e-mail: wxy03@ssap.cn

L. Shen
e-mail: wxy03@ssap.cn

X. Li and C. An (eds.), *The Changing World and Africa*,
https://doi.org/10.1007/978-981-16-4983-7_6

Portuguese-speaking African countries (Portuguese: Países Africanos de Língua Oficial Portuguesa, PALOP) gained their independence in 1975, due to its historical connection with its former colonies, Portugal has maintained unique relations with them. In the twenty-first century, after the global financial crisis and the European debt crisis, Portugal has attached increasing importance to its ties with PALOP countries in the arenas of politics, the economy, and socio-cultural development. Meanwhile, their economic and social engagement has increased, especially through the use of the Forum for Economic and Trade Cooperation between China and Portuguese-speaking Countries (Macao) (also known as Forum Macao) and the Community of Portuguese Language Countries (Portuguese: Comunidade dos Países de Língua Portuguesa, CPLP).

1 Portugal–PALOP Economic Relations Based on Two-Way Demand

Portugal has become more committed to enhancing its relations with the PALOP countries since the impacts of the global financial and debt crises hit, continuously expanding bilateral commodity trade and promoting economic exchanges. Compared with Portugal's trade with members of the European Union (EU), its trade with the PALOP countries is not that large; however, Portugal has indeed taken note of the significance of commodity trade with these former colonies amid the overall economic slump among European countries.

Import and Export of Commodities

Due to their geographic proximity, the EU countries—including Spain, Germany, France, Italy, and the Netherlands—are Portugal's most important import and export partners. However, Portugal's GDP shrank by 17% between 2008 and 2012, beginning with the global financial crisis of 2008, while the economies of Angola and Mozambique in Africa have witnessed robust growth, with the average growth rate of the two countries reaching 11% and 7.3%, respectively, from 2009 to 2013. Changes were observed in the trends of imports and exports between Portugal and Angola. Angola accounted for 7.4% of Portugal's total exports in 2009 and 3.4% in 2017, while it accounted for 4.1% of Portugal's total imports in 2009 and 9.8% in 2017. Nevertheless, Portugal remains Angola's main import partner, making up 11.8% of Angola's total imports from Portugal

(Table 2). In addition, Angola is also the main export destination of Portuguese products outside the EU, as well as Portugal's largest trading partner in Africa (Table 1), enjoying a constant improvement in its trading status.

According to Table 2, Portugal is one of the main exporters for a host of PALOP countries. It is the source of 54.5%, 39.5%, and 34.4% of the total imports of São Tomé and Príncipe, Cape Verde, and Guinea-Bissau, respectively. Although it only accounts for 11.8% of Angola's total imports, Angola is Portugal's largest importer among the PALOP countries in terms of the total volume. It is worth noting that Mozambique has witnessed medium–high rates of economic growth over the past decade and a gradual expansion of economic and trade cooperation with Portugal. According to the *Statistical Yearbook 2018*, published by Mozambique's National Institute of Statistics, 3.5% of its total imports (230 million USD) were from Portugal, its seventh largest source of imports.

As the purchasing power of PALOP continues to increase, Portugal should pay more attention to products that can be exported to these

Table 1 Statistics of Portugal's Imports from PALOP Countries in 2018 (Unit: thousand USD)

Country	*Portugal's Import from PALOP (1)*	*PALOP's Total Exports (2)*	*Portugal's Total Imports (3)*	*Proportion of each Country's Export to Portugal in its Total Exports (4) = (1)/(2)*	*Proportion of Portugal's Import from each Country in its Total Imports (5) = (1)/(3) (5) = (1)/(3)*
Angola	1,173,908	42,021,981	95,629,299	2.8%	1.2%
Cape Verde	18,166	75,271	95,629,299	24.1%	0.02%
Mozambique	49,328	5,160,904	95,629,299	1.0%	0.05%
Guinea Bissau	658	150,109	95,629,299	0.4%	–
São Tomé and Príncipe	815	12,087	95,629,299	6.7%	–
Equatorial Guinea	640,657	6,543,443	95,629,299	9.8%	0.7%

Source Authors' own calculation from TRADE MAP. https://www.trademap.org/Index.aspx

Table 2 Statistics of PALOP Countries' Imports from Portugal in 2018 (Unit: thousand USD)

Country	*Imports from Portugal (1)*	*PALOP's Total Imports (2)*	*Portugal's Total Exports (3)*	*Proportion of each Country's Import from Portugal in its Total Imports (4) = (1)/(2)*	*Proportion of Portugal's Export to each Country in its Total Exports (5) = (1)/(3)*
Angola	1,927,911	16,385,608	74,135,906	11.8%	2.6%
Cape Verde	321,589	814,605	74,135,906	39.5%	0.4%
Mozambique	236,260	6,819,155	74,135,906	3.5%	0.3%
Guinea Bissau	120,064	349,326	74,135,906	34.4%	0.2%
São Tomé and Príncipe	75,493	148,306	74,135,906	54.5%	0.1%
Equatorial Guinea	14,701	1,021,484	74,135,906	1.4%	0.02%

Source Authors' own calculation from TRADE MAP. https://www.trademap.org/Index.aspx

countries so as to alleviate its own current trade deficit. For example, Portugal can reduce its reliance on natural resources from the PALOP countries by producing export products made from resources needed by those countries. Angola is one of the world's top ten exporters of crude oil, for example, with a daily export of approximately 1.7 million barrels. However, it also needs to import a large amount of refined oil from overseas. The same is also true for Mozambique and most other PALOP countries. By contrast, crude oil is the largest import commodity for Portugal, while refined oil is its largest export commodity.

Therefore, bilateral trade ties should be promoted and diversified by docking Portugal's demand for the importation of natural resources from the PALOP countries with their demand for importing finished products from Portugal by implementing the following steps. First among these is to develop bilateral cooperative trade driven by oil. For Angola, oil accounted for 99.2% of its total exports to Portugal in 2010. This staggering figure highlights the importance of the oil industry to Portugal–Angola economic relations. Energy companies in Angola have already established partnerships with their counterparts in Portugal. Portugal also encourages more energy companies to partner with those in Angola in an effort to boost bilateral trade relations.

Second, Portugal should develop the bilateral trade of automobiles and auto parts. Auto parts, one of Portugal's main exports, are major imports for Angola and Mozambique. With its abundant mineral resources, Mozambique's largest export commodity and one of Angola's top five exports are aluminum and iron, respectively, both of which are essentials to the automobile manufacturing industry. Therefore, curtailing restrictions on the supply and demand between the countries can indeed increase bilateral trade in the field of automobiles and auto parts.

Third, develop bilateral trade in timber and paper. Portugal is hailed as the world's largest producer of cork products. Making pulp and paper is one of Portugal's central industries. However, due to domestic shortages, it needs to import a large amount of timber from other countries. In this case, the arable land of the PALOP countries is the most suitable source for raw materials.

Fourth, establish cooperation in tourism. Amid the boom of the global tourism industry, the superior geographical locations and pleasant climate of Portugal and the PALOP countries create perfect conditions to develop tourism bilaterally or even multilaterally.

The goods exported by Portugal exactly meet the import needs of the PALOP countries, and the natural resources of the latter also fully fit the demands of the former. If Portugal and the PALOP countries can provide specific resources based on their own developmental needs, it will not only promote bilateral economic exchanges and diversify traded commodities, but also strengthen bilateral friendship.

Two-Way Investment

As Portugal-PALOP trade cooperation continues to grow and deepen, two-way investment has become the highlight of their commercial collaboration. Portugal invested heavily in the PALOP countries before the global economic crisis of 2008, especially in the banking and construction industries of Angola. However, this investment pattern has gradually begun to change due to the rapid economic development of the PALOP countries after Portugal's economic crisis. Driven by the booming oil economy, Angola has been investing in other countries instead of relying on foreign direct investment (FDI) to support its own economy, which marks an investment turn from a net inflow to a net outflow. According

Table 3 Stock of FDI from PALOP Countries to Portugal (2009–2018) (Unit: million USD)

Country Year	*Angola*	*Guinea Bissau*	*Equatorial Guinea*	*Cape Verde*	*Mozambique*	*São Tomé and Príncipe*
2009	327.0	–	–	4.3	2.9	–
2010	809.7	4.0	–	14.7	6.7	–
2011	538.9	3.9	–	22.0	5.2	– 1.3
2012	1,378.6	5.3	–	23.7	2.6	–
2013	2,254.9	2.8	59.3	22.1	–11.0	5.5
2014	2,442.8	3.6	74.1	27.9	63.1	– 2.4
2015	2,280.9	4.4	–56.6	26.1	52.3	27.2
2016	1,541.1	4.2	–157.1	15.8	17.9	30.6
2017	2,374.7	4.8	–152.3	32.4	15.6	33.6
2018	2,504.0	4.6	–170.6	34.3	56.1	32.1

Source Authors' own calculation from OECD.Stat. https://stats.oecd.org

to Table 3, PALOP investment in Portugal grew out of nothing from 2009 to 2018, with an overall rise in the direct investment inflows from these countries, with the exception of Equatorial Guinea. As for Mozambique, the investment it receives has mushroomed since 2010, owning to its abundant natural and human resources. Portugal's stock of FDI inflows also increased slightly after the economic crisis.

Despite the rise in Angola's outbound investments, Portugal remains its most important investor. According to Table 4, Portugal's investment in Angola has climbed since 2009 and reached its peak in 2014. Regardless of the drop afterward, the annual investment from 2015 to 2018 was still higher than that of 2009. Portugal's investment in other African countries also grew during the same period amid some fluctuations. Statistics show that the Agency for Private Investment and Promotion of Exports of Angola (Portuguese: Agência de Investimento Privado e Promoção das Exportações de Angola, AIPEX) registered 155 investment projects altogether from June 26, 2018 to June 30, 2019, involving a total amount of 2.489 billion USD. China and Portugal were the two largest sources of FDI for Angola in the past 12 months, with Chinese companies investing 175 million USD in 15 projects and Portuguese companies 5.8 million USD in 12 projects, said Licínio Vaz Contreiras, Chairman of the APIEX Board of Directors.

Table 4 Stock of FDI from Portugal to PALOP Countries (2009–2018) (Unit: million US)

Country Year	*Angola*	*Guinea Bissau*	*Equatorial Guinea*	*Cape Verde*	*Mozambique*	*São Tomé and Príncipe*
2009	2,782.0	8.6	7.2	191.6	734.4	21.6
2010	3,991.2	13.4	42.8	326.0	583.9	25.4
2011	5,019.4	13.0	101.0	322.5	905.4	25.9
2012	5,800.86	17.2	228.2	353.6	1,000.0	30.3
2013	6,048.8	29.0	205.5	416.5	1,278.4	37.2
2014	6,769.8	15.8	151.8	433.4	1,545.5	49.8
2015	5,575.4	20.7	143.7	365.8	1,400.1	44.6
2016	4,428.2	21.1	53.7	357.3	1,100.5	52.7
2017	5,404.2	30.0	58.8	415.0	1,538.7	87.6
2018	3,442.9	21.8	61.8	379.0	1,514.81	24.0

Source Authors' own calculation from OECD.Statistics. https://stats.oecd.org

According to a statement issued by Portugal's Securities Market Commission (Portuguese: Comissão do Mercado de Valores Mobiliários, CMVM), Mozambique's Bank of Commercial Investment (Portuguese: Banco Comercial e de Investimentos, BCI) contributed 18.7 million euros to the Portuguese Investment Bank (Portuguese: Banco Português de Investimento, BPI) in 2019, 11.1% lower than the 21.1 million euros it invested in 2018. By contrast, the Angola Development Bank (Portuguese: Banco de Fomento Angola, BFA) saw its contribution to the BPI increasing by 8%, from 73.2 million euros in 2018 to a total of 78.9 million euros in 2019. The PALOP countries have begun to invest in Portugal and have benefitted from doing so. In addition, Portugal's deep-rooted ties with the PALOP countries give it many unique and exclusive advantages. Portuguese companies can participate in the fast-paced market economy through two-way investment than they could otherwise. Meanwhile, successful African companies can also be introduced to promote Portugal's economic development.

As for the methods used to boost foreign investment, Portugal has already established a package of incentives for FDI, but these are not designed solely for the PALOP countries. For example, the Golden Visa program, one of its major incentives, grants foreigner entities the right to apply for permanent residence in Portugal once they meet certain investment requirements. Most attractive to Chinese investors, this plan did not make a splash among the PALOP countries. If Portugal can introduce targeted policies for these African countries, it will attract more investment to stimulate its own economy while promoting bilateral economic relations. Moreover, it is necessary for Portugal to take measures to reduce tariffs in its trade with the PALOP companies, since the EU's common external tariff that it has adopted stands in the way of the PALOP countries being able to increase investment.

Cooperation in Infrastructure and the Banking Industry

Portugal accounts for a decreasing proportion of the PALOP countries' external trade volume and FDI, due to increasing competitiveness among countries. However, it can maintain its economic competitiveness by virtue of its unique relationship with these countries and, in doing so, strengthen cooperation in the fields of infrastructure, banking, and education so as to increase bilateral economic exchanges and enhance trade cooperation.

In the field of infrastructure, Portugal boasts more advanced transportation and information and communications technologies than the PALOP countries. It has deep water ports, four international railways, and many high-quality airports and roads. It is among the top ten EU countries in terms of the speed and cost of broadband service, with digital subscriber lines (DSL) accessible to 95% of the Portuguese population. In Angola and other PALOP countries, however, infrastructure like roads and buildings has largely been destroyed by long-term civil wars. As most of these countries depend on agriculture, a safer and faster system of commodity transportation will be instrumental to enhancing their trade competitiveness, which means a large amount of money is needed for infrastructure construction. For instance, Angola has spent 3.3 billion USD on railways alone since 2005, and has allocated 1.25 billion USD to repair the port of Lobito. Between 2013 and early 2015, Mozambique invested 17 billion USD in transportation projects, mainly to improve the transportation of natural resources. Therefore, increasing investment in constructing the infrastructure of the PALOP countries will not only bring about considerable returns for Portugal, but also promote the trade of natural resources between the two sides.

The banking sector also boasts huge potential for Portugal–PALOP cooperation. For example, Angola's banking industry is weak, despite the rapid development of its financial sector since the end of its civil war. The number of its banks increased from nine in 2003 to 23 in 2011, with total assets jumping from 3 billion USD to more than 57 billion USD. Yet the spike in bank assets is dwarfed by the absence of a stock market and limited access to financial services. Its banks also face uncertainties in providing loans to credit-worthy borrowers because of deficient infrastructure and data management. Similarly, banks in other PALOP countries are also confronted with many problems in both software and hardware, which calls for further cooperation in the establishment and management of the banking industry. Portuguese banks can accordingly design a number of programs to facilitate financial transactions and charge their PALOP counterparts to increase revenue. In this process, the banking industry in these countries can be better structured. In addition, such cooperation will enable the PALOP countries to enter the global capital market by virtue of Portugal's economic links.

Economic Aid

Aid plays a significant role in Portugal maintaining friendly relations with the PALOP countries. In terms of financial aid, Portugal intended to provide Guinea-Bissau with various forms of aid totaling about 40 million euros from 2015 to 2020, according to a five-year plan on bilateral cooperation signed by Portuguese Prime Minister Pedro Passos Coelho during his visit to Guinea-Bissau in July 2015. At the end of 2016, São Tomé and Príncipe signed a five-year cooperation agreement worth 57.5 million euros (62 million USD) with Portugal, its former suzerain, laying the foundation for its overall economic and financial development in the next few years. In 2017, Portugal provided 11.8 million USD in aid to Guinea-Bissau. After the fourth Portugal–Mozambique Summit in Lisbon in 2019, representatives of the two countries signed 13 agreements. According to one of them, Portugal would provide Mozambique with a credit line of 400 million euros, the repayment of which was to be March 2020. As for Angola, its foreign aid is mainly provided by the US, Japan, the Netherlands, Norway, Sweden, and Portugal.

In terms of cooperative agreements, Portugal has signed many deals with Angola, boosting bilateral economic and trade cooperation. To be specific, the foreign ministers of Angola and Portugal signed eleven cooperative agreements in Luanda on March 6, 2019, including the "Memorandum on the Training of Technical and Diplomatic Officers," the agreement between Portugal's Agency for Competitiveness and Innovation and Angola's Institute for Support to Micro, Small, and Medium Enterprises, the agreement between Angola's Institute of Asset Management and State Stakeholdings and its Portuguese counterpart Parpública, Participações Publicas de Portugal, etc. Following Mozambique and Cape Verde, Angola signed the "Lusophone Compact" with Portugal and the African Development Bank in order to promote its economy and sustainable development. Portugal also signed 13 agreements with Mozambique to strengthen cooperation.

2 Portugal–PALOP Social and Cultural Relations Based on Language and Culture

Two-Way Flow of People as a Prominent Feature

The historical connection between Portugal and the PALOP countries has underlain the characteristics of the flow of people, a connection that is particularly reflected in the areas of finance, labor, and society. Some Portuguese immigrate to the PALOP countries because of the same language, similar lifestyle, and similar foods. These immigrants aim to earn more money through hard work, which means economic independence for young people and raising families for the middle-aged. Therefore, most of them need access first and foremost to the labor market. Only in this way can they gain stable income, the same as for every immigrant around the world.

The two-way flow of people is a prominent feature of the interactions between Portugal and the PALOP countries. In 2008, Portugal witnessed a surging number of citizens immigrating to other countries for better opportunities after the global economic crisis broke out. As the economies of some PALOP countries were improving following years of development alongside a rising demand for labor, Angola and Mozambique became the first and second biggest African destinations for Portuguese immigrants. Statistics show that the number of Portuguese immigrants in Angola exceeded 33,300 in 2007 and rose to 113,000 in 2012, an increase of 240% in just five years. As of December 2019, there were approximately 200,000 Portuguese immigrants in Angola. Attracted by

Angola's robust economic growth, Portuguese immigrants are seeking more job opportunities in Angola.

At the same time, Portugal also accepts a large number of immigrants from the PALOP countries. There is a remarkable two-way flow of people between Portugal and these countries. Portugal believes that immigration plays an important role in the country's economic development by promoting growth and reducing the unemployment rate. Therefore, Portugal does not worry about or oppose foreign immigrants. According to Portugal's Immigration and Borders Service (Portuguese: Serviço de Estrangeiros e Fronteiras, SEF), Portugal witnessed an increasing number of applicants for Portuguese citizenship from 2014 to 2018. In 2018, the SEF registered 41,324 applications for Portuguese citizenship, a significant year-on-year increase of 10.9%. Among these applications, 11,586 were from Brazil, 4,259 from Cape Verde, 1,953 from Angola, and 1,550 from Guinea-Bissau. Most of them were from Portuguese-speaking countries.

In addition, Portugal and other Portuguese-speaking countries are working hard to achieve a completely free flow of people among the Portuguese-speaking countries. In 2019, although the CPLP member states did not reach an agreement on the "the maximum free movement of people," Portugal and other member states indicated that they would still promote the free movement of people among them, and hope to settle the issue at the CPLP conference to be held in 2020 in Luanda, capital of Angola.

Camões Institute's Promotion of Educational Cooperation and Development

The Camões Institute is an official Portuguese institution for the promotion of the Portuguese language and culture. It promotes language and culture in many PALOP countries and other countries around the world through Portuguese language majors and lecturers in universities, so as to strengthen bilateral cooperation and exchanges in education, science, technology, culture, sports, youth programs, and mass media. In order to vigorously promote education in the Portuguese language, the Camões Institute has established Portuguese language centers in the higher polytechnic institutes (Portuguese: Instituto Superior Politécnico) in Angola and Guinea-Bissau, and in the Maputo University in Mozambique, helping them develop education and promote the Portuguese language. It also forged partnerships with local universities. In addition, the Camões Institute has sent language teachers to four private Portuguese language schools in Angola's capital, Luanda, and in Guinea-Bissau. What's more, about 200 young Portuguese language teachers were assigned to Cape Verde, Guinea-Bissau, and São Tomé and Príncipe.

Portugal Attaches Great Importance to Improving Human Resources in PALOP

The PALOP countries need to improve their education levels to be able to promote their development, while Portugal is able to strengthen cooperation in this field through its existing advantages. The PALOP countries generally lack skilled labor forces, making it hard for both local and foreign companies. According to the PwC's 15th Annual Global CEO Survey, one-third of CEOs said that a lack of well-educated workers would have a negative impact on their business. As part of the dialogue between Angola and the EU, the *Joint Way Forward* focuses on addressing global issues of common concern and outlines the importance of improving education and skills of the Angolan people to achieve sustainable economic growth. Strengthening Portugal–PALOP educational cooperation is not only important to Portugal's political agenda, but also conducive to the development of all countries involved.

3 Portugal–PALOP Development Cooperation Under Multilateral Frameworks

Forum Macao Promotes Multilateral Exchanges

Portugal has actively participated in the Forum for Economic and Trade Cooperation between China and Portuguese-speaking Countries (Macao) (also known as Forum Macao) to increase friendly exchanges with other Portuguese-speaking countries. Forum Macao was established in October 2003 in Macao, China, with the participation of eight countries, including Portugal, five PALOP countries, Brazil, and Timor-Leste. Equatorial Guinea may become the ninth member. Since its establishment, Forum Macao has successfully held five ministerial meetings in Macao in October 2003, September 2006, November 2010, November 2013, and October 2016, with a plan of actions passed at each meeting aimed at strengthening economic and trade exchanges between China and Portuguese-speaking countries.

Seizing the development opportunities offered by Forum Macao, Portugal has taken an active part in the forum activities. José Duarte, the then Assistant Minister of the Portuguese Prime Minister, attended the first ministerial meeting in October 2003. Mário Lino, the then Portuguese Minister of Public Works, Transportation and Communications, attended the second meeting in September 2006. José Sócrates, the then Portuguese Prime Minister, attended the third meeting in November 2010. Paulo Portas, the then Portuguese Deputy Prime Minister, attended the fourth meeting in November 2013. António Costa, Portuguese Prime Minister, attended the fifth meeting in October 2016. Through this platform, Portugal has not only accelerated economic and trade exchanges with the PALOP countries, but has also promoted multilateral diplomacy.

Through their cooperation with China, the PALOP and other member states of the Forum have been expanding and deepening their relationships. The first ministerial meeting only witnessed seven fields for cooperation, while the number increased to 18 at the fifth meeting, with extensive and in-depth cooperation in government affairs, investment, trade, agriculture, forestry, animal husbandry, fishery, enterprises, provinces and cities, finance, human resources, natural resources, production capacity, infrastructure, tourism, transportation, communications, culture, broadcasting, sports, public health, marine affairs, other traditional and emerging fields.

Making full use of Forum Macao, Portugal has strengthened its ties with the Chinese mainland and Macao, promoting and spreading the Portuguese language together with the PALOP countries. Many colleges and universities in Macao, such as the Institute for Research on Portuguese-Speaking Countries at the City University of Macau, the Department of Portuguese at the University of Macau, and the annual summer training courses for Portuguese language teachers held by the Macao Polytechnic Institute, offer a Portuguese major or courses. What's more, the number of universities in Chinese mainland offering Portuguese majors and courses has now increased from ten to more than forty. In addition, the number of students enrolled in Portuguese-related majors in Macao rose sharply from 759 in the academic year of 2014–2015 to 1,490 in the academic year of 2018–2019, an increase of nearly 97%. Based on the platform, Portugal and the PALOP countries have trained a large number of bilingual talents for the Chinese mainland and Macao. At the same time, they have also promoted the Portuguese language and improved its international recognition.

CPLP Accelerates Cooperation and Development

Due to Portugal's historical colonization of African countries, it has continued its historical, political, economic, and cultural relations with Guinea-Bissau, Cape Verde, Equatorial Guinea, Angola, São Tomé and Príncipe, and Mozambique today. Initiated by Portugal and Brazil, the CPLP was formally established on July 17, 1996, with its headquarters in Lisbon. Its member states include Portugal, Angola, Mozambique, Cape Verde, Guinea-Bissau, São Tomé and Príncipe, Brazil, and Timor-Leste, with the accession of Equatorial Guinea in 2014 as the ninth member state. The CPLP aims to deepen mutual friendship and cooperation among the member states and establish a closer, more practical, and more comprehensive relationship linked by the Portuguese language.

Portugal regards the CPLP as one of the three pillars of its diplomacy and attaches great importance to its development. It has continued expanding cooperative efforts with Portuguese-speaking countries during its rotating presidency of the CPLP from July 2008 to July 2010. In July 2010, Portuguese President Aníbal Cavaco Silva paid a state visit to Angola and attended the 8th CPLP Summit, which adopted the *Luanda Declaration* and the *Brasilia Plan of Actions for the Promotion, Dissemination and Projection of the Portuguese Language*. In February 2012, the new headquarters of the CPLP was inaugurated in Lisbon. In July, Portuguese President Aníbal Cavaco Silva, Prime Minister Pedro Passos Coelho, and Minister of State and Foreign Affairs Paulo Portas attended the CPLP Summit held in Maputo, capital of Mozambique.

In terms of international collaboration, the CPLP serves as an important bridge for Portugal and the PALOP countries to be able to expand international diplomacy and develop relations with other countries and regions. The CPLP, as indicated in its own name, has signed nearly 30 cooperation agreements, conventions, memorandums of understanding, resolutions, or protocols with other international organizations, indicating its member states' will to speak with one voice in international governance and cooperation in related fields. Besides, Portugal, the PALOP countries, and other members have been recognized by the UN and its subsidiary agencies, international organizations, and regional organizations.

In terms of economic, trade, and financial cooperation, Portugal has reached a number of substantive cooperative agreements with other CPLP member states. According to data quoted by financial news agency Bloomberg, from the United Nations Conference on Trade and Development, Angola, Mozambique, and São Tomé and Príncipe are among the African countries that have attracted the most foreign investment in 2013. Among them, Mozambique was one of the Portuguese-speaking countries that received the most foreign investment, with a per capita foreign investment of 248 USD. In 2013, José Eduardo Carvalho, President of the Association of Portuguese Industries, stated that Angola had become Portugal's second largest foreign trade market. Every year, about 8,000 companies in Portugal export products to Angola, with a total export volume of up to 3 billion euros. Angola is also a large investor in Portugal. According to "Africa Intelligence," Angola's state oil company Sonangol was once a major shareholder of Millennium BCP (Portuguese Commercial Bank). As of December 31, 2018, it still owned 19.49% of the latter.

In terms of diversified cooperation, Portugal's trade cooperation with other CPLP member states encompasses many fields. In 2012, the CPLP launched a series of plans aimed at creating better cooperation conditions for the member countries by focusing on cooperation in knowledge (research and development), emerging technologies, agricultural and rural development, infrastructure construction, ocean and natural resources, energy, and tourism, in order to promote the further growth of trade and investment. In July 2014, the tenth CPLP Summit held in Dili, capital of Timor-Leste, adopted the *Dili Declaration*, which confirmed the political and diplomatic coordination of the member states in the energy field. Up till now, Portugal's cooperation with other member states has covered fields like education, public health, science, technology, national defense, agriculture, public administration, communications, justice, public safety, culture, sports, and media.

In terms of language promotion, Portugal and other CPLP member states have jointly promoted the use of Portuguese and enhanced the soft power of its international status. The fifth CPLP Summit, held in São Tomé and Príncipe in July 2004, emphasized that member states should coordinate their positions on major international issues, strengthen their cooperation in defense, poverty eradication, AIDS, and malaria prevention, and make efforts to standardize the Portuguese language. At the seventh CPLP Summit, held in July 2008 in Lisbon, capital of Portugal, with the theme of "The Portuguese Language: A Common Heritage, A Global Future," the leaders of the member states jointly signed the *Lisbon Declaration* and expressed the necessity of improving the value of Portuguese and popularizing it internationally. Through their efforts, the Portuguese language is developing rapidly, with a large number of speakers in South America, Europe, Africa, and Asia. Portuguese is now the sixth most widely spoken language in the world.

4 Reflections on and Prospects for Portugal–PALOP Relations

First, political exchange is the guarantor of a relationship. Since the independence of the PALOP countries in 1975, diplomatic relations between Portugal and these nations have changed from a state of "hot north and cold south" to a collective warming, enjoying sound development. In the 1970s and 1980s, Portugal had relatively tense relations with Mozambique and Angola while maintaining relatively stable relations with Guinea-Bissau, Cape Verde, Equatorial Guinea, and São Tomé and Príncipe in northern Africa. Since the 1990s, and especially in the twenty-first century, Portugal's economy has been sluggish due to domestic instability and the stagnation of the eurozone. Given its historical connections and extant pressures, the Portuguese government changed its diplomatic strategy, re-examined its relations with the PALOP countries, and increased its diplomatic contact with them. As a result, they have enjoyed a much more active bilateral political interaction and a higher degree of political warmth. Such a change is not only due to the potential need for and inevitable result of their own development, but also a natural and inevitable trend in changes to the international environment. Sound political relations have provided a solid guarantee for the close economic and social relations between Portugal and the PALOP countries.

Second, their shared language promotes communication. The Portuguese language is the cultural foundation that connects Portugal and the PALOP countries. Throughout the history of Portugal's 500-year colonization, culture and education have had the biggest influence on these countries, with Portuguese still being their official language. Sharing the same language and a similar culture has removed the major obstacles hindering their communication and cooperation.

Third, international organizations promote further cooperation. Both Portugal and the PALOP countries are members of the Forum Macao and the CPLP. These international arrangements guarantee the cooperation and interests of their member states and enhance their international status. In addition, Portugal can skillfully use its platform in Macao to expand exchanges with China in the fields of science, technology, education, and culture, strengthening bilateral cooperation in the financial field, and maintaining friendly bilateral relations. Therefore, participation in the same international organizations is also a catalyst for a more long-term and stable development of Portugal's relations with the PALOP and other countries.

Fourth, a complete two-way flow of people remains to be realized. Portugal supports a two-way flow of people between itself and the PALOP countries. A high-level two-way flow of people is important for Portugal and these countries to be able to solve unemployment and develop their economies.

Fifth, strengthening cooperation in key areas is the direction these countries must take. It is Portugal's future goal to increase trade, two-way investment, and assistance with Angola and Mozambique. In terms of the economic strength of the PALOP countries, Angola and Mozambique are two large economies experiencing rapid economic growth. Portugal is expected to increase its trade cooperation with these two countries. At the same time, as the economic strength of these two countries continues to grow, characteristics of two-way investment and assistance will become more obvious. However, to what are considered economically backward PALOP countries, as a major country, Portugal is supposed to offer major assistance to help them overcome difficulties, so as to enhance their bilateral friendship. In conclusion, in the face of economic recession and vicissitudes in the international situation, it will be a significant path forward for Portugal to maintain friendly exchanges with the PALOP countries in its foreign policy.

CHAPTER 7

Japan's Diplomacy Toward Africa: From Pragmatistic Balance to Strategic Importance

Yongpeng Zhang

Beginning in the early twentieth century and up through the twenty-first century, Japan's foreign policy toward Africa has evolved from an "economy-oriented" pragmatic balance to one of strategic emphasis. "Pragmatism" is used in this article only in its general sense, which is different from that in the classical philosophical tradition of pragmatism. It refers to the policy or concept that focuses on short-term interests and prioritizes actual benefits, similar to the philosophical concept of "utilitarianism." "Pragmatic balance" in this article refers to Japan's principle of obtaining political and economic interests practically in a certain period as the central priority of its diplomacy with Africa, under the condition of

This article was originally published in Chinese in the fifth issue of *West Asia and Africa* in 2018 and was revised when included into this collection. This paper is translated by Liu Shuang, Cao Daqin, and Wang Weigang.

Y. Zhang (✉)
Institute of West-Asian and African Studies, Chinese Academy of Social Sciences, Beijing, China
e-mail: wxy03@ssap.cn

X. Li and C. An (eds.), *The Changing World and Africa*,
https://doi.org/10.1007/978-981-16-4983-7_7

lacking or ignoring major strategic principles. Due to the constraints of its own national power and those imposed by the United States (US), Japan has lacked an independent diplomatic strategy since the early twentieth century, during the Cold War and toward the early 1990s. Japan often tried to maintain a diplomatic balance between "black Africa" and "white Africa" to gain economic interests through technical adaptations, avoiding the strategic constraints of major powers and ideological confrontation in Africa. Therefore, Japan's foreign policy toward Africa during the Cold War was characterized by pragmatism centered on short-term interests, utilizing a strategy of balanced diplomacy. The end of the Cold War enabled Japan to get rid of, to some extent, the constraints of ideological confrontation between major powers, and to pursue its own strategic interests and goals. Besides, the birth of democracy in South Africa in the early 1990s terminated the confrontation between black Africa and white Africa. These put an end to Japan's pragmatic and balanced foreign policy toward Africa. With the opening of the Tokyo International Conference on African Development (TICAD) in 1993 and the establishment of a Japan–Africa cooperation mechanism, Japan has entered a new era of strategic emphasis in terms of its foreign policy toward Africa. Since the beginning of the twenty-first century, Japan has constantly strengthened its strategic emphasis on Africa under the framework of the TICAD, which can be seen in the implementation of its strategic goals as well as its political, economic, security, and geostrategic planning, corresponding with notable strategic competition with China in Africa.

1 The Historical Basis of Japan's Foreign Policy of Pragmatic Balance Toward Africa

The relationship between Japan and Africa in modern times can be theoretically traced back to the end of the nineteenth century and the beginning of the twentieth century. During the Meiji Restoration in the second half of the nineteenth century, Japan carried out profound, comprehensive reforms in politics, the economy, society, the military, and other fields. At the same time, the Japanese elite's unprecedentedly vigorous pursuit of overseas territories and interests led the country into a new era of external aggression and expansion. Japan's political elite then set their eyes on Africa, which was then being carved up by various Western colonizers. At that time, only Liberia and Ethiopia were relatively independent, declaring their independence in 1847 and 1896, respectively.

From the end of the Russo-Japanese War (1904–1905) to the end of World War I (WWI), Japan primarily developed its relations with countries or regions such as South Africa, Egypt, and Ethiopia. WWI greatly promoted the rapid development of Japan's capitalist industry and economy. "Because European products were withdrawing from the Asian market at this time, a large number of military orders and demand for daily necessities caused a rapid increase in the total exports of Japanese products," stimulating Japan's demand for overseas strategic resources. At that time, the main motivation for and goal of Japan's foreign policy toward Africa was to obtain the mineral resources and overseas markets needed by its industry and to lay a foundation for its diplomatic and international development outside Asia. Rich in resources, South Africa served as the bridgehead for Japan to expand its trade and markets in the early years of its expansion in Africa.

In 1926, Japan's Osaka Shosen Kaisha (OSK), a shipping company, started a commercial maritime route from Japan's Kobe to South Africa's Durban via Kenya's Mombasa, which marked the formal establishment of Japan–South Africa economic relations. The trade volume between Japan and South Africa was 521,000 yen in 1913 and rose to 38.46 million yen in 1941, an increase of more than 700%. During that period, bilateral trade volume reached its peak at 143 million yen in 1937, when the two countries formally established diplomatic relations.

In addition to South Africa, Japan also held political and economic ties directly or indirectly with a few other African regions during the aforementioned period. In 1919, South West Africa (now Namibia) was declared a League of Nations (LON) mandate territory, with South Africa being responsible for its administration. Japan was one of the five major countries that formulated relevant LON agreements. In this sense, safeguarding the interests it achieved in Asia and the Pacific by cooperating with the Western powers to carve up Africa was the core of Japan's policy on South West Africa and still stands as part of modern Japan's foreign policy toward Africa. In addition, Japan developed relations with Ethiopia in the late 1920s and early 1930s, and established the legation in Addis Ababa in 1936.

Due to the fact that Africa was carved up by Western powers and reduced to colonies, the relations between Japan and Africa before World War II (WWII), or even in modern times, have been relatively simple.

During WWII, Japan's contact with Africa was mainly aimed at the war against the United Kingdom (UK), and Japan's foreign policy toward Africa focused more on increasing economic and trade benefits. Nevertheless, as a trading partner and overseas market during Japan's early expansion in Africa, South Africa was the initial base for Japan to practice its foreign policy toward Africa. From the end of WWII to the early 1990s, South Africa, Zimbabwe, Zambia, Malawi, Namibia, and other regions in Africa were all still UK colonies or governed by white UK people, though beginning to declare their independence one after another. South Africa, in particular, remained as a fortress of the white regime. Afterward, other black African countries rich in natural resources became independent successively, and differences in ideology and institution emerged between the white African and black African countries. The former nations were Japan's traditional partners and ideological allies in Africa, while the latter were resource-rich countries or regions with ideological differences or opposition to the white regime. This posed a dilemma for Japan, which needed natural resources to develop its economy, yet was under the strategic containment of major powers in the post-war period. Therefore, a pragmatic foreign policy that prioritized interests became the main option for Japan to take toward Africa, and the best way to achieve this goal was to maintain a diplomatic balance between the black and white African countries.

2 Japan's Foreign Policy of Pragmatic Balance Toward Africa During the Cold War

Japan's foreign policy toward Africa from the post-war era to the 1950s was the basis of its foreign policy in modern times. With the gradual formation of the Eastern and Western Blocs, as well as the forthcoming Cold War, Japan naturally followed the Western Bloc led by the US in diplomacy. The international environment during that period determined the basic principles of Japan's foreign policy toward Africa in the early post-war period: adhering to an anti-socialist and anti-communist position, as well as expanding its foreign trade and overseas markets. In fact, the diplomatic practice throughout the Cold War showed that Japan's foreign policy toward Africa placed more emphasis on economic interests compared with the political confrontation that characterized the Cold War, and as such was consistent with Japan's national policy prioritizing the economy—or what is known as an "Economic Emphasis."

The Formation and Development of Japan's Foreign Policy of Pragmatic Balance Toward Africa

The first Asian-African Conference (Bandung Conference) in 1955 provided a platform for Japan to have its first official contact with African countries at the international, multilateral level after WWII, marking an official start to post-war Japan–Africa diplomacy. In order to prevent the so-called "red" threats and possible proposals that "attack the West" at the meeting, the US suggested that its allies, Japan and the Philippines in particular, participate in the Asian-African Conference. The final result of the Conference, however, didn't meet Japan's expectations. Nevertheless, this initial contact between post-war Japan and Africa provided certain experience and served as a reference for the formulation of its subsequent foreign policy toward Africa.

The Japan–South Africa relationship has always been an important part of Japan–Africa relations after the war. The traditional relations between Japan and South Africa were re-established after the war through the "Treaty of Peace with Japan" at the San Francisco Peace Conference held in 1951. South Africa, present at the conference, gave its support to the US and Japan. In 1952, Japan set up a consulate general in Pretoria, South Africa, and then a consulate in Rhodesia (now Zimbabwe). In addition, the trade relationship between Japan and South Africa was restored as early as 1948, when the National Party won the election and began to implement apartheid in South Africa—a situation that would last for more than 40 years.

Since only a few black African countries became independent in the 1950s, Japan began to restore economic ties with the Eastern and Western African countries that established connections with Japan before WWII. For example, Japan had particularly active diplomatic contacts with Ghana and Ethiopia, establishing its embassy in Ghana in 1959. In 1957, Emperor Haile Selassie of Ethiopia paid a visit to Japan, making him the first state guest to be received by the post-war Japanese government. Besides political contacts, Japan also developed its economic and trade ties with the black African countries. During the second half of the 1950s, some black African countries or regions became important export markets for Japanese textiles.

The period from the 1960s to the mid-1970s witnessed a further intensification of the Cold War between the East and the West, as well as the rapid development of the Japanese economy. Since the black African countries generally opposed and condemned the white regime in South Africa, Japan faced a dilemma in its foreign policy toward Africa. On the one hand, as the white regime followed the Western Bloc during the Cold War, Japan had to maintain close ties with South Africa due to the strategic struggle with the Eastern Bloc being led by the Union of Soviet Socialist Republics (USSR). On the other hand, the rapid development of the Japanese economy required natural resources. Such demand stood higher in the face of the 1973 oil crisis, which forced Japan to value the resources of black Africa. As South Africa and black African countries both boasted strategic resources desired by Japan, it could afford to offend neither, but maintained a balance between the two. Therefore, Japan's policy toward white Africa during this period was to expand ties with white regimes and strengthen its economic support for them to prevent them from diplomatic and social isolation. Japan's support reaped rewards from South Africa. In April 1961, the white South African regime officially declared that Japan and the Japanese in South Africa could enjoy the privileges of the white population, which provided favorable conditions for Japan–South Africa exchanges in various fields such as the economy, culture, and education. In the same year, Japan and South Africa sent an ambassador to each other, and Japan listed South Africa as a country with diplomatic relations.

As for the economy, despite the continuous calls by the Organization of African Unity (OAU) and the United Nations (UN) for economic sanctions against South Africa, the Japan–South Africa trade relationship was still developing rapidly. South Africa ranked third in Japan's export partners in 1975, following the UK and the European Economic Community. Based on the pragmatism of its "Economic Emphasis," Japan began to attach greater importance to and substantively develop its relations with the black African countries from the 1960s to the mid-1970s while continuing to develop political and economic ties with South Africa. At the end of the 1950s, Japan only had embassies in Ethiopia and Ghana. However, from 1960 to 1981, it established embassies in 13 black African countries.

During the same period, Japan's foreign policy of striking a pragmatic balance toward Africa was also reflected in its attitude and practice toward the decolonization movement taking place there. After WWII, it was obviously difficult for US-occupied Japan to detach itself from the US in terms of policy and attitudes toward African decolonization in the 1950s. At that time, Japan opposed the process of decolonization in Africa and the realization of national self-determination through revolutionary means. In the late 1950s, with the relative reduction in US diplomatic restrictions on Japan, Japan was able to state its position on the African decolonization movement. In 1957, the Japanese Foreign Minister expressed Japan's respect for the African National Liberation Movement at the UN General Assembly. However, Japan took a cautious attitude or tried to balance the conflicts with regards to decolonization or colonization faced by Africa. In 1966, when the UN Security Council voted on the proposal submitted by Mali, Nigeria, and Uganda to deploy all means to end the white minority regime in Southern Rhodesia, Japan, together with the US and the UK, abstained from voting.

Japan's relevant policy and stand changed in the 1970s as the momentum of international anti-colonialism and anti-racial discrimination increased. For example, in the late 1970s Japan voted in the UN General Assembly for Resolution 33/38 B, which condemned the white regime in South Africa, indicating that Japan had begun to lean toward the African countries in terms of policy. However, it was undoubtedly difficult for Japan to truly separate itself from the white regime in South Africa in practice. Japanese scholars believe that Japan pursued a "dual-track" foreign policy toward Africa at that time; that is, while supporting the white regime in South Africa, Japan tried to quell the dissatisfaction of other African countries and at the same time developed relations with both black and white Africa. By supporting the South African government, Japan was able to obtain the natural resources needed for economic development; and by balancing relations with the black African countries, Japan could silence the criticism of Japan–South Africa relations.

Japan's Two Adjustments to Its Balanced Foreign Policy Toward Africa: Economic Goals of Pragmatism

From the 1960s to the mid-1970s, Japan had two important adjustments to its foreign policy toward Africa. The first was the adjustment from a politics-oriented diplomacy to an economy-oriented diplomacy in the early 1960s. At that time, Japanese Prime Minister Hayato Ikeda proposed a ten-year "Income Doubling Plan" as the domestic economy developed rapidly, which established new goals for Japan's economic development. Under the principle of this economy-oriented diplomacy, Japan's foreign policy toward Africa focused completely on utilitarian and practical economic goals.

The second adjustment was seen during the period around the first oil crisis that occurred in the first half of the 1970s. The specific measures included strengthening mining cooperation with African countries. In February 1970, Japan sent a delegation of entrepreneurs headed by the vice chairman of Mitsubishi Heavy Industries to discuss further cooperation in terms of the resources of nine African countries, including Ethiopia, Kenya, and Tanzania. By co-founding companies with Zairian companies, Japanese companies became deeply involved in copper mining and product transportation to Japan. The measures also included increasing development assistance to Africa. In 1969, Japan provided 340 million USD of Official Development Assistance (ODA) to black Africa, a share of only 1.1% in its total foreign ODA. In 1975, it rose sharply to 6.9% of its total ODA, about 850 million USD. In the early and mid-1970s, three factors led to a further adjustment of Japan's foreign policy toward Africa. First, although the Japanese economy came to the end of its period of rapid growth, the growth rate remained at a high level of more than 7%, which continued to fuel its demand for African resources. Second, the 1973 oil crisis caused Japan to realize how fragile its ability to acquire resources was, and it had to reconsider diversifying sources of resources and energy. Third, an increasing number of African countries adopted the socialist system from the late 1960s to the early 1970s, and movements of anti-colonial rule and anti-racial discrimination in Rhodesia, Namibia, and South Africa were in full swing, which struck Japan and troubled its diplomacy in Africa.

Under the global context of ideological confrontation during the Cold War, Japan's active adjustment of its policy toward the black African countries featured pragmatism by prioritizing actual effects.

The Continuation and Conclusion of Japan's Foreign Policy of Pragmatic Balance Toward Africa from the Late 1970s to the Early 1990s

From the late 1970s to the late 1980s, as the USSR gradually caught up with the US, global competition between the two superpowers became increasingly fierce. During this period, the USSR signed the "Treaty of Friendship and Cooperation" with Ethiopia, intervened actively in the Angolan civil war, and became involved in the wars between Somalia and Ethiopia. Its aggressive struggle for hegemony put the US under great pressure. As Japan became industrialized in the 1970s, the US intended to take advantage of Japan to resist the USSR threats in Africa.

In line with the US strategy in Africa, Japan began to reinforce its support for the white regimes in Southern Africa that were against communism. Emphasizing both pragmatism and balance, Japan appeased the black African countries in five aspects while continuing to support South Africa, all with the purpose of maintaining its political and economic interests: showing sympathy and support for the liberation movement of the African countries, strengthening high-level exchanges with them, increasing development assistance to them, publicly announcing sanctions against South Africa, and starting dialogues with the African National Congress (ANC) of South Africa.

In the 1990s, both Africa and the world at large witnessed tremendous changes following the end of the Cold War and the ideological confrontation between the Eastern Bloc and the Western Bloc. As for Africa, as the target of much condemnation and criticism, the South African white regime finally announced the abolition of apartheid in 1990. In 1991, ANC leader Nelson Mandela signed a peace agreement with President Frederik Willem de Klerk, and in April 1994 witnessed the birth of the new South Africa. Hence, Japan's foreign policy toward Africa was free from the restrictions of the Cold War. With the end of the divisions and contradictions between white Africa and black Africa, there was no need for Japan to maintain a balance of interests between the two. Japan's foreign policy of pragmatic balance toward Africa was put to an end.

3 TICAD: The Start of Japan's Post-War Strategic Emphasis on Africa and Its Manifestation

The 1990s was a decade marking rapid progress in economic globalization and a turn to economic development in countries around the world. The African countries also adopted a number of measures to develop their economies and become involved in globalization at the same time that regional organizations in Africa accelerated the process of regional economic integration. Meanwhile, the OAU was committed to establishing the African Economic Community (AEC) and an African Common Market, and resumed the relevant agenda for African political, economic, and social development. As a result, the organizational and social foundations gradually came into being for the establishment of the African Union. The efforts of the African countries to strive for economic development provided the world with a huge potential market. More importantly, the new landscape of international forces after the Cold War provided new opportunities for the African countries to play a role in the international arena. These factors, coupled with its rich natural resources, made Africa central to the strategies of major countries.

As for Japan, the end of the Cold War also offered Japan new opportunities for diplomacy. It began to have greater ambitions again, with its main strategic goals of becoming a political power and a permanent member of the UN Security Council. Exploring Japan's diplomatic strategies—as well as realizing the significance of Africa's vitality to Japan's political and economic development—Japan held the first TICAD in 1993, which marked the start of the institutionalized and comprehensive development of its diplomacy with Africa.

TICAD: Opening and Deepening of Institutionalized and Comprehensive Strategic Cooperation with Africa

Most Japanese scholars of African studies deny that Japan had diplomatic strategies for Africa, believing that it has always ignored Africa. In fact, as we review the past today, it is clear that Japan has been gradually strengthening its strategic leverage over Africa, which is demonstrated by the continuous development of the TICAD scale, the upgrading of cooperation, the proposal of the "Asia-Africa Growth Corridor" (AAGC), and Japan's active deployment of the geostrategic concept of the "Indo-Pacific." All of these developments should be credited to the convening of the first TICAD and its institutionalization.

The greatest significance of the first TICAD at the policy level was that regular meetings have institutionalized and normalized Japan–Africa relations, making Japan's cooperation with Africa a long-term strategy. This was the biggest change in Japan's foreign policy toward Africa in the early 1990s. According to the speech delivered by then Japanese Prime Minister Morihiro Hosokawa at the 1993 TICAD, Japan's future foreign policy toward Africa lay in four parts. First, Japan would continue to support the political and economic reforms of the African countries and play a greater role in Africa based on its cooperation with the UN and other relevant organizations. Second, Japan would increase its development assistance to Africa by providing 70 to 75 billion USD in aid to Africa within five years, starting in 1993. Third, Japan would actively support Africa in human resource development, strengthen exchanges among the young people between the two sides, and pay attention to environmental issues in Africa. Fourth, Japan hoped to establish "long-term and reliable friendly relations" with the African countries beyond those of "donor" and "recipients." The first TICAD served as a platform dedicated to topics on Japan-Africa cooperation, with leaders or representatives from 48 African countries, twelve European donor countries, and eight UN organizations. It held broad international significance in terms of such coverage, far beyond the scale of Japan–Africa relations during the Cold War period, with its political influence bearing obvious strategic importance across the globe.

The following TICADs continuously deepened and expanded Japan's cooperation with Africa. The sixth TICAD, held in Nairobi, capital of Kenya, in August 2016 unveiled new features. First, it was held on the largest-scale ever. It is said that more than 11,000 people, including entrepreneurs, participated in the conference. Besides representatives from Japan, the conference was attended by heads of state and government from 54 African countries, as well as representatives from 52 partner countries and 74 international and regional organizations, such as the UN Development Program, the World Bank, and the African Union. Second, Japan's cooperation with Africa was broadened and intensified. The meeting identified a wide range of areas of cooperation, which could be grouped into the following three aspects of "boosting African economic diversification and industrialization, promoting public health, as well as advancing social stability and peacekeeping." It covered political, economic, cultural, and security topics such as agriculture, the blue economy, human resource development, government governance, finance, anti-terrorism, anti-extremism, global issues, climate change, and food safety. In his opening speech at the sixth Conference, Japanese Prime Minister Shinzo Abe spoke of the UN reform, saying that Japan supported Africa obtaining a permanent seat on the Security Council by 2023 at the latest. He also emphasized that the reformation of the UN Security Council was the common goal of both Japan and Africa and that the two sides should join efforts to this end. The plan of action released after the conference specifically mentioned maritime security, international law, and the reformation of the UN Security Council.

With the theme of "Advancing Africa's Development through People, Technology and Innovation," the seventh Conference, held in Japan's Yokohama in August 2019, was comparable in scale and topics to the previous one. With three major cooperative areas comprising the economy, society, and peace and stability, it focused on the following agenda: economic transformation and improvements in the business environment and institutions through private investment and innovation; the promotion of a resilient and sustainable society for human security; peace and stability (including support for Africa's own proactive efforts). Prime Minister Abe announced at the conference that Japan would step up its cooperation with Africa in three years in terms of human resource (including the implementation of the "ABE Initiative 3.0"), promotion of Universal Health Coverage in Africa, the New Approach for Peace and Stability in Africa, and increased private investment in Africa. The topics and results of the conference showed that Japan's strategic cooperation with Africa was further advanced, especially in maritime security, UN reform, active confirmation of the "Free and Open Indo-Pacific Strategy," and other strategic issues.

An Overall Deepening of Economic and Trade Cooperation with Africa

Japan's economic cooperation with Africa lies mainly in three interconnected parts: trade, investment, and assistance, which constitute the economic and political foundation of Japan's cooperation with Africa.

First, Japan–African trade has continued to develop yet has seen relatively modest growth. Instead of a constant, steady rate of growth, Japan's trade with Africa witnessed ups and downs from 2007 to 2016. Japan's total trade volume with Africa was 26.37 billion USD in 2007, and remained above 30 billion USD from 2011 to 2013. In 2018, it was 17.13 billion USD, a year-on-year increase of 23.4%, with Japan's export to Africa of 8.15 billion USD and import of 8.97 billion USD. South Africa accounted for a significant proportion of Japan's trade with Africa. From 2007 to 2018, Japan's trade with South Africa basically drew abreast of its trade with Africa. South Africa accounted for more than 30% of its trade with Africa in each year during this period, and even came close to or exceeded 50% in many years.

Second, Japan's investment in Africa has fluctuated significantly in recent years. Its investment in Africa fluctuated distinctly from 2007 to 2018, a period during which South Africa was Japan's main investment destination. With an investment of 1.52 billion USD in 2008, it showed negative growth for two consecutive years, from 2009 to 2010, being significantly affected by the financial crisis. In 2016, it slumped to −431 million USD. However, in terms of stock, Japan's investment in Africa has shown an overall upward trend, from 3.895 billion USD in 2007 to 99.92 billion in 2016, or an increase of 15.7%. During the same period, the stock of Japanese investment in South Africa increased from 852 million USD in 2007 to 8.208 billion in 2016, increasing 86.3%. In 2018, Japan's direct investment in Africa reached 1.575 billion USD, among which its investment in South Africa was 1.253 billion USD. Meanwhile, the stock of Japan's investment in Africa was 8.776 billion USD, among which that in South Africa reached 6.937 billion USD. This demonstrates a seriously unbalanced distribution of Japan's investment in Africa.

Third, Japan's assistance to Africa has continued to develop over the years. Japan has expanded its assistance to Africa in every conceivable way. Japan's assistance to Africa could be dated back to 1954. Since the 1990s—and especially the beginning of the twenty-first century—economic assistance has increasingly become an important tool for Japan to seek its political and economic interests in Africa, corresponding with the improvement of Africa's strategic status and Japan's need to become a political power and explore emerging markets and resources. Consequently, Japan increased its assistance to Africa gradually. Its assistance to Africa was 90 million USD per year from 2003 to 2007, increasing sharply to 1.75 billion USD in 2008. Despite the international financial crisis, the number reached 1.68 billion USD in 2009. The total assistance, announced at the fifth TICAD in 2013, soared to 32 billion USD in that same year, greatly exceeding the volume of the fourth session. Most of its assistance has been provided in the form of aid *gratis*, followed by technical assistance, concessional loans, and grants to the African Development Bank (ADB). It was announced at the sixth TICAD in 2016 that Japan would provide 30 billion USD in aid to Africa over the next three years, with an average of 10 billion per year. In 2016, the total amount of assistance projects implemented by the Japan International Cooperation Agency (JICA) in 48 African countries was 115.696 billion yen (approximately 1.154 billion USD at that time). Japan announced at the seventh TICAD in 2019 that it would put forth every effort to surpass its private investment of 20 billion USD, and support Africa's development in the three main areas of the economy, social development, and peace and security.

It is worth noting that Japan has made important progress in supporting Africa in cross-regional infrastructure construction. It has participated in such projects in many African countries, including the

Mediterranean coastal road across the Maghreb region, as well as nine trans-African highways such as Cairo-Dakar, Algiers-Lagos, Lagos-Mombasa, Tripoli-Windhoek-Cape Town, and N'Djamena-Djibouti. From 2016 to 2018, the total amount of infrastructure projects implemented by the JICA was 125.136 billion yen (approximately 1.101 billion USD as of August 2018), including ODA loans of 120 billion yen and grants of 30 billion yen. Japan ranks third in terms of support to Africa in infrastructure construction, following China and the European Union.

Increased Participation in African Security Affairs

Japan has participated in African security affairs at three levels: UN peacekeeping operations, bilateral cooperation projects, and strategic planning. As for UN peacekeeping, Japan was a member of the United Nations Mission for the Referendum in Western Sahara as early as 1991. Beginning in the twenty-first century, the Japan Self-Defense Forces (JSDF) have participated in UN peacekeeping operations mainly by sending engineers or coordinators to the UN Mission in Liberia (2003), the UN Operation in Côte d'Ivoire (2004), UN–African Union Mission in Darfur (2007), the UN Organization Stabilization Mission in the Democratic Republic of the Congo (2010), the UN Multidimensional Integrated Stabilization Mission in Mali (2013), etc. In March 2017, the Japanese cabinet decided to extend the mission of JSDF engineers in South Sudan for another year. However, among all the UN peacekeeping operations in Africa, only two Japanese individuals were in the UN Mission in South Sudan by the end of 2017.

At the bilateral level, Japan's security cooperation with Africa has mainly been carried out under the framework of the TICAD. After the Cold War, Japan put forward the concept of "human security," an outward security outlook. In August 2003, Japan revised its ODA program to take "human security" as a basic policy as well as a framework for its assistance to Africa. From January 2013 to March 2016, Japan provided 16.28 million USD through the UN Development Program to support twelve African countries in establishing peacekeeping operation training centers. During the same period, Japan sent experts 15 times to participate in peace and stability operations aimed at eliminating the roots of instability, covering regions such as the Sahel, Great Lakes, Sudan, South Sudan, the Horn of Africa, North Africa, etc. In addition, Japan also provided financial support to the African Union for peace-building projects initiated by the African Union and African regional organizations. At the seventh TICAD in 2019 with "peace and security" as one of the three main themes, Japan intended to support the establishment of conflict mediation and arbitration mechanisms for the African continent, participate in UN peacekeeping in Africa, and train 60,000 judicial officers, police, and security personnel for the African countries.

As for strategic planning, military security cooperation between Japan and Djibouti was an important factor. Since February 2009, Japan has continued to send destroyers and anti-submarine patrol aircrafts to participate in international anti-piracy operations in the Gulf of Aden, and has provided assistance in the form of technology and patrol vessel to the African Union and the Djibouti Coast Guard under the framework of maritime security cooperation. In August 2016, Japanese Defense Minister Tomomi Inada, who was visiting Djibouti, stated that Japan was considering expanding the "function" of its military base in Djibouti. In October, the spokesperson of the Japanese Foreign Ministry confirmed that Japan intended to further expand the area of the Japanese military base in Djibouti to the east of the base, and was considering deploying C-130 transport aircraft and more military personnel at the base.

Japan's impact on African security involves strategic aspects such as military deployment, peacekeeping, training, financial support, and regional cooperation, and covers most countries and regions in Africa. Japan's security strategy in Africa has four goals: to protect Japan's growing economic interests and investments in Africa, to respond to China's presence and development in Africa, to add to Japan's profile as an "active pacifist," and to include Africa in Japan's strategic framework.

Strategic Competition with China in Africa

Japan's strategic competition with China in Africa is a strategic extension of its response to China's rise in East Asia, the East China Sea, and the South China Sea. It is also a reflection of China–Japan relations and China–Japan strategic competition at a broader geopolitical level. Japan's competition with China in the field of diplomacy with Africa lies primarily in its cooperation platform, public opinion image, and strategy.

Competition for Cooperative Platforms with Africa from the Third to the Seventh TICAD

Japan's foreign policy toward Africa has gradually shown signs of competition with China, a characteristic of Japan's diplomacy with Africa since the beginning of the twenty-first century. First, Japan's foreign policy toward Africa has been covering more areas, making Japan and China competitors in terms of foreign policy toward Africa. For example, the third and fourth Forum on China–Africa Cooperation (FOCAC) held in 2006 and 2009, respectively, proposed "eight measures" for cooperation with Africa. China–Africa cooperation covered almost all areas that could be involved at that time, including agriculture, environmental protection, infrastructure construction, investment, trade, debt reduction, climate change, science, technology, medical care, education, people-to-people exchanges, and cultural exchanges. In comparison, the "Yokohama Action Plan" of the fourth TICAD in 2008 also covered aspects such as economic growth, realization of human security, sound governance, environmental and climate change, infrastructure construction, security, trade, investment, tourism, agriculture, and the realization of the UN Millennium Development Goals. What's more, the fifth, sixth, and seventh TICADs placed more emphasis on Africa's needs and cooperation in infrastructure, trade, and investment. Second, Japan's competition with China was also reflected in the arrangement of the TICAD mechanism. With previous TICADs all held in Japan every five years, the sixth and following conferences were held every three years in an African country and in Japan in turn, which is exactly the same as the FOCAC.

Competition in Public Opinion Image

Since 2010, Japanese leaders have deliberately downplayed China's influence and continued to belittle the quality of Chinese products and its national image in collaboration with Japanese media, in order to compete with China for political and economic interests and for a better national image in Africa. For example, after Chinese Foreign Minister Wang Yi began his visit to Africa on January 6, 2014, Shinzo Abe began his visits to Côte d'Ivoire, Mozambique, and Ethiopia on the 9th of that same month, arousing speculations about his intention to downplay China's influence. At the same time, Japanese media also deliberately reinforced the reports on the confrontation between China and Japan. TV TOKYO said Japan "will fight against China who has already invested in Africa first." Experts invited by Japan's Fuji TV claimed on the program that, "we must make Africa understand that unlike China, Japan's assistance is for the future development of Africa while China's ultimate goal is for their untapped resources." At the sixth TICAD in 2016, Japanese Prime Minister Shinzo Abe deliberately emphasized the quality of Japanese products in his opening speech, highlighting the spirit of Japanese companies to pursue quality, Japan's high-quality workers, and the importance of Japanese product quality to the economic and social development of Africa. It was easy to tell he was implying that the quality of Chinese products and the quality of workers were not as good as those of Japan. The documents of the sixth TICAD mentioned multiple times that Japan would help Africa build "high-quality infrastructure," which was a clear attack on the infrastructure China had already built in Africa.

Geostrategic Competition

Shinzo Abe was elected Prime Minister in 2012, when the US was engaged in an important stage of the Asia–Pacific Rebalancing strategy. The US repeatedly encouraged Japan to strengthen its own defense capabilities in order to act as a pawn for the US in its efforts to contain China. Its strategy coincided with Abe's own strategic intention to deny Japan's aggression in WWII, confront China's rise, and become a true political power. Therefore, Abe proposed a series of strategies, including "overlooking the globe" diplomacy, framing the concept of an expanded "New Asia" together with the US, Australia, and India, and actively implementing the so-called "Free and Open Indo-Pacific strategy" that strategically considered the Indian Ocean and the Pacific Ocean as a whole.

The so-called "Indo-Pacific Strategy" was derived from the concept the "Indo-Pacific," which originally referred to the geographic region from the Indian Ocean to the Western Pacific Ocean and the central Pacific Ocean. However, since 2011, it has been used to refer to geopolitical regions covering the coast of East Africa, West Asia, the Indian Ocean, South Asia, the Western Pacific Ocean, and East Asia. To increase the number of its allies and contain China, Japan has been particularly active in promoting the concept the "Indo-Pacific." At the ministerial meeting of the seventh TICAD, held in June 2018, Japan's Foreign Minister, Taro Kono, described the link between the Indo-Pacific Strategy and Africa as one intended "to connect Africa all the way to the coast of North and South America through Indian Ocean and Pacific Ocean." The Indo-Pacific Strategy was confirmed in the TICAD "Yokohama Declaration" in

August 2019. This was the first time that the Indo-Pacific Strategy had been confirmed in the official documents of the Japan–Africa Summit.

The core of Japan's so-called Free and Open Indo-Pacific Strategy is to contain China, in view of the strategic competition between promoters of the Indo-Pacific Strategy and China, as well as the considerable overlap between the coverage of this strategy and that of the 21st Century Maritime Silk Road of China's "Belt and Road" Initiative (BRI).

In addition, another vision of Japan forcing Africa into its strategic framework was that represented by the so-called "Asia-Africa Growth Corridor" (AAGC). During Indian Prime Minister Narendra Modi's visit to Japan in 2016, Japan and India mentioned the AAGC in their joint statement. At the 52nd AfDB Annual Meeting, held in India in May 2017, India officially announced the launch of the AAGC. Indian officials said that the AAGC would provide financial support for the development of Africa and would work to connect Africa and Asia and help increase the connection between the people of the two continents. Japan and India also discussed their cooperation with Africa. According to a vision document on AAGC issued by think tanks of India, Japan, and Indonesia, the establishment of the AAGC was based on the fact that both India and Japan are highly competent in their cooperative efforts with Africa, especially with the TICAD initiated by Japan and the Indo-Africa Forum Summit (IAFS) launched by India; the "Special Strategic and Global Partnership" between India and Japan adds further value to this vision. According to the document, the AAGC framework mainly covers four aspects of Japan–India–Africa cooperation, namely, enhancing capacities and skills, quality infrastructure and institutional connectivity, development and cooperation projects, and people-to-people partnership. The AAGC additionally deliberates on these four aspects: the existing mechanisms for cooperation between Asia and Africa (i.e., the existing Japan–Africa and India–Africa cooperation platforms), the broad-based agenda for the synchronized growth of Asia–Africa aimed at sustainable and innovative development, the establishment of optimum linkages and cooperation among the subregions of Asia and Africa, and the establishment of an industrial corridor and industrial network. The Japan Bank for International Cooperation (JBIC) decided to support projects of Japanese and Indian companies in Africa as a form of financial follow-up support.

The AAGC is a new framework for cooperation with Africa proposed by Japan and India based on their existing cooperation mechanisms with Africa. Its essence is to include Africa into a larger strategic framework, and to work in concert with the Indo-Pacific Strategy of Japan, the US, and India. Its ultimate goal is to form a complete geostrategic cooperative regional belt, with the Indo-Pacific Strategy as the political framework and the AAGC as the economic cooperation framework. As the economic framework of the Indo-Pacific Strategy, the AAGC lacks actual actions and projects, but it has been exploring methods and ideas for further cooperation, many of which are similar to China's BRI. Neither India nor Japan participates in the BRI, and they regard it as a threat. With a common competitor (or rival) in mind, Japan and India have come together and to launch their own strategy of cooperation with Africa. It is self-evident that there will be confrontation and competition between the Indo-Pacific Strategy, the AAGC, and China's BRI.

4 Conclusion

Generally speaking, Japan's diplomacy with Africa has developed from a balance of interests during the Cold War to a stance characterized by a more strategic emphasis since the latter half of the 1990s. Its diplomacy with Africa during the Cold War lay in two parts. On the one hand, it firmly followed the Western Bloc led by the US and its ideology and developed its relations with traditional white African partners represented by South Africa. On the other hand, it actively developed its relations with black African countries or regions in order to have access to resources and economic interests, successfully realizing its economic take-off. Due to the eventual containment of US and Western strategies, Japan's diplomacy with Africa during the Cold War focused on short-term utility, and was theoretically characterized by utilitarianism and pragmatism.

As part of the new international environment, Japan started to place strategic emphasis on Africa as part of its diplomacy beginning with the first TICAD in 1993. This is reflected in its cooperation platforms with Africa, bilateral economic and trade relations, cooperation in security, and its guarded and strategic competition with China. Japan's strategic emphasis on Africa stems from its realistic understanding of the importance of Africa. Japan's *Diplomatic Bluebook 2012* clearly states the importance of Sub-Saharan Africa to Japan. First, it is conducive for Japan to earn the trust of the international community by helping Africa tackle its own problems. Second, Japan could gain economic benefits from Africa's potentially huge market, abundant natural resources, positive economic growth, and growing population. Third, cooperation with the African countries is an important part of the UN Security Council reform. The *Diplomatic Bluebook 2017* greatly increased the content about Sub-Saharan Africa to an unprecedented coverage of 13 pages, and specifically mentioned Abe's proposal of a "Free and Open Indo-Pacific Strategy," a strategic vision for improving the connectivity between Asia and Africa. This indicates that the "Free and Open Indo-Pacific Strategy" has officially become a foreign policy expression of Japan. In fact, there is only one ultimate goal for Japan to place strategic emphasis on Africa: to make Japan a political power or a true world power commensurate with its economic strength. There are five basic methods or approaches to achieving this goal. First, to take an active part in the containment of or confrontation with China, so as to eliminate or reduce the influence of China's rise on Japan as much as possible. Second, to actively respond to the US strategy of containing China, so as to increase Japan's power and influence. Third, to establish a larger-scale geostrategic cooperation framework, so as to improve Japan's international status and influence. Fourth, to win the support of African countries, so as to obtain a permanent seat on the UN Security Council. And fifth, to enjoy stable gains from Africa's natural resources and potential market, so as to maintain a sustained growth of the Japanese economy. These are the main components and characteristics of Japan's diplomacy today.

CHAPTER 8

The Status Quo and Trends in the Development of Relations Between ROK and Africa

Rui Guo and Lichang Liang

With the introduction and implementation of its "Medium-Power" strategy, the Republic of Korea's achievements in strengthening economic diplomacy, expanding foreign trade, and obtaining energy resources have drawn much attention. It is well known that the African region is widely regarded as being simultaneously rich in resources and the largest untapped market in the world today. It is not only a long-term concern for large Western countries and emerging countries, but also highly valued

The original version of this chapter was revised: Chapter author "Rui Guo's" affiliation has been updated. The correction to this chapter is available at https://doi.org/10.1007/978-981-16-4983-7_23

R. Guo (✉)
School of Public Administration, Jilin University, Changchun, China
e-mail: wxy03@ssap.cn

L. Liang
Department of Politics, Huaibei Normal University, Huaibei, China
e-mail: wxy03@ssap.cn

X. Li and C. An (eds.), *The Changing World and Africa*,
https://doi.org/10.1007/978-981-16-4983-7_8

by powers like ROK. As a global strong trading power and a major energy importer, the status of diplomacy with Africa in ROK's overall diplomacy has been rising and its function has been strengthened in recent years. Against this background, the development of ROK–Africa relations has rapidly entered a new stage and reached new heights. ROK is committed to initiating new levels of cooperation with Africa.

1 The History and Status Quo of the Relationship Between ROK and Africa

Since the beginning of the twenty-first century, ROK and Africa's energy-based economic and trade relations have accelerated toward a new situation of all-round cooperation under the impetus of successive South Korean governments. After more than 20 years of exploration, ROK and Africa have achieved several important cooperative plans in many priority areas, such as political diplomacy, economic and trade cooperation, and regional governance, which is connected with ROK's various practical needs. Since the 1960s, ROK has paid attention to the development of relations with Africa.

Political and Diplomatic Relations

In 1961, under the leadership of President Park Chung-hee, ROK established diplomatic relations with Niger, Benin, Cameroon, Chad, and Côte d'Ivoire. According to the statistics, ROK established relations with 24 countries in Africa in the 1960s, opening the prelude to the development of ROK–Africa relations.[1] By 2006, ROK had established relations with 53 countries in Africa.[2] In 2014, it set up 23 permanent embassies and consulates in Africa. According to the 2019 ROK's Diplomacy White Paper, currently there are 18 African embassies stationed in ROK.

After the end of the Cold War, with the emergence and strengthening of the concept of non-pragmatic diplomacy, ROK and Africa's political and diplomatic relations have consistently developed based on close economic cooperative relations. However, this does not mean that

[1] Ministry of Foreign Affairs, Republic of Korea: 60 years of ROK's Diplomacy (2009, p. 101).

[2] African and Middle Eastern Affairs Bureau: *African Handbook* (2014, p. 11).

ROK pays more attention to the development of relations with Africa. During this period, due to the transformation of ROK's policy toward Africa, the diplomatic situation became chaotic, and ROK's diplomatic emphasis on Africa declined. Since the new century, the political dialogue and diplomatic interaction between ROK and Africa have become more frequent. President Roh Moo-hyun's visit to Africa in 2006 was regarded as a significant shift in ROK's diplomacy with Africa. The concept of an "emphasis on Africa" was extended to a certain degree during the periods of Li Myung-bak and Park Geun-hye. Up to May 2016, four of ROK's presidents have visited African countries.

In 2008, in order to strengthen ROK's energy diplomacy with Africa, the Li Myung-bak government sent staff from the reorganized and streamlined Ministry of Foreign Affairs to several key energy countries in Africa and set up new embassies in Cameroon and the Democratic Republic of the Congo. Meanwhile, business with Africa taking place through the African and Middle Eastern Affairs Bureau of the Ministry of Foreign Affairs was downsized to the only African section.[3] For the purpose of strengthening diplomacy with Africa and achieving the multilateralization of diplomacy, the Moon Jae-in government reorganized the diplomatic department of Africa in January 2020.[4] Since then, focus has been on expanding the Africa section of the Africa and Middle East Bureau of the Ministry of Foreign Affairs to two departments, increase staffing levels, and enrich business functions, aiming to enhance substantive cooperation with Africa from various fields such as economics, culture, and regional development.

With the help of multilateral international conferences and bilateral dialogue and cooperation mechanisms, ROK has continuously increased its diplomacy with Africa in recent years. Not only has ROK attached great importance to developing relations with the African Union (AU) for the long term, it is also committed to strengthening relations with different individual countries in Africa. In considering ROK's diplomatic architecture with Africa, South Africa, the five countries of North Africa (Egypt,

[3] "韓 외교 '아프리카' 강화...외교부, 11년만에 담당과 확대", https://news.mt.co.kr/mtview.php?no=2020010511237630480, January 30, 2019.

[4] *ROK's government adjusts foreign officials to strengthen energy diplomacy*, [ROK], Korea JoongAng Daily, June 27, 2008.

Morocco, Libya, Algeria, and Tunisia),[5] West African countries (Senegal, Ghana, Côte d'Ivoire), East African countries (Ethiopia, Kenya, Mozambique), and Central African countries (Rwanda, Democratic Republic of the Congo, Zimbabwe, Cameroon) all occupy important positions and have close relations with each other.

Economic and Trade Cooperation Relations

Since the statistics were released in 1988, ROK-Africa trade volume has shown an increasing trend. Compared with the Cold War period, ROK–Africa trade volume has increased rapidly and significantly since the end of the Cold War. After President Roh Moo-hyun's visit to Africa in 2006, the trade volume between ROK and Africa greatly increased again. Under the efforts of the two governments of Li Myung-bak and Park Geun-hye, ROK's imports and exports to Africa maintained an apparent growth trend. In 2011, ROK–Africa trade volume broke its previous historical record, reaching a total value of 20 billion US dollars. In that year alone, ROK established the largest trade surplus with Africa, reaching 8.8 billion US dollars.[6] After 2016, ROK–Africa trade volume began to decline. ROK's exports to Africa have slowed down significantly, and its imports from Africa have also shown negative growth.

From the country distribution,[7] ROK's exports to Sub-Saharan African countries fell by 15% compared with the previous year, with the total amount reduced to 6.4 billion dollars. Among them, ROK's exports to Nigeria, Angola, and Ghana fell the most, at 56.7, 46.7, and 9.8%, respectively. ROK's exports to seven countries including Tanzania (55.5%), Senegal (51.9%), Kenya (16.7%), and South Africa (5.5%) continued to show strong growth momentum. Among them, ROK's exports to Tanzania's computers and electronic accessories, and Senegal's ships, exports of vehicle accessories, and consumer goods to South Africa have increased greatly, which has driven this trend.

[5] Since North African countries belong to the Middle East in ROK'S overall diplomacy, the relevant data are generally not included in ROK's diplomacy with Africa.

[6] Korea International Trade Association: "the statistics of ROK's trade", http://stat.kita.net/stat/kts/rel/RelColligationList.screen, January 31, 2020.

[7] Korea International Trade Association: "the statistics of ROK's trade", http://stat.kita.net/stat/kts/rel/RelColligationList.screen, January 31, 2020.

From the perspective of export commodities,[8] among the ten largest export commodities, excluding ships (−52.1%) and petroleum products (-1.3%), the export value of eight items including machinery, chemicals, and consumer goods all increased. In terms of reductions, ROK's ship exports to Africa, which accounted for 53% of its total exports (4.7 billion US dollars) in 2016, fell sharply from 40% in 2017 (3 billion US dollars) to 22% in 2018 (1.4 billion US dollars). ROK's exports of non-petroleum products greatly increased by 33% in 2017, but in 2018 there was a slight decrease of − 1.3%. From the perspective of growth, ROK's automobile (4.6%), synthetic resin (2.6%), other petrochemicals (27.1%), computers (481.3%), precision chemical raw materials (21.7%), dry battery and storage battery (10.3%), and other export items maintained their growth momentum, averaging 77%.

It is widely known that ROK is completely dependent on imports to sustain its energy resources like oil and natural gas. According to the statistics, 82.4% of ROK's 2017 oil imports and 45% of its natural gas imports came from oil-producing countries in the Middle East.[9] To ensure a stable supply of energy and maintain national energy security, ROK has been working hard to promote the diversification of energy imports. In 2016, ROK imported crude oil from seven African countries, including Equatorial Guinea, Gabon, Angola, Congo (DRC), The Republic of Congo, Nigeria, and Chad, whose imports value accounted for 2.5% of the total ROK crude oil imports that year, amounting to 1.12 US billion dollars. In 2017, ROK imported crude oil from four African countries, including Equatorial Guinea, Algeria, Gabon, and Cameroon, with its imports value accounting for 1.6% of ROK's total crude oil imports that year, amounting to 970 million US dollars, or a decline of 13.4% year-on-year.

From the perspective of investments, ROK's investment projects in African countries continue to increase in the fields of energy resources, logistics, and infrastructure construction. In 2006, ROK officially started the process of direct investment in Africa. Since 2007, ROK's companies have maintained more than 200 million US dollars in investments in Africa per year in energy resource development and infrastructure

[8] Korea International Trade Association: "the statistics of ROK's trade", http://stat.kita.net/stat/kts/rel/RelColligationList.screen, January 31, 2020.

[9] Export–Import Bank of Korea: *ROK's Handbook* (2018, p. 32).

construction. By the end of the first half of 2018, ROK's foreign direct investment in Africa (FDI) totaled 3.3 billion US dollars, but it accounted for less than 1% of ROK's FDI during the same period. From the perspective of investment fields, ROK's foreign direct investment in Africa is concentrated in manufacturing, construction, wholesale, and retail, as well as mining industries, accounting for 68.8% of its total foreign direct investment in Africa. Direct investment in non-manufacturing industries, specifically, has continued to increase yearly. From the perspective of investment distribution, by the first half of 2018, ROK's investment in Madagascar totaled 1.8 billion US dollars. In addition, ROK's investment in South Africa, Nigeria, Sudan, Ghana, and other African countries continues to increase, and its investment in Equatorial Guinea is at a relatively high level.[10]

Assistance to Africa and Regional Governance

In order to promote African policy and deepen relations with Africa, participation in African regional governance has become an important part of ROK's diplomacy to demonstrate its commitment to public welfare. Since 2006, ROK has continued to increase its assistance to Africa, and it has continuously expanded the fields and channels of its assistance to the continent. According to the statistics of the Export–Import Bank of Korea (KEXIM), ROK's total aid to Africa was only 402.4 million US dollars in 2005, exceeding the 100 million US dollars in 2008 and reaching 107.1 million US dollars, exceeding the 200 million US dollars in 2012 and reaching 260 million US dollars, exceeding the 300 million US dollars in 2014 and reaching 340 million US dollars in 2014, exceeding the 400 million US dollars in 2017 and reaching 420 million US dollars.[11] ROK's assistance to Africa has increased rapidly every year. During the 12 years of statistics collected here, its total annual amount of assistance to Africa has increased nearly 10 times over.

From 2006 to 2017, the ROK's total grant assistance to Africa reached 1.48 billion US dollars, benign loans amounted to 840 million US

[10] Export–Import Bank of Korea: "*Statistics of investment in Africa 2006–2018*", https://stats.koreaexim.go.kr/, February 1, 2020.

[11] Export–Import Bank of Korea: "*Statistics of investment in Africa 2006–2018*", https://stats.koreaexim.go.kr/odastats.html, February 1, 2020.

dollars, and loans relief was 13 million US dollars. From the perspective of regional distribution, ROK's total caliber assistance to Sub-Saharan African countries reached 2.46 billion US dollars, accounting for 90.3% of its total caliber assistance to Africa, with aid to other African regions totaling more than 200 million US dollars. From the 13 categories of ROK's aid to Africa, eight categories exceed 100 million US dollars, including logistics (1.04 billion US dollars), health care (750 million US dollars), education (650 million US dollars), public administration and civil society (410 million US dollars), agriculture (410 million US dollars), drinking water and sanitation (380 million US dollars), multi-sector (160 million US dollars), and communications (140 million US dollars).[12]

In eliminating humanitarian disasters and resolving hunger problems in Africa, ROK reached an agreement with the World Food Program (WFP) in February 2018 to invest at least 46 billion Korean won to provide approximately 50,000 tons of rice to Africa per year. Prior to February 2017, the ROK formally joined the "Food Assistance Convention" (FAC) to remove policy obstacles for launching food assistance programs to Africa. In May 2019, the ROK government provided 50,000 tons of rice assistance to the Middle East and African regions for the first time through WFP. Among them, Ethiopia received 16,000 tons, Kenya received 10,000 tons, and Uganda received 0.5 million tons.[13]This is the largest food aid provided to WFP by ROK in 1996 after having changed their status from "recipient country" to the status of "donor country."

For the purpose of maintaining stability in the African region and undertaking further international responsibilities, the ROK has also actively participated in the United Nations peacekeeping operations in Africa. In July 1993, the ROK participated in the UN peacekeeping operation for the first time. At that time, it dispatched 250 engineering troops and 30 ceasefire observers to participate in the UN peacekeeping operation in Somalia, which opened the prelude of ROK's participation in the UN peacekeeping operation after the end of the Cold War.[14] By the end

[12] Export–Import Bank of Korea: *"Statistics of investment in Africa 2006–2018"*, https://stats.koreaexim.go.kr/odastats.html, February 1, 2020.

[13] *ROK has provided 50,000 tons of rice to Yemen and other four countries through the World Food Program for the second time*, Yonhap News Agency, May 10, 2019.

[14] Zhao Jianming: An Analysis of ROK's UN Peacekeeping Operations after the Cold War, International Forum, 2010(1), p. 21.

of 2013, the ROK had participated in 13 UN peacekeeping operations in eight different countries in the African region.[15] Since 2010, ROK had followed Western great powers to actively participate in the Libyan reconstruction process. ROK consistently supports UN Security Council resolutions to resolve the Libya issue and seeks a political solution led by the Libyan people based on the Geneva Declaration (2012), hoping to provide assistance to Libya in vocational education, health care, and economic development. In order to participate in the Libyan power plant project, ROK dispatched official and private delegations to conduct field visits three times in 2017. According to the statistics, from 2012 to 2018, ROK provided 63.45 million US dollars in humanitarian assistance to the Libya reconstruction project.[16]

In order to establish long-term, stable dialogue and cooperation with the African Union (AU), the ROK has continued to increase its diplomacy with the African Union in recent years. In April 2005, ROK officially obtained the necessary AU Observer State qualification. From 2004 to 2009, ROK provided continuous financial support to the AU based on specific matters, totaling 1.2 million US dollars. In June 2009, ROK and the African Union entered into a partnership, becoming the fifth country after China, Japan, Turkey, and India to establish a long-term cooperative relationship with all countries on the African continent.[17] In December 2013, ROK formally established the "Korea-African Union Cooperation Fund." From 2013 to 2015, ROK assisted the African Union with 500,000 US dollars each year, and in 2016, it provided support to the African Union with 1 million US dollars each year.[18] According to the plan laid out in the "ROK–Africa Comprehensive Blueprint for Strengthening Comprehensive Cooperation" (2016) announced by President Park Geun-hye during her visit to Africa, ROK provided a peace fund ranging from 2 million US dollars to 4 million US dollars per year from 2016 to 2018.

[15] nhn?docId = 929,183&cid = 43,667&categoryId = 43,667, February 1, 2020.

[16] Ministry of Foreign Affairs, Republic of Korea: Diplomatic White Paper (2019, p. 136).

[17] [Korea] Rui Rongjun, Xu Chengyu: "Looking for opportunities in the last engine of Africa", [ROK], Korea JoongAng Daily, November 24, 2009.

[18] Ministry of Foreign Affairs, Republic of Korea: An Introduction to the African Union (2018, p. 56).

2 ROK's Policy Objectives and Major Initiatives Toward Africa

In order to promote long-term policies toward Africa and build a closer relationship between the ROK and Africa, the ROK has introduced a series of policy measures and is committed to exploring new ideas and practices. This not only reflects the distinctive characteristics of ROK's policy toward Africa and its unique aspirations as a "medium power," but also shows that ROK's diplomacy with Africa is becoming more mature and flexible.

Strengthening the Construction of a Pivot Country

With the recovery of ROK's economy and the expansion of its foreign trade, ROK once again intensified its economic diplomacy in order to ensure the supply security of overseas energy resources during the presidencies of Roh Moo-hyun and Lee Myung-bak. Subsequently, during the period of Park Geun-hye and Moon Jae-in, ROK has maintained a strong sense of momentum in its diplomacy toward Africa. Coordinating with the comprehensive development of its relations with Africa, ROK has given priority to some specific African countries, namely, African pivot countries.

By strengthening its energy diplomacy with Africa, ROK's Africa policy reflects the distinctive characteristics of the "strong economy." Under the guidance of the economic pragmatism approach to Africa, the ROK has gradually strengthened the construction and overall layout of African pivot countries. And according to the principle of "centralized selection," Kenya, Nigeria, the Democratic Republic of the Congo, Ghana, and South Africa have all become important pivot countries for ROK's layout in Africa, while South Africa, Angola, Nigeria, Libya, and Algeria have been prioritized for developing and strengthening the ROK's energy diplomacy with Africa.[19] Since the Roh Moo-hyun period, South Korea has re-established or newly established embassies in Angola (2007), Cameroon (2008), Democratic Congo (2008), Uganda

[19] 이호영:《한국의 대 아프리카 외교의 동학과 대응방안》,《국제관계연구》, 2011, 제 16 권, 제 2 호, p. 156.

(2011), Equatorial Guinea (2012), Mozambique (2013), and Madagascar (2016).[20]

According to the expansion of cooperative fields and changes in the foreign policy, the criteria used by ROK to select the pivot countries for African policy are not stable. Judging from the size of embassy staff in Africa, South Africa, Kenya, Uganda, and Nigeria are undoubtedly ROK's most significant partners in Africa. However, taking the particularity of African regional organization diplomacy into consideration, Ethiopia is also an important pivot country for ROK's African policy. On the whole, these countries are not only important destinations frequently visited by Korean leaders and government officials, but also the most important trade and investment objectives of Korean enterprises. It should be noted that from Park Geun-hye's presidency on, the ROK's diplomatic demands toward Africa have become increasingly diversified. During the period of Moon Jae-in, ROK further defined its Africa policy with some countries as pivots, and intended to promote a more comprehensive partnership between ROK and Africa.

Sharing the Development Pattern of ROK

Providing free aid and technical assistance is a consistent way for ROK to promote the development of ROK–African relations. After the end of the Cold War, ROK not only provided financial assistance to Africa through the Economic Development Cooperation Fund (EDCF), but also tried to make African countries copy the development pattern of ROK. In 2004, Roh Moo-hyun's government launched the Economic Development Experience Sharing Project (KSP) to help developing countries. The project is a government project based on the ROK's own experience growing into a global trade power, which it has subsequently put forward as a series of policy suggestions to developing countries that are tantamount to guiding and instructing other countries to copy the development pattern of the ROK itself.[21]

[20] 김용빈, 안광석, 김원녕:《우리 정부의 대아프리카 외교 인프라 현황과 문제점 및 강화 방안 》, 개발마케팅연구소최종보고서, 2017.10, p. 9.

[21] [Han] Xu Jingxu:" Do you know the economic development experience sharing project?" [ROK], Korea JoongAng Daily, September 3, 2010.

From the period of Lee Myung-bak on, the ROK paid more attention to spreading and expanding the experience of the "Saemaeul Movement" to Africa. In 2009, ROK and the United Nations (UN) jointly launched the "South Korean Millennium Saemaeul" project,[22] which aimed to explore the development pattern of backward countries based on the "Saemaeul Movement," itself aimed at helping African countries achieve poverty alleviation and self-reliance.[23] During her visit to Africa in May 2016, President Park Geun-hye said ROK was willing to share the development experience accumulated in the past with African countries and the spirit of the "Saemaeul Movement," providing education and training opportunities for the youth and women of African countries. In addition, President Park Geun-hye introduced the new concept of the "South Korea aid project" to some African leaders for their comprehensive development and cooperation, and indicated that the ROK was dedicated to setting up a new development cooperation pattern with African countries.[24]

As a whole, the ROK's development assistance program to Africa has focused on two aspects in recent years: one is aimed at the backward, rural areas of African countries; the other is for the economic development planning projects of African countries. In order to achieve these goals, ROK has formulated a specific program for discussing assistance to Africa projects through the "South Korean-African Forum" and the Korean-African Ministerial dialogue mechanism. The ROK's measures include expanding the scale of economic development cooperation and financial

[22] The "Korean Millennium Village" project targets four villages in the Mbora region in central Tanzania and the Luxiila region in southern Uganda. The main contents include the establishment and repair of sanitation centers, popularization of mosquito nets, improvement of drinking water supply, and school facilities, distribution of malaria drugs, and provision of anthelmintic, etc. See "The 'New Village Movement' Will Spread to Poor Areas in Africa", Yonhap News Agency Seoul, July 22, 2009, https://cn.yna.co.kr/view/ACK20090722001200881?section=search, February 9, 2020.

[23] "The government plans to teach the 'Saemaeul Movement' to developing countries", Yonhap News Agency, Seoul, May 22, 2011, https://cn.yna.co.kr/view/ACK20110522001100881?section=search, February 9, 2020.

[24] "Park Geun-hye speaks at the head of the African Union and emphasizes that ROK and Africa are partners in mutually beneficial cooperation", February 9, 2020, Yonhap News Agency, Addis Ababa, May 27, 2016, https://cn.yna.co.kr/view/ACK20160527005000881?section=search, February 10, 2020.

aid to Africa, helping African countries cultivate human resources, identifying priority areas for cooperation, helping African countries complete developmental project planning, strengthening dialogue and cooperation between governments and enterprises, and jointly promoting the implementation of relevant cooperation projects with international institutions in Africa.

Enhancing Civil Society Communications

After the end of the Cold War, the ROK paid more attention to the development of overseas assistance. In Africa, the ROK's civil society is currently active in poverty eradication, public health, and many other fields. For example, the non-governmental organization of "friendly neighbors" was established in 1991, and has been officially carrying out public welfare undertakings to eliminate parasites in Africa since 2005. In the Mwanza Region on Lake Victoria in Africa, for example, the organization not only conducts parasite inspection and drug popularization for local residents, but also ensures that local residents can access clean drinking water by digging wells and installing water purification devices.

In terms of promoting personnel exchanges between the ROK and Africa, the voluntary activities of Korean non-governmental organizations are the main parts. At the government level, the ROK mainly promotes and strengthens the non-governmental interaction between the ROK and Africa by inviting research and sending volunteer groups. In fact, in addition to trade personnel exchanges, personnel exchanges between the ROK and Africa that are mainly focused on social and cultural exchanges are not common. According to the statistics of overseas compatriots' residence data in 2019, although Korean people lived in more than 180 countries across the world, and their overseas population reached more than 7 million, the number of overseas Korean nationals living in Africa was very small, with a total of no more than 10,000, accounting for 0.15% of the total number of overseas Korean nationals.[25] In recent years, the ROK has been intending to increase its strength in this regard, so as to promote deeper exchanges between Korean and African personnel.

In order to strengthen its public diplomacy, the ROK not only provides free Korean TV series to African countries, but also sets up "Korean

[25] Ministry of Foreign Affairs, Republic of Korea: *Diplomatic White Paper* (2019, p. 284).

Culture Institutes." Through cultural exchanges and cooperation, it not only enhances the understanding and trust between the ROK and Africa, but also provides opportunities for Korean enterprises to enter these countries. In addition, with the promotion of non-governmental organizations (NGO) at home and abroad, Korean stars' African volunteer activities have become an important way for Korean non-governmental organizations to communicate with Africa. As Lee Gen, chairman of The Korean Foundation (TKF), said, "Hallyu is the 'leader' of ROK's soft power and an important asset for ROK to carry out public diplomacy."[26]

The case most worthy of attention is that African countries not only hope to get help from the ROK in terms of poverty eradication and development, but also urgently hope to copy some successful practices already enacted by the ROK in science, technology, and education. For example, the National University of Ethiopia not only appointed a Korean as the president of the university, but also actively introduced the "Silicon Valley" pattern to change the operation pattern of the university. With the expansion and strengthening of academic exchanges with ROK, Ethiopia has started a "Hallyu of education" that has since spread to the entire African region.[27] ROK's actress Kim Yeon'er also opened a school for free in South Sudan to provide private support for African education.

Constructing a Multi-Dialogue Mechanism

During his March 2006 visit to Africa, President Roh Moohyun published *The South Korea Initiatives for African Development*. As an important achievement and follow-up measure of the initiative, the ROK and African countries now hold the "South Korean-African Forum" every three years. The second "South Korea-Africa Forum" held since 2006 has been upgraded to a formal ministerial agreement between the ROK and the African Union (AU) since 2009. The third and fourth "South Korea-Africa Forum" was held in Seoul, ROK, and the African Union headquarters in Ethiopia, respectively. In order to promote the implementation of a consensus on economic cooperation between the ROK

[26] *"Statistics: The Number of Hallyu Fans Worldwide Has Increased to Nearly 100 Million"*, Yonhap News Agency, Seoul, January 14, 2020, https://cn.yna.co.kr/view/ACK20200114004700881?section=search, February 12, 2020.

[27] [Korea] Li Yuanzhen: "The First Foreigner to Draw the Blueprint of Ethiopia", [ROK], Korea JoongAng Daily, September 17, 2012.

and Africa, the Ministry of Planning and the Ministry of Finance of the Republic of Korea, the Export and Import Bank of Korea, and the African Development Bank (AFDB) decided to jointly hold the Korean-African Ministerial Economic Cooperation Conference (KOAFEC). Since the first meeting of the KOAFEC in April 2006, a total of six meetings have been held as of 2018. This is an opportunity to further strengthen the cooperation between ROK and African countries.

Since 2008, the policy dialogue between ROK and Africa has maintained normal operations. Through these policy dialogue mechanisms, ROK and Africa have not only strengthened high-level exchanges and economic cooperation, but also carried out frequent dialogue and cooperation on international cooperation, economic development, regional situation, ecological environment, and other related issues. In the process of the institutionalization of policies toward Africa, Moon Jae-in's government extended the practice of the first three governments while also putting forward some new measures to strengthen South Korea's policies toward Africa, such as the establishment of the "South Korean-African Consortium," the Middle East Africa Environmental Forum, the Korean-African energy cooperation seminar, the Korean-African business forum, the Seoul Africa dialogue, and the Korean-African youth forum. The fifth "South Korean-African forum" will be held in 2021. On July 14, 2019, ROK's foreign minister Kyung-wha Kang presided over the meeting of heads of ROK's diplomatic agencies in Africa in South Africa and attended the "South Korea Africa Economic Forum." It can be said that attaching more importance to Africa and strengthening its institutionalized diplomacy to Africa has become an important part of the ROK's "medium-power" diplomacy in the new era.

3 Factors Influencing the Implementation of ROK's African Policy

As the ROK attaches great importance to the development of its relations with Africa and strengthens the implementation of its policy toward Africa, influencing factors such as competition among major powers outside the region, Africa's political, economic, and social development, and the ROK's own capabilities and demands cannot be underestimated. These will become essential factors that will affect the implementation of ROK's Africa policy and the development of ROK–Africa relations in the future.

Factors Related to Major Powers Outside the Region

Whether it was during or after the end of the Cold War, the ROK's diplomacy has never been able to get rid of the influence of great power politics and great power diplomacy. In terms of diplomacy toward Africa, the ROK faces strong competition from major powers at different levels and in different fields, especially due to the obvious influence of major powers such as China, the US, Europe, Japan, and the regional blocs.

First among these is the Chinese factor. Entering the new century, China's diplomatic relations with Africa have made great efforts again. With the Forum on China–Africa Cooperation functioning as its institutionalized platform, China's all-around cooperation with Africa has been strengthened, and China–Africa relations have entered a new stage. With the pragmatic advancement of the "Belt and Road Initiative," Africa has become an emerging destination for China's foreign aid and investment in the new era. By the end of 2017, China's investment stock in Africa had exceeded 100 billion dollars, covering almost every country in Africa.[28] According to these statistics, China's investment is primarily in projects in nine countries including Nigeria, South Africa, Malawi, Mozambique, the Democratic Republic of Congo, Sudan, Egypt, Kenya, and Tanzania and numbers close to 11 billion US dollars.[29] China-Africa trade volume has broken through the 200 billion US dollars mark. At the 2018 China-Africa Beijing Summit, President Xi Jinping proposed that China and Africa cooperate closely on the basis of advancing the "Ten Major China-Africa Cooperation Plans," with the "Eight Major Actions" to be implemented as key works in the next three years. The continuous strengthening and deepening of Sino-African relations and China's increasing influence in African countries have drawn the attention of Western countries, including the ROK, which has increasingly become an important factor for the ROK to formulate and adjust its Africa's policy and schedule the development of its relations with Africa.

Second is the American factor. According to the statistics, American direct investment in Africa has reached 745.87 billion US dollars over

[28] "Ministry of Commerce: Africa Has Become an Emerging Destination for China's Foreign Investment", People's Daily Online, August 28, 2018, http://www.dzwww.com/xinwen/guoneixinwen/201808/t20180828_17778525.htm, February 15, 2020.

[29] KOTRA: *2019 세계시장 진출전략II*, December, 2018, p. 219.

the past two decades. Among these investments, American direct investment in Africa reached 69.03 billion US dollars in 2014 and dropped to 47.8 billion US dollars in 2018.[30] Compared with the continuously increasing trend of Chinese investment in Africa, American investment in Africa has declined significantly in recent years. However, the US still has a strong influence on Africa's policies in the political, military, and security fields. In October 2018, the US Congress passed a 60 billion US dollar financing bill aiming at expanding overseas infrastructure investment. In November of the same year, the United States Agency for International Development (USAID) released an upgraded version of the "Power Africa Initiative"–*Power Africa Gas Roadmap to 2030*, delineating Africa's top ten cross-border power trade opportunities, and identified 47 projects among them, including 18 priority projects.[31] In December of the same year, the Trump administration announced a new version of its investment strategy in Africa, indicating "the United States' unilateralism considerations and its tendency of great powers competition mindset in its relations with Africa."[32] The US adjusted its Africa policy with need for maintaining its hegemony and global strategy, and adopted targeted strategies for specific countries in the African region, which intensified the strategic game of major powers in the region and brought multiple unpredictable results to the ability of the ROK to stably develop its relations with Africa.

Third are European and Japanese factors. In recent years, political entities such as Europe and Japan have stepped up their investments in Africa. According to the statistics, as early as 2017, the bilateral trade volume between the European Union and African countries exceeded 300 billion US dollars. At the Fifth EU–Africa Summit held in 2017, the EU pledged to invest 54 billion US dollars in sustainable development to African countries by 2020. In 2018, the European Union and the United Kingdom pledged to invest 40 billion euros and 5 billion

[30] *"The Capital Competition of U.S., Europe, and Japan in Africa: U.S. Direct Investment of 745.87 Billion US Dollars in 20 Years"*, China Business News, September 4, 2019, https://baijiahao.baidu.com/s?id=1643734886608163338&wfr=spider&for=pc, February 15, 2020.

[31] Zhang Rui: *"China's Power Investment in Africa: Opportunities and Challenges under the Belt and Road Initiative"*, *International Economic Cooperation*, Issue 2, 2019.

[32] Song Wei: *"The Trump Administration's African Strategy and Its Impact"*, *Modern International Relations*, Issue 3, 2019, p. 54.

euros in African countries, respectively.[33] At the Seventh African Development Conference held in August 2019, Japan proposed that private investment in Africa in the next three years would exceed the sum of the previous three years, which means it is projected to break through the 20 billion US dollar mark.[34] European countries and Japan have strengthened their deployment in Africa, which has brought about complex and subtle influences on ROK's policy toward Africa. Especially in the context of the current deterioration of Japan–ROK relations, it is inevitable that the two countries will compete in Africa.

At present, although ROK-African relations have developed rapidly in recent years, ROK's policy toward Africa cannot cast off the influence of major powers, and the development of its relations with Africa lags behind the above-mentioned countries. The reasons for this are multiple: first, the ROK has a weak historical connection with African countries, and the ROK–Africa economic dialogue and cooperation mechanism started late; second, the scale of ROK's investment, aid, and trade in Africa cannot be compared with that of major countries such as China, the US, Europe, and Japan; third, the investment of ROK's enterprises in Africa is concentrated in the fields of raw materials and infrastructure construction, and its breadth and depth are obviously insufficient. Given that, in order to improve the effectiveness of ROK's African policy, the Moon Jae-in government intended to promote more dialogue and policy coordination between the ROK and China, the US, Japan, Britain, France, and other countries on African affairs.

Africa's Own Factors

Since the beginning of the new century, the African region has maintained an average annual economic growth rate of 5%, and its economic growth rate has exceeded the global average, which is about 3%. Africa is rich in natural resources, and its oil and natural gas reserves rank third and fourth in the world, respectively. Additionally, the African region is still a huge consumer market with rapid development and growth. According

[33] KOTRA: 2019 세계시장 진출전략II, December 2018, p. 220.

[34] "*The Capital Competition of U.S., Europe, and Japan in Africa: U.S. Direct Investment of 745.87 Billion US Dollars in 20 Years*", China Business News, September 4, 2019, https://baijiahao.baidu.com/s?id=1643734886660816333&wfr=spider&for=pc, February 15, 2020.

Table 1 Visions of the African Union (AU) and African Development Bank (AFDB)

AU's seven visions	*The high 5s of AFDB*
Inclusive growth and sustainable development	Light up and power Africa
Pan-Africanism	Feed Africa
Achieving democracy and embody justice	Industrialize Africa
Peace and security	Integrate Africa
Cultural community	Improve the quality of life for the people of Africa
Women, teenager, and children's welfare	
Being strong and united in the international community	

Material Resources Websites of African Union and African Development Bank

to the statistics, the annual population growth rate in Africa is 3%, of which nearly half of the population is under 15 years old.[35] In addition to the rapid development of African economic integration in recent years, these have brought a lot of opportunities for the ROK, prompting it to accelerate its entry into the African market, especially the Sub-Saharan region.

From the perspective of the political environment, African countries have achieved national independence, but some countries in the region are still in a state of long-term political turmoil. In addition, the international community has imposed economic sanctions and other severe measures to some African countries, which have become an obstacle for the region to be able to develop cooperation with other countries. However, for the purpose of achieving political stability and economic and social development, African countries such as South Africa, Kenya, Ethiopia, and regional international institutions such as the African Union and the African Development Bank have formulated and proposed various reform plans and visions, as shown in Table 1. In view of this, in recent years, ROK has continuously promoted the implementation of its Africa policy through strengthening bilateral policy dialogues and high-level

[35] [Korea] Jin Jinqing, Kim Ki-hwan: *ROK's Small and Medium-Sized Enterprises Successfully Marching into the Africa Called "Land of Opportunity"*, [Korea] Korea JoongAng Daily, April 18, 2011.

exchanges of visits, with a view to building a new type of partnership between ROK and Africa.

From the macro policy environment, the accelerated process of African economic integration has brought potential opportunities for Korean enterprises to be able to achieve new development. In January 2015, the African Union passed a new plan for the long-term development of Africa Agenda 2063. In the first ten-year implementation plan of the agenda, the key areas of African development and the expected results and targets are clearly set out. In response to this, the ROK has specifically proposed cooperation project ideas that are aligned with this agenda, as shown in Table 2. What's more, the industrial diversification policy, the expansion of foreign investment, and the carrying out of the "African Continental Free Trade Area Agreement" also provide many opportunities for the ROK to speed up its entry into the African region. From the perspective of the industrial environment, core issues which ROK has focused on include: first, the huge value brought about by the rich energy resources and untapped markets in Africa; second, the industrial structure

Table 2 ROK-aligned cooperative version of *Agenda 2063*

Main areas	*Objective version*
Building integrated high-speed rail networks to connect all capitals and major commercial centers in Africa	Promoting the movement of personnel, materials, and services, reducing logistics costs, and eliminating stagnant logistics systems
Constructing the African E-University	Use long-distance online learning to greatly improve African students' access to high school and educational services
Building a free trade zone on the African continent	By 2022, achieving the goal of doubling the scale of trade between among countries
Democratic Republic of the Congo's Grand Inga Hydropower Construction Project	The power plant can output a maximum of 43200 MW of electricity, and neighboring countries can obtain clean energy supply
Building a pan-African electronic network	Through the construction of broadband base stations and wired communication infrastructure among African countries, it will promote the entry of African countries into the "electronic society."

Material Resources Export–Import Bank of Korea: ROK's *Handbook* (2018, p. 39)

dominated by the primary industry, undeveloped agriculture, and manufacturing of African countries, which presents opportunities for Korean enterprises to achieve new development; third, the fact that there is a huge gap in public infrastructure construction in African countries and the ROK hopes to find opportunities for cooperation between the two sides in the fields of public health care, education, and others; and fourth, that the rapid growth of the youth population and the constant advancement of urbanization have enabled the continuous expansion of the African consumer market and brought new opportunities for cooperation.

ROK's Domestic Factors

After the end of the Cold War, ROK's foreign policy concept has undergone a shift toward "de-politicization" and the so-called "utilitarian diplomacy" of pursuing economic interests.[36] As the world's tenth largest trading country, the ROK's inexhaustible impetus for Africa's economic diplomacy comes from two aspects: one is the practical demand for the supply of raw materials such as oil and minerals, and the other is the great expectations the country holds for a huge export market in the future.[37] In this context, the focus of the development of ROK-African relations has subsequently shifted to the field of economic cooperation.

After entering the new century, diversifying the supply of raw materials such as energy resources, getting rid of excessive dependence on oil and gas resources in the Middle East, ensuring the security of the energy supply, and achieving stable development for the country have become the top priorities of each successive Korean government. As early as 2005, ROK was already the world's fourth largest crude oil importer and seventh largest oil consumer. Its crude oil imports amounted to 87.5 billion US dollars, accounting for 33.5% of the total imports that year. Among them, crude oil imports account for 63.7% of ROK's total energy imports, and more than 80% of its crude oil imports come from Middle

[36] Ministry of Foreign Affairs, Republic of Korea: *Diplomatic White Paper* (1990, p. 80).

[37] Ministry of Foreign Affairs, Republic of Korea: *Diplomatic White Paper* (1991, p. 41).

Eastern countries.[38] It can be seen that ROK's energy structure and energy supply security are extremely fragile. As a result, ROK puts its focus on Africa, which has abundant energy resources, and looks forward to enhancing energy cooperation between ROK and Africa to handle its energy supply dilemma in the future. In this sense, the ROK indicates a strong policy willingness to strengthen cooperation with African countries in specific areas.

The great success of ROK's economic development largely depends on the support of the huge markets that already exist in China, the US, and Japan. In recent years, China has deepened its "supply-side" reforms, coupled with the impact of the "Terminal High Altitude Area Defense (THAAD)" incident on China–ROK relations. This is coupled with the slowdown of the US economy and the Trump administration's "anti-globalization," "trade bullying," "America First," and other paranoid actions and policies. The intensification of disputes over historical issues between ROK and Japan has triggered Japan's escalation of retaliation against ROK. All of these have made the ROK realize that over-reliance on the above-mentioned traditional markets carries great uncertainty and high risks. ROK's capital, technology, goods, and services urgently need to find new market resources, and Africa, which is regarded as a potential huge consumer market, is undoubtedly the best choice.

Due to the deterioration of the North Korean nuclear issue during the Lee Myung-bak and Park Geun-hye period, ROK's conservative government began to regard African countries as important dialogue partners for the implementation of the UN Security Council's sanctions against North Korea. Especially after the fourth nuclear test conducted by North Korea in 2016, President Park Geun-hye took advantage of her visit to Africa to strengthen the cooperation between ROK and Africa on the North Korean nuclear issue. Obviously, this visit was "not only for economic and development cooperation, but also for pressure on North Korea. It was North Korea's nuclear and missile provocations and a concentrated strategy against North Korea's allies to pressure and isolate the DPRK."[39] According to a report by Japan's Yomiuri Shimbun

[38] Ministry of Foreign Affairs, Republic of Korea: *Diplomatic White Paper* (2006, pp. 185–186).

[39] [Korea] Liu Wisdom: *President Park's Visit to Africa to Launch Strategies Against North Korea's Friendly States Uganda and Ethiopia* [Korea] Korea JoongAng Daily, May 25, 2016.

on May 30, 2016, after the ROK-Ukrainian summit, Ugandan leaders were "instructed to sever cooperation with North Korea in the security, military, and police fields, and faithfully implement UN Security Council sanctions at the request of ROK.[40] Previously, ROK's Ministry of National Defense designated Uganda and Ethiopia as key countries for defense industry cooperation and, in 2015 sent military attachés to the Ugandan Embassy in an effort to increase pressure on North Korea through non-military diplomacy. It can be said that ROK's constantly increasing diplomatic activities in Africa have obvious pragmatic considerations for the continued repression and isolation of North Korea. In the coming period, cooperation in areas such as military security and the defense industry will become a new focus area for the development of ROK–Africa relations.

4 Studying and Judging the Development Trends of the Korea-African Relationship

From the perspective of the main policy, cooperation field, and objective environment, the future relationship between the ROK and Africa will develop more strongly and deeply. ROK's Africa policy system will continue to improve, and its cooperation field with Africa will continue to expand. Under the guidance of the economic cooperation relationship based on investment and trade, the ROK and Africa are expected to form a comprehensive partnership, but the conditions for the establishment of a strategic partnership between the two parties are not yet mature.

Improving the African Policy System and Expanding the Cooperation Fields with Africa

After President Roh Moo-hyun's visit to Africa in 2006, the ROK's energy diplomacy that took Africa as its axis once again promoted the development of relations between the ROK and Africa. In July 2007, the Africa and Middle East Bureau of the Ministry of Foreign Affairs of the ROK was adjusted to contain four departments, focusing on strengthening the diplomatic layout of Southeast and Northwest African

[40] [Japan]Yomiuri Shinbun, May 30, 2016, quoted from *Foreign Media: Park Geun-hye Visits Three African Countries to Seek to Block North Korean Military Cooperation*, http://top.chinadaily.com.cn/2016-05/31/content_25554727.htm.

countries, thus inaugurating a new process for the ROK's diplomacy with Africa. Nevertheless, under the banner of the so-called creative pragmatic diplomacy during Lee Myung-bak's period, the Ministry of Foreign Affairs of the ROK reduced the non-business departments into one department organized around Africa. This diplomatic system was retained during the Park Geun-hye period. Although this does not mean that ROK ignores the development of relations with Africa, it reflects the imperfection and instability of ROK's policy system toward Africa from this one aspect.

As an important approach to strengthening its diplomacy with Africa and achieving the multilateralization of that same diplomacy, ROK established the "Korea-Africa Consortium" in June 2018 and promoted the establishment of a framework for ROK-Africa cooperation in the new era, as shown in Table 3. In March and October 2019, the Ministry of Foreign Affairs of the ROK hosted two inter-departmental coordination meetings on African business, mainly to discuss the issue of Korean enterprises entering the African market and strengthening foreign policy toward Africa. In January 2020, the Moon Jae-in government once again reorganized its diplomatic departments with Africa, aiming to strengthen its diplomatic functions with the continent. However, ROK's African policy has not changed on the whole because of economic factors, which means that ROK's diplomacy with Africa will continue to follow a pragmatic route in the future. Since the Roh Moo-hyun presidency and under the promotion of successive Korean governments, the cooperative fields between ROK and Africa have been continuously expanded, and cooperation between the two sides has deepened. Due to the importance of the African energy resource market and the development of the African economic integration process, the ROK will increasingly regard the African region as a place to realize its dream of economic growth in the future. Judging from the existing important consensus and agreements reached between the ROK and the African Union, with the continuous improvement of ROK's policy system toward Africa, the cooperation fields between ROK and Africa will be further expanded and the cooperative relationship will exhibit a trend of deepening development.

Table 3 Cooperative framework between ROK and Africa for the new period

Key aspects		*Cooperation fields*	*Projects conceived*
Economic cooperation	1.1	Sustainable and inclusive growth, added value	Projects of industrial and manufacturing hub development related to international value chains
			Projects of achieving the African industrial "Korea–Africa Business Forum"
	1.2	Trade, industrial development, and investment	Projects of commodity exchange in Africa
			Projects of Korea–Africa Trade Investment Network
	1.3	Natural resources, inventory, and exploitation	Projects of earth sciences, mining information systems, value addition, and other projects related African institutions strengthening
	1.4	Agricultural production	Projects of grain harvest reduction and enterprise-type agricultural investment supporting
			Projects of improving the production capacity of fragmented farmers, cultivating self-reliance, using African grain seeds, and realizing sustainable production
	1.5	Blue economy, harbor, and aquatic transportation	Projects of transportation industry, Maritime University, Excellent Research Center Support
			Projects of African Transport Industry Development

(continued)

Table 3 (continued)

Key aspects		*Cooperation fields*	*Projects conceived*
			Projects of "Blue Economy" Commercial Development
	1.6	Public health	Projects of emergency Surgery Center Support
			Projects of Africa Center for Disease Control and Prevention Support
			Projects of strengthening the public health information system
			Projects of strengthening the health responsibility mechanism for mothers, children, and adolescents
2. Human resource development and scientific technology	2.1	Human resource development	Projects of the pan-African University and Korean University Network
			Projects of Korean-African Scholarship
	2.2	Scientific technology	Projects of achieving technological innovation through the "Korea-African Entrepreneurs Forum"
			Projects of Korean-African Universe Education, Training and Research Network
3. Youth development and female empowerment	3.1	Youth development	Projects of youth technology development, entrepreneurship cultivation, exchange business
	3.2	Female empowerment	Projects of women's education: supporting girls to complete school

(continued)

Table 3 (continued)

Key aspects		*Cooperation fields*	*Projects conceived*
4. Peace and security	4.1		Projects of the new African Anti-Terrorism Education Center
	4.2		Mali Secondary Medical Facilities aiding project
5. Monitoring and evaluation	5.1	Monitoring, evaluation	Projects of Korea-Africa joint supervision, evaluation, and expert meeting

Material Resources Ministry of Foreign Affairs, Republic of Korea: An Introduction to the African Union (2018, pp. 61–62)

The Accelerating Tendency of ROK's Capital and Technology to the African Market

In contrast to South Korea's excellent performance in Asian, European, and American markets, the advantages and potential for South Korea's capital and technology have not been fully released in Africa. According to regional investment statistics from 2006 to 2019, South Korea's investment in Africa is far lower than that in Asia and North America. With a few exceptions, its investment in Africa is generally lower than that in the Middle East. During this period, South Korea's investment in Africa ranged from a low of just $150 million (2015) to a high of no more than $530 million (2019). In the same year, South Korea's investment in Asia reached US $10.7 billion and US $14.4 billion, respectively.[41] According to statistics, in 2019, South Korea's investment in Africa only accounted for 1.2% of its total foreign investment in the current year which has remained basically unchanged for more than a decade.

From the perspective of regional trade statistics, the trade volume between Korea and Africa is much lower than that of other regions, except that the trade volume with Oceania is slightly higher in some years. Taking the trade data of December 2019 as an example, South Korea's exports and imports to Africa were US $6.3 billion and US $4.9 billion, respectively, while South Korea's exports and imports to Central and South

[41] Website of Export–Import Bank of Korea, https://stats.koreaexim.go.kr/sub/interregionalStatistics.do.

America during the same period were as high as US $20.6 billion and US $9.7 billion respectively. Affected by the global economic slowdown, the trade between South Korea and Africa has experienced negative growth in recent years, but it is lower than the negative growth of its trade with other regions.[42] At present, Korea–Africa economic relations are much less close than Korea's relations with other regions. This is due to the low economic and social level of Africa and a weak overall consumption capacity, which makes the advantageous industries and products of South Korea incapable of entering the African market on a large scale. However, the accelerating urbanization process and the gradual release and expansion of the consumer market in Africa provide great opportunities for ROK's capital and technology to participate in African regional development in the future.

In terms of the expectations of Korean companies, there are two aspects that are most immediately pressing. The first is the expansion of the African consumer market. Through a comprehensive analysis of the size of the African consumer market, its urbanization process, income level, and other factors, Korean companies' organizations have identified African countries such as South Africa, Nigeria, Kenya, Ghana, and Mozambique as having great market potential. By using Hallyu culture and products (K-pop and K-beauty) and combining African cultural industries with Korean consumer products, it is possible to create a new Hallyu marketing model targeting the young generation of African consumers. At the same time, by overcoming the geographical limitations between South Korea and African countries through online sales (K-shop), the potential ability of Korean SMEs to enter the African market can be increased. The second pressing concern is how to strengthen industrial cooperation with African countries. Currently, African countries are promoting industrialization in three major areas: the cultivation of manufacturing, modernization of agricultural, livestock, and mining industries, and infrastructure investment. African countries are quite interested in Korea's development model based on manufacturing, and they hope to strengthen cooperation with Korea, which will help Korean enterprises to export machinery and equipment, raw materials, parts, and other products. Korea's advantageous industries such as electronics and automobiles are speeding up their entry into the African market, and it can be

[42] Website of Korean International Trade Association, http://stat.kita.net/stat/kts/rel/RelColligationList.screen#none.

said that the door to the African market has been opened to Korean enterprises. However, compared with countries such as the EU, China and Japan, South Korea lacks competitiveness in infrastructural investment in Africa due to restrictions on investment scales and concerns about cost recovery.

There is Still a Long Way to Go to Build a Comprehensive Cooperative Partnership Between ROK and Africa

Compared to other countries, ROK is a veritable "middle power" in terms of economic size, technological strength, and military capability. From the perspective of economic factors alone, ROK has been a member of the Organization for Economic Cooperation and Development (OECD), a club of developed countries, since the 1990s, and it occupies an important position in the global trade system. According to statistics, in 2019, Korea's exports ranked seventh in the world at 2.9% of the world total, imports ranked ninth in the world at 2.6% of the world total, and trade volume accounted ninth in the world at 2.8% of the world total.[43] These favorable factors will undoubtedly help promote the development of ROK–Africa relations in the new era.

However, in contrast to moderate policy expectations, the shortcomings of having an insufficiently comprehensive development of ROK–Africa relations has not yet been overcome, and they both have not yet been able to talk about building a comprehensive strategic partnership. The reason for this is that ROK lacks its own hard power. Compared with China, the US, Japan, and other global economic powers, the scale of trade between ROK's and African countries is significantly smaller, and the proportion of ROK–Africa trade to the ROK's total trade is also low. Second, the structure of economic and social development in the African region constrains ROK's capital and technology input. Judging from the scale of governmental economic cooperation funds, the proportion of development assistance expenditures, and regional distribution, African countries are not the key targets of ROK's foreign economic cooperation. In addition, the main model of economic and social development in Africa is government-led investment in infrastructure projects, but these projects

[43] Website of Korean International Trade Association, http://stat.kita.net/main.screen.

are heavily dependent on high international loans and international development assistance that give global economic powers, including China, a significant advantage. For ROK capital, which is strong in manufacturing (semiconductors, automobiles, shipbuilding, chemicals, etc.) and cultural soft power, the African region is merely a market with great potential for cooperation that is waiting to be developed. Third, the imbalance in the development of ROK–Africa relation is difficult to change, especially as the exchanges and cooperation between the two sides in non-economic fields have just begun. Since the 1990s, when ROK participated in UN-led peacekeeping operations in Africa, it has continued to participate in African regional governance under the name of "contribution diplomacy," thereby improving its national image and international status. In the future, the ROK will take more substantial measures to strengthen economic sanctions against the DPRK, increasing military aid and arms sales to Africa. Yet the future of socio-cultural exchanges between Korea and Africa is still greatly limited due to the huge differences in culture and religion and the influence of geographical factors.

It can be said that the ROK's long-standing pragmatism toward Africa will not enable it to bring the ROK–Africa relationship to a more benign stage of comprehensive and balanced development in a short period of time. In the future, the development of ROK–Africa relations will not have the necessary mature conditions for upgrading to a comprehensive strategic partnership. In terms of developmental trends, although the ROK–Africa partnership lacks a solid strategic significance, the framework for the new era of comprehensive partnership is becoming clearer, with political dialogue as a guide, economic cooperation as a priority, and social and cultural exchanges as a complement. By learning from the successful experiences and mature practices of China, the US, Japan, and Europe in the development of their relations with Africa, the ROK is making efforts to compensate for its lack of hard power. In addition, the ROK is actively using soft power tools such as cultural industries and development assistance to explore African markets and tap the potential of ROK–Africa relations, aiming to strengthen the development of friendly and cooperative relations between the ROK and Africa.

5 Conclusion

The ROK's diplomacy to Africa is an important part of its overall diplomacy and represents the primary aspect of its regional diplomacy. It is an important means of promoting and safeguarding ROK's national interests, as well as an important territory for realizing the diversification of energy and resources, as well as developing emerging markets. In the new century, ROK's diplomacy with Africa has gradually changed from weak to strong, from single to three-dimensional. In the process of changing from government led to non-governmental, multi-agental participation, the development of Korean-African relation has also entered the fast track. Against a backdrop of fierce competition among the world's major powers, "smart diplomacy" in Africa by "middle powers" such as Korea has played a central role. From ROK's point of view, African countries do not want the great powers in the region engaged in fierce competition and excessive interference in their domestic and foreign affairs. As a moderately developed country, ROK's involvement in African affairs and the strengthening of its relations with Africa will not only avoid arousing the resentment of African countries and the caution of some major powers, but will also help ROK to use its advantages to promote a new, future-oriented ROK–Africa partnership that is commensurate with its national strength.

Although ROK still lacks systematic and overall strategic planning for its African policy, through a process of strengthening and deepening the mechanisms in place for policy dialogues and cooperation between ROK and Africa, especially through ROK's conscious dissemination of the "ROK experience," promotion of the "ROK model," and deepening of internal social governance system of African countries. The ROK's approach, such as the social governance system, shows that the ROK attempts to concentrate its advantageous resources and limited diplomatic power on economic cooperation, deepening the livelihood of the people, and integrating mutual interests in a way that leads to the new development of ROK–Africa relations. Of course, there are both positive and negative factors that facilitate and constrain the formulation and implementation of ROK's Africa policy. Among these factors are both structural factors and normal problems, as well as linkage conflicts and potential risks. It can be said that the future development of ROK–Africa relations into a comprehensive partnership still has a long way to go and that the two sides are not yet ready to build a comprehensive strategic partnership.

CHAPTER 9

The Evolution of Australia's African Policy and Its Value Orientation

Fei Chen

"In the history of Australia's foreign relations, Africa has been marginal relative to Australia's engagement with a number of other world regions and major powers."[1] Although Australia has more or less "ignored" Africa throughout the period from the twentieth century to the early twenty-first century, Africa has not completely disappeared from Australia's radar. For one thing, Australia has developed limited ties with African countries within the framework of the United Nations and the Commonwealth, while on the other hand it maintains a certain degree of political

[1] David Mickler and Nikola Pijovic, "Engaging an Elephant in the Room? Locating Africa in Australian Foreign Policy", *Australian Journal of Politics and History*, Vol. 61, No. 1, 2015, p. 100.

This article was originally published in the 6th issue of "West Asia and Africa" in 2016. There are minor revisions when collecting this book.

F. Chen (✉)
School of Politics and International Studies, Central China Normal University, Wuhan, China
e-mail: wxy03@ssap.cn

X. Li and C. An (eds.), *The Changing World and Africa*,
https://doi.org/10.1007/978-981-16-4983-7_9

and economic cooperation with South Africa and other Commonwealth member countries. It wasn't until the Rudd and Gillard administrations that Australia was first identified as a medium-power country with global interests, and a creative medium-power diplomacy replaced the more traditional medium-sized power country-oriented diplomacy in the Asia–Pacific region, promoting changes in Australia's foreign policy and Africa policy alike. Australia, which now pursues a greater degree of international influence and voice, no longer ignores Africa and, in fact, promotes the rapid development of Australia–Africa relations through diversified policy instruments and tools. Entering the period of the Abbott and Turnbull governments, the focus of Australia's foreign policy has once again returned to the Asia–Pacific region, but it is increasingly difficult for Australia and African countries to ignore each other. The breadth and depth of bilateral cooperation remain at a relatively high level. Observing and analyzing the development of Australia's African policy will help us to fully understand the interaction between medium-power countries and African countries, as well as provide inspiration and reference for the sustainability of China–Africa cooperation.

1 Twentieth Century to Early Twenty-First Century: "Ignoring" Africa

The history of Australia–Africa relations can be traced back to the connection and interactions that existed between the sovereign state and the colonies during the British colonial rule. But before the establishment of the Commonwealth of Australia in 1901, there were few official links between Australia and the African continent. In 1885, Australian New South Wales sent a military delegation to support the British Empire's military expansion into Sudan.[2] After the outbreak of the Boer War in South Africa, Australia sent volunteers to support the British sovereign. These two cases are considered to be important cases of official contact between Australia and the African continent. During World War I and World War II, the Australian Armed Forces maintained synergy with the British army in North Africa and jointly carried out military operations, which also enabled the Australian government to have a certain degree of

[2] Malcolm Saunders, "A Case Study of Historical Inaccuracy: New South Wales and the Sudan Campaign of 1885", *Journal of Australian Studies*, Vol. 8, Issue 14, 1984, pp. 29–38.

interaction with North African countries. However, since there was little autonomous diplomacy in Australia before World War II, Nikola Pijovic of the Australian National University once pointed out, "As Australia had no substantive independent foreign policy until after World War II, its foreign policy engagement with African issues prior to that time came solely within the confines of British foreign policy."[3]

After World War II, under the framework of the United Nations and the Commonwealth, the Australian Menzies government (1949–1972) actively used multilateral platforms to maintain healthy interactions with various African countries. Australia uses the Commonwealth Prime Ministers' Meeting as a channel to deepen intergovernmental communication with the African member states of the Commonwealth. However, during the 1956 Suez Canal crisis, Australia misjudged the international influence that Britain and France's invasion of Egypt would exert globally and the possible response of the US to the crisis, choosing to adopt a firm supportive attitude toward the UK, which led to the decline of Australia's national image in Africa. In addition, because the Australian government has historically shown a certain degree of sympathy for the Afrikaner government and the system of apartheid in South African, and has also prevented South Africa's racial issues from being discussed within the Commonwealth framework, Australia is regarded as a racist country by most African countries. Australian Prime Minister Robert Menzies did not actually support apartheid in South Africa, however. In 1961, Menzies' speech in the Australian Parliament clearly stated that he opposed apartheid.[4] Australia's support for South Africa is based on safeguarding Australia's own interests, and Australia is concerned that the issue of immigration within its country will be increasingly internationalized. In a letter to then-British Prime Minister Macmillan, Menzies pointed out that South Africa's precedent meant that such issues would have been "quite legitimate" to discuss in the Commonwealth.[5]

[3] Nikola Pijovic, "The Commonwealth: Australia's Traditional 'Window' into Africa", *The Round Table*, Vol. 103, No. 4, 2014, p. 385.

[4] Robert Menzies, "House of Representatives Question International Affairs Speech", http://parlinfo.aph.gov.au/parlInfo/search/display/display.w3p;query=Id%3A%22hansard80%2Fhansardr80%2F1961-04-11%2F0134%22, accessed March 10, 2020.

[5] Menzies to Macmillan, 15 January 1962, Menzies Papers, MS4936, Series 1, Box 22, Folder 187, National Library of Australia, in David Goldsworthy, "Australian External

In the context of the intensification of Cold War confrontations and the rise of the wave of decolonization in Africa, Australia has also had to constantly adjust its policies toward Africa. "In particular, there was a concern in Australia that newly independent but weak African states would fall into the hands of the Eastern bloc, threatening security in the Indian Ocean region and potentially cutting off strategic transportation, communication and trade routes between Australia and Europe."[6] Therefore, on the one hand, Australia has prevented the spillover of regional conflicts from affecting its own strategic interests, and on the other hand, it has developed bilateral relations with African countries and the Commonwealth countries as breakthroughs. From the end of World War II to the mid-1960s, Australia successively established consulates in South Africa (1946), Egypt (1950), Ghana (1957), Nigeria (1960), Tanzania (1962), and Kenya (1965). Aside from Egypt, these African countries are all members of the Commonwealth.

In the 1970s, Australia's foreign policy, especially its foreign policy toward Africa, saw a major turning point. The foreign policy of the Labor Party's Whitlam government (1972–1975) differed from the previous Liberal government, with "Three of the major declared principles of the Whitlam Government's foreign policy are anti-racism, anti-colonialism and respect for UN resolutions."[7] These three principles have given Australia a new impetus for diplomacy with Africa. At the Commonwealth Prime Ministers' Meeting, Nigerian leader Yakub Goon commented that Whitlam's government's new African policy was "very pragmatic and very helpful." Other African countries commented that the Whitlam government's African policy was "very progressive ... if we can say so without being offensive or patronizing."[8] The representative of the General Assembly of the Whitlam government voted in favor of the anti-racial discrimination bill. The government closed the Rhodesia Intelligence Center serving the illegal Smith regime in South Africa, promulgating a ban on wheat exports to Rhodesia. And athletes exhibiting racial discrimination are not allowed to come to Australia or go to other countries

Policy and the End of Britain's Empire", *Australian Journal of Politics and History*, Vol. 51, No. 1, 2005, pp. 26–27.

[6] David Mickler and Nikola Pijovic, op. cit., p. 103.

[7] David Goldsworthy, "Australia and Africa: New Relationships?", *The Australian Quarterly*, Vol. 45, No. 4, 1973, p. 69.

[8] David Goldsworthy, op. cit., p. 59.

through Australia.[9] These measures will help Australia establish closer diplomatic relations with most African countries.

Since then, the Fraser government (1975–1983) continued the Whitlam government's African policy. At the 1977 Commonwealth Heads of Government meeting, Fraser publicly condemned racism, criticized racism for poisoning international politics, and for the first time publicly called Rhodesia Zimbabwe.[10] In 1979, the Commonwealth Government Heads' Meeting was held in Lusaka. This was an important meeting dedicated to the transformation of the Rhodesian ruling system. Fraser was the key coordinator, uniting other Commonwealth countries to collectively pressure the newly appointed British Prime Minister Margaret Thatcher, and prompting Mrs. Thatcher to sign the Zimbabwe Bulletin of the Commonwealth Government. Australian scholars David Mickler and Nikola Pijovic believe that, "during Fraser's tenure, Australia's most prominent foreign policy initiative regarding African issues was his role in mediating the 1979 Commonwealth Lusaka Agreement, which eventually (in 1980) paved the way for majority rule in Zimbabwe."[11] After the Lusaka Agreement, Australia's image and prestige in African countries improved to a large extent, an improvement that is conducive to Australia's economic and trade investment environment in Africa.

During the period of the Hawke government and the Keating government (1983–1996), Australia's foreign policy showed an increasingly obvious Asia–Pacific orientation, while African policy took a more marginal position. At this stage, Australia's African policy still focused on the issue of apartheid in South Africa. As Australia firmly opposed apartheid policies in South Africa, it supported economic sanctions against South Africa. This also enabled Australia to gain more support from third world countries and became a non-permanent member of the Security Council in January 1985. In April 1994, South Africa held its first national election after abolishing the apartheid policy. Taking this key

[9] Gan Zhenjun, "Australia's Policy towards South Africa after the Second World War", https://www.library.uq.edu.au/ojs/index.php/asc/article/view/629/806, accessed March 10, 2020, accessed March 10, 2020.

[10] Malcolm Fraser, "Prime Minister's Address to the Royal Commonwealth Society", June 3, 1977, http://pmtranscripts.dpmc.gov.au/release/transcript-4414, accessed March 10, 2020.

[11] David Mickler and Nikola Pijovic, op. cit., p. 104.

event as an important point of departure, the focus of Australia's African policy also began to shift, and apartheid or racial discrimination policies in South Africa were no longer a core part of Australia's African policy from that point on. By the time of the Howard government (1996–2007), Australia–Africa relations remained relatively flat, but Australia had significantly increased its official development assistance to Africa. However, the subtle relationship between Australia and Africa changed in 2002. At that time, Australian Prime Minister Howard served as the chairman of the Commonwealth Government Summit. At the same time, Howard was also the leader of the "three carriages" in the Commonwealth against Zimbabwe. The other two were South African President Thabo Mbeki and Nigerian President Olusegun Obasanjo. However, Howard's tough stance on the Zimbabwe issue not only led to the deterioration of Australia's relations with Zimbabwe, but also led to cracks in the "troika."

In short, from the establishment of the Commonwealth of Australia to the beginning of the twenty-first century, Australia lacked a clear, comprehensive, and unified policy toward Africa. Australia's African policy at this stage had three characteristics. First, based on the needs of strategic interests, Australia maintained a relatively low level of cooperation with Africa. Second, Australia's cooperation with Africa used the United Nations and the Commonwealth Framework as its main communication channels. Third, South Africa and other Commonwealth member countries were the main target countries for Australia's cooperation with Africa. However, changes in the international political and economic situation since the twenty-first century and the prominence of the African region's own development potential have caused major powers and medium-sized powers to turn their attention to the African region, often for the first time. Compared with the past, the political and economic situation among most African countries has been relatively stable since the twenty-first century. The peace dividend is one important source of economic growth in Africa. The prices of various resource-based primary products continued to rise globally from 2000 to 2012, and resource prosperity led to sustained economic growth in some African countries. According to data provided by the International Monetary Fund, the average GDP growth rate in Africa from 2000 to 2010 was 5.5%, much higher than the global average. Over the previous 20 years, Africa's average annual growth rate was only 2.4%.[12] Australia now hopes to promote business opportunities in Africa. The Howard government reopened the High Commission in Ghana in 2004 after almost twenty years of absence and established the Australian Trade Commission office in Libya in 2005.[13] Despite slight changes in specific measures, the Howard government has not proposed an integrated and comprehensive policy on Africa. In fact, the shift in Australia's African policy actually started with the Rudd government.

2 2007–2013: "New Engagement" Policy

During the Rudd and Gillard administrations (2007–2013), Australia's interest in the African continent increased significantly, and the political, economic, and cultural ties between them have continued to increase. Africa has become increasingly important to Australia's foreign relations and international agenda. The new adjustments and changes to the African policies of the Rudd and Gillard governments are also known as Australia's "new engagement" with Africa. During this period, the specific contents of the Australian government's "new engagement policy" for Africa include the following:

Firstly, increase Australia's diplomatic cooperation with Africa. The Rudd and Gillard governments have already expanded and deepened relations between Australia and African countries at multiple levels.

To do so necessitates strengthening high-level diplomacy with African countries. The high frequency of high-level official visits is a direct manifestation of Australia's increased diplomacy with Africa. In January 2009, Australian Foreign Minister Stephen Smith visited Ethiopia, attended the African Union Executive Council Meeting, and delivered a speech.[14] Smith also became the first Australian foreign minister to attend the African Union meeting in history. From March to April 2009, Australian

[12] "Statement by H. E Ambassador Regina Maria Cordeiro Dunlop Deputy Permanent Representative Mission of Brazil to the United Nations", New York, 11 October 2011, quoted from Xu Guoqing, "The Evolution and Characteristics of Brazil-Africa Relations", "West Asia and Africa", Issue 6, 2012, p. 140.

[13] David Mickler and Nikola Pijovic, op. cit., p. 106.

[14] Stephen Smith, "Visit to Ethiopia and the African Union", Media Release, http://foreignminister.gov.au/releases/2009/fa-s008_09.html, accessed on March 10, 2020.

Governor Quentin Bryce visited nine African countries across 19 consecutive days.[15] With the visits of Smith and Bryce, leaders of African countries have also returned to visit Australia. Nikola Pijovic of the Australian National University pointed out that from May to June 2009, the Tanzanian Foreign Minister, Kenyan Foreign Minister, Rwandan Foreign Minister, and Botswana Foreign Minister visited Australia. In the space of around one month, Australia had been visited by more high-level African ministers than during the whole four terms of the Howard government.[16]

Second, establish formal diplomatic relations with more African countries. In 2007, Australia established diplomatic relations with 41 African countries. By the end of 2013, Australia had established diplomatic relations with 54 African countries.

This requires attaching greater importance to the establishment of diplomatic institutions for African countries. In 2011, Australia reopened the Addis Ababa Embassy in Ethiopia. Addis Ababa is also the headquarters of the African Union, and Australia has strengthened ties with the African Union through its embassy in Ethiopia. On May 9, 2012, Australian Foreign Minister Bob Carr announced that he would build a new Australian embassy in Dakar, Senegal. This is the first time Australia has expressed its intention to build an embassy in French-speaking Africa.[17] Although the number of Australian embassies and consulates in Africa lags far behind the US, France, China, and Brazil, and despite the obstacles to the construction of the Australian embassy in Senegal, during the Rudd and Gillard administrations, Australia paid an increasing amount of attention to the settings and layout of its diplomatic institutions in Africa.

At last, Australia must pay greater attention to the role of African international organizations and multilateral institutions. In September 2010, Australian Foreign Minister Kevin Rudd signed a Memorandum of Understanding with the Chairperson of the AU Commission, Jean Ping, establishing "close cooperation" with the AU in the areas of trade and

[15] "Governor-General's Visit to Africa", Media Release, https://www.gg.gov.au/media-release/governor-general%E2%80%99s-visit-africa, accessed on March 10, 2020.

[16] Nikola Pijovic, "The Liberal National Coalition, Australian Labor Party and Africa: Two Decades of Partisanship in Australia's Foreign Policy", *Australian Journal of International Affairs*, 2016, p. 12.

[17] Bob Carr, "Opening of New Embassy in Senegal", Media Release, http://foreignminister.gov.au/releases/2012/bc_mr_120509.html, accessed on March 11, 2020.

investment, peace and security, the Millennium Development Goals, agriculture and food security, democracy, governance and human rights, and climate change.[18] Both Smith and Rudd visited the African Union headquarters as foreign ministers and delivered speeches. During the Rudd and Gillard administrations, Australia applied to join the African Development Bank and was willing to inject 161 million Australian dollars in advance for the African Development Fund. The strengthening of diplomatic cooperation between Australia and Africa is due to the outstanding performance of the Rudd and Gillard governments' "new engagement policy" with Africa.

The Australian government's "new engagement policy" for Africa also involves increasing official development assistance to Africa. For a long time, the main objectives of Australia's foreign aid were concentrated in Asian and Pacific countries, and African countries were not the main targets of Australia's foreign aid. However, during the Rudd and Gillard administrations, Australia continued to increase its assistance to African countries, especially those in Eastern and Southern Africa. In this regard, some scholars even pointed out that the aid to Africa during the Rudd and Gillard administrations has almost achieved "unprecedented growth."[19] Australia's aid to sub-Saharan Africa increased from A$83.849 million (2007–2008) to A$436 million (2011–2012).[20] As Kevin Rudd wrote in 2013, "Australia may have been a latecomer to Africa's rise, but as my friend David Mickler notes, we were there when it mattered for South Africa and Zimbabwe—and more recently Australia has been there for South Sudan and for the people living in the Horn of Africa."[21]

There are often three explanations for Australia's motivation for increasing aid to Africa. The first one is based on humanitarian considerations; the second is based on strategic interests; and the third is based on

[18] David Mickler and Nikola Pijovic, op. cit., p. 110.

[19] Nikola Pijovic, "The Liberal National Coalition, Australian Labor Party and Africa: Two Decades of Partisanship in Australia's Foreign Policy", *Australian Journal of International Affairs*, latest article, 2016, p. 13.

[20] "Table 6.Australian Official Development Assistance, Partner Country", http://dfat.gov.au/about-us/publications/Documents/2013-14-std-time-series-table-6.xlsx, accessed on March 11, 2020.

[21] Kevin Rudd, "New Engagement: Contemporary Australian Foreign Policy Towards Africa", pp. ixx–xxi, in David Mickler and Tanya Lyons, *New Engagement: Contemporary Australian Foreign Policy towards Africa*, Melbourne: Melbourne University Press, 2013.

economic interests. Joel Negin of the University of Sydney believes that although geopolitical, cultural, and economic interests have prompted Australia to increase its aid to Africa, the most important reason for Australia's aid to Africa is to reduce the humanitarian needs of global poverty.[22] "In addressing the humanitarian needs in Africa, the Australian government has focused on areas that fall within four MDGs: agriculture and food security (MDG 1); water and sanitation (MDG 7); the improvement of maternal health (MDG 4); and a reduction in child mortality (MDG 5)."[23] The Australian government uses its expertise and skills in these areas to address Africa's practical needs in a targeted manner. However, there are still some with the view that Australia increased its aid to Africa in order to seek non-permanent seats on the Security Council. After achieving this strategic goal, Australia's aid to sub-Saharan Africa was reduced to A$395 million in the 2012–2013 financial year. In the 2013–2014 fiscal year, it was further reduced to 254 million Australian dollars.

The third aspect, that of economic interests, promotes Australian mining enterprises to invest in African ore resources. The economic interaction between Australia and Africa is mainly concentrated in the mining field. Although Australia is also a traditional source of iron ore, the cheaper prices of African iron ore have prompted Australian mining companies to actively invest in African ore resources. Between 2007 and 2013, the number of Australian companies investing in ore resources in Africa continued to increase. In February 2012, Greg Hull, Australia's Senior Trade Commissioner for Africa, pointed out, "At least 230 Australian companies have invested in Africa. They hold approximately 650 projects of mining exploration, mining exploitation and mining processing in 42 African countries."[24] Some data show that "over 40% of Australian overseas mining projects are invested in Africa."[25]

[22] Joel Negin, "Australian Development Policy in Africa", in David Mickler and Tanya Lyons, *New Engagement: Contemporary Australian Foreign Policy towards Africa*, Melbourne: Melbourne University Press, 2013, p. 134.

[23] Samuel Makinda, op. cit. p. 59.

[24] Australia High Commission Pretoria, "Australian Mining Investment in Africa Booming", *Media Release*, February 6, 2012.

[25] Raymond Da Silva Rosa, "Out of West Perth, Mine Skills for Africa", https://www.businessnews.com.au/article/Out-of-West-Perth-mine-skills-for-Africa, accessed March 10, 2020.

"Research shows that in terms of the number of transactions, India and Australia (not China) are the most merged and acquired Asian and Oceanian countries in Africa,"[26] and M&A in Australia is mainly concentrated in the field of ore resources.

There are three main reasons why Australian mining companies flock to Africa: African ore resources are abundant and their prices are relatively low; the ruling governments of Rudd and Gillard are in a "resource boom period;" and the Rudd and Gillard governments take measures to continue to encourage and support Australian mining companies to invest in African ore resources. In October 2011, the Commonwealth Heads of Government Meeting was held in Perth, a city in Western Australia. At this Commonwealth Government Heads of State Meeting, Australian Prime Minister Gillard proposed the Mining Development Initiative, which aims to help developing countries use their natural resources to improve the economy in a sustainable manner. With the support of the Australian government, the University of Western Australia and the University of Queensland jointly formed the International Mining for Development Centre to provide developing countries with Australian professional experience in achieving sustainable mining development. At the Africa Down Under mining conference held in Perth in August 2012, Australian Foreign Minister Bob Carr emphasized, "Resources must be effectively managed through legislation, governance frameworks, sustainable environmental protection measures, and skilled labor. Australia has more than 100 years of experience using natural resources to build the economic future. Australia is committed to sharing this experience with Africa."[27] Australia provides African countries with proactive skills and experience in specific practices, which can help deeper cooperation with African countries in mineral resources.

The cross-border economic activities of the Australian mining sector have also allowed the Australian government to establish a series of supporting measures to serve domestic enterprises. Various types of Australian mining companies have entered Africa to invest and operate. Mining giants such as Rio Tinto and BHP Billiton, even without the

[26] Anjali Piramal: "India and Australia are the Asian Countries with the Most M&A Transactions in Africa", compiled by Wang Hongfeng, published in "China Mining News" on March 30, 2013.

[27] Bob Carr, "Speech by the Foreign Minister Given at the Africa Down Under Conference", Africa Down Under, Perth, August 29, 2002.

special support or assistance of the Australian diplomatic service, are able to independently handle their business in Africa. However, the operation of some small mining companies in Africa sometimes requires the assistance of the diplomatic service. The Australian government is also actively providing various services for these small mining companies investing in Africa. Small mining companies have become an emerging force in terms of investing in African ore resources. Systemic guarantees and regulatory support can effectively ravel out the concerns of these small enterprises.

The "new engagement policy" also aims to promote cooperation between Australia and various African countries in the fields of education and culture. Educational assistance and educational cooperation are the most prominent features of the Rudd and Gillard governments' "new engagement policy" to Africa. The Australian government promotes the realization of its strategic and economic interests through educational exports. In the area of education, the number of Australian scholarships to Africa increased from 100 in 2007–2008 to 1000 in 2012–2013.[28] In July 2012, Australia established the "Australia–Africa Universities Network." Dennis Richardson, Secretary-General of the Australian Department of Foreign Affairs and Trade, said in his address at the launch ceremony of the network, "This new network will provide a valuable opportunity to improve links between academics both here and in Africa, further strengthening both our research potential and our people-to-people links."[29] In fact, the establishment of the "Australia–Africa Universities Network" has far-reaching strategic significance. On the one hand, the network can function as an intelligence and consulting portal for the Australian government and provide Australian companies with the expertise necessary to enter Africa. On the other hand, the network can promote Australia's soft power in Africa. According to some statistics, 90% of the scholarships received by African students will still be used in Australia, which also helps to stimulate the Australian economy to a certain extent. Moreover, African students studying in Australia may continue to play the role of bridge between Australia and Africa in the future.

[28] Samuel Makinda, op. cit., p. 59.

[29] Raymond Da Silva Rosa, "Out of West Perth, Mine Skills for Africa", https://www.businessnews.com.au/article/Out-of-West-Perth-mine-skills-for-Africa, accessed on March 10, 2020.

The "new engagement policy" of the Rudd and Gillard governments is not only the result of an active adjustment of Australia's diplomatic strategy, but also the result of the more general prosperity of global resources and the sustainable development of the African economy. Specifically, when Rudd and Gillard were in power, the Australian government's implementation of the "new engagement policy" for Africa was mainly based on the following two strategic considerations. First, the support of African countries is the basis for the country's aggressive diplomacy. The foreign policies of the Rudd and Gillard governments are more active and aggressive than any of the previous governments. In the 1990s, former Australian Foreign Minister Gareth Evans once pointed out that Australia is a medium-power diplomacy oriented in the Asia–Pacific region. The iconic method typical of Australia's medium-power diplomacy is to establish alliances with like-minded countries.[30] Although Australia is also considered to be a medium-power member, itself, Australia's foreign policy has obvious geographical characteristics, and Africa, Latin America, and the Middle East are not areas of concern for Australia's foreign relations. But Kevin Rudd believes that Australia is "a middle power with global interests,"[31] and "the creative middle power diplomacy principle is the best way to promote Australia's national interests."[32] David Santoro of Macquarie University in Australia also pointed out that "Creative medium power diplomacy is based on the concept that followers and leaders are not mutually exclusive. Medium powers such as Australia are destined to follow big powers, but they can and must actively share ideas and participate in the future."[33] Some scholars believe that "a proactive and creative diplomacy makes Australia a more attractive

[30] Gareth Evans, "Australia's Middle Power Diplomacy", http://www.gevans.org/speeches/old/1993/101193_fm_australiasmiddlepowerdiplo.pdf, accessed on March 10, 2020.

[31] Hon Kevin Rudd, "Australia's Foreign Policy Looking West", Speech Given at the University of Western Australia, 12 November 2010, *Journal of the Indian Ocean Region*, Vol. 7, Issue 1, 2011, p. 131.

[32] Kevin Rudd, "Address to the East Asia Forum in Conjunction with the Australian National University, Advancing Australia's Global and Regional Economic Interests", https://pmtranscripts.dpmc.gov.au/release/transcript-15823, accessed on March 10, 2020.

[33] David Santoro, "No Australian Uranium to India: 'Creative Middle Power Diplomacy' in Action?", PacNet No. 16, *Pacific Forum CSIS*, March 6, 2008.

partner internationally."[34] Australia hopes to further enhance its international image and, in doing so, seeks to play a more constructive role and even be the leader in certain regional and international affairs. The proactive adjustment of Australia's diplomatic strategy has led it to reach countries in the Third World (including countries in the African region) more broadly, and the support of African countries is the therefore the basis for Australia's aggressive diplomacy.

At the same time that Australia is implementing its "new engagement" with African countries and continuously strengthening its cooperation with African countries, it also aims to increase Australia's influence on the international stage, promotes Australia's more comprehensive participation in global governance, and facilitates the ability of Australia to play a more active role in the formulation of international rules. On the other hand, Australia also hopes to gain a relatively favorable position in the big power game. Through close cooperation with African countries, Australia can gain more weight to play against the big powers. Australia's increasing interaction with African countries has achieved more prominent political clout, and most African countries support Australia's greater role in the international arena. The 67th UN General Assembly was held on October 18, 2012, voting to elect non-permanent members of the Security Council for 2013–2014. Australia was successfully elected with high votes in the first round. At the time of its election, it had been 18 years since the last time that Australia was elected in January 1985.

Some people in Australia hold the view that Australia values the development of relations with African countries in its foreign policy only because Australia wants to strive for the 2013–2014 non-permanent member of the Security Council position.[35] However, Australia's closer interaction with African countries is not just organized around seeking the status of a non-permanent member of the Security Council, but is based on a more aggressive diplomatic strategy. African countries not only

[34] Gürol Baba and Taylan Özgür Kaya, "Testing the Creativity of Kevin Rudd's Middle Power Diplomacy: EU–Australia Partnership Framework Versus the Asia–Pacific Community", *Journal of International Relations of the Asia–Pacific*, Vol. 14, No. 2, 2014, p. 263.

[35] Samuel Makinda, "Between Jakarta and Geneva: Why Abbott Needs to View Africa as a Great Opportunity", *Australian Journal of International Affairs*, Vol. 69, No. 1, 2015, p. 54.

have the value of votes in Australia's diplomatic territory, but also play an important part of Australia's diplomatic strategy.

The rapid development of the African economy is itself the new impetus for the Australian economy. From an economic perspective, the Rudd and Gillard governments' "new engagement" with Africa was in order to seize new opportunities for African economic development. Africa is the new pole of world economic growth. Many economies regard Africa as the new frontier for foreign economic development. From 2000 to 2010, six of the ten fastest-growing countries in the world came from Africa. The African Development Bank's 2013 report also showed that around one-third of African countries grew by more than 6%.[36]

As a result, Australia's emphasis on Africa's economic development continues to increase. In 2012, Bishop also claimed, "I contend that while China, Japan, South Korea, Indonesia, India and other nations in our region will be primary drivers of economic activity for many years to come, we should not focus too narrowly on the potential of the Asia Century. We need to be outward looking and seek opportunities wherever they exist."[37] Africa has become a key area for Australia to "look out," and the continuous development of the African economy may become the driving force for Australia's own economic growth again. Especially from 2000 to 2012, the prices of various primary products in the world continued to rise, and the periodic resource boom led to an increase in the interest of African mining and exploration companies. The Rudd and Gillard governments' economic "new engagement" with African was mainly reflected in the cooperation between resource companies of Australia and Africa. In addition, Australian resource companies have invested heavily in Western and Southern African countries such as Senegal, Ghana, Namibia, South Africa, and Zambia, which will likely trigger continued attention from the Australian government in the future.

Overall, in the context of the "new engagement," Australia has further deepened its political, economic, and cultural interactions with African

[36] "Annual Development Effectiveness Review 2013", African Development Bank Group, 2013, p. 1.

[37] Julie Bishop, "Success This Century Demands People and Trade Pacts Beyond Asia", *Sydney Morning Herald*, http://www.smh.com.au/federal-politics/political-opinion/success-this-century-demands-people-and-trade-pacts-beyond-asia-20120607-1zyyk.html, accessed on March 10, 2020.

countries. The "new engagement policy" has promoted the continuous improvement of the level of political interaction between Australia and Africa. Australia–Africa high-level exchanges have become more frequent, mutual political trust has continued to increase, and bilateral and multilateral contacts have become increasingly institutionalized. Australia–Africa economic and trade cooperation is also booming, driving Australian mining companies to invest heavily in African ore resources. In addition, the Rudd and Gillard governments have greatly increased Australia's assistance to Africa, and actively promoted the cultural and educational cooperation between Australia and Africa. These measures are also conducive to the construction of Australia's national image and soft power. Based on the strong support of African countries, Australia was successfully elected as a non-permanent member of the UN Security Council. The "new engagement policy" of the Rudd and Gillard governments is the most proactive and positive period in the history of Australia–Africa diplomacy. And the legacy of the "new engagement policy" may permanently promote the further in-depth development of Australia–Africa cooperation.

3 Since 2013: "Here and Faraway" with Africa

After five years and nine months of rule by the Labor Party government, the ruling coalition based on the Liberal Party won the general election and won the opportunity to rule again. In terms of foreign policy, the Liberal Party's Tony Abbott, Malcolm Turnbull, and Scott Morrison did not interact with Rudd and Gillard. China's foreign policy has been completely broken, but a certain degree of adjustment has taken place.

Abbot's questioning and criticism of the foreign policies of Rudd and Gillard's governments mainly comes from two aspects: first, Rudd's foreign policy did not match Australia's strength and positioning. During the campaign, Mr. Abbott criticized Mr. Rudd for being overly ambitious on the global stage.[38] Hugh White of the Australian National University also expressed a similar point of view. He pointed out that the reason why Rudd's foreign policy failed was that, "There were too many initiatives, too little preparation and too little follow-through, turbocharged by a slightly deranged egocentricity that fed an illusion that he

[38] Michael Fullilove, "Caught Between the U.S. and China", *The New York Times*, September 5, 2013.

belonged at centre stage on every issue, no matter what he actually had to contribute."[39] The Abbott government's slogan, "more Jakarta and less Geneva,"[40] is understood not only as calling for increased focus on bilateral diplomacy over multilateral diplomacy, but also as a focus on regional issues over gaining global influence. The biggest difference with Rudd's foreign policy is that Abbott's foreign policy clearly emphasized the abandonment of "idealistic and misguided global initiatives."[41]

Second, Australian diplomacy should pay more attention to core areas rather than marginal areas. Abbott and Turnbull disagreed with Rudd's excessive promises to African countries in terms of their winning non-permanent seats on the Security Council and believed that sub-Saharan Africa should not be the focus of Australia's official development assistance. In September 2012, Abbott, who was then the leader of the opposition party, once proposed that, "Prime Minister Julia Gillard should not be 'swanning' around in New York 'talking to Africans,' rather she should be in Jakarta talking to the Indonesian head of state about border security."[42] During the election stage, Abbott also promised that he would choose Asian countries for his first visit as prime minister.

Compared with the idealistic Rudd, Abbott and Turnbull's foreign policy is more pragmatic. Andrew Phillips of the University of Queensland believes that, "For the Abbott government, deals — not ideals — would be the lodestar of its foreign policy."[43] Abbott and Turnbull's foreign policy is marked by pragmatism, abandoning the idealistic feelings of Rudd and Gillard's government. Although Abbott's foreign policy philosophy has a strong Anglo tendency and believes that Australia should strengthen its relations with traditional allies such as the US and the UK, Australia's geographical location makes it equally impossible to ignore cooperation with Asia. The foreign policies of the Abbott and Turnbull

[39] Hugh White, "Foreign Policy Is Kevin Rudd's Forte—Isn't It?", *Strategic Insights*, No. 65, https://www.aspi.org.au/publications/strategic-insights-65-keep-calm-and-carry-on-reflections-on-the-anglosphere/SI65_Anglosphere-2.pdf, accessed on March 10, 2020.

[40] Lenore Taylor, "Ten Things to Know about Foreign Policy under Julie Bishop and Tony Abbott", *Guardian*, 3 June 2013.

[41] David Mickler and Nikola Pijovic, op. cit., p. 112.

[42] Richard Willingham, "Focus on Boats, not UN, Abbott tells PM", *The Sydney Morning Herald*, September 24, 2012.

[43] Andrew Phillips, "Issues in Australian Foreign Policy July to December 2014", *Australian Journal of Politics and History*, Vol. 61, Issue 2, 2015, p. 265.

governments have once again returned to the traditional two pillars of diplomacy: security and trade. In terms of security, Australia has strengthened its military cooperation with the US and its allies in the Asia–Pacific region. In terms of trade, Australia has established closer economic and trade ties with Asia–Pacific countries.

It seems as if Australia's interest in African affairs is declining, and Africa is not even the key to Australia's cooperation with developing countries. Throughout the Abbott government, Australia hoped to invest 90% of official development assistance in the Indo-Pacific region, especially in Pacific Island countries and Southeast Asia. Among them, Indonesia and Papua New Guinea are the two countries that have received the most official development assistance from Australia. Before the election in September 2013, the "Russian Union Foreign Policy" issued by the ruling coalition also clearly identified the "Indo-Pacific" region as a priority geostrategic region, emphasizing the establishment of a strong and effective relationship.[44] "Africa" appeared only once in the entire "Foreign Policy of the Ruling Union," which also heralded the trend and direction of Abbott's policy toward Africa.

After Abbott took office, Australia's declining interest in Africa was based on three iconic cases. First, in December 2013, Abbott attended the funeral of former South African President Nelson Mandela, but Abbott did not publish any guidelines or routes for the new Australian government's policies on Africa while in South Africa. Second, in May 2012, then Australian Foreign Minister Carl promised to establish an Australian embassy in Senegal, but the Abbott government temporarily suspended the proposal. Third, Australia proactively proposed the concept of the Indo-Pacific, but the main reason for this concept was "the rise of India and the increasing importance of the Indian Ocean."[45] "As for the definition of the Indo-Pacific system, Australian domestic officials and scholars are basically the same, but they differ on whether to extend to the east coast of Africa."[46] In fact, the main target of "Indo" in the concept of

[44] "2013 Coalition Foreign Affairs Statement – The Coalition's Policy for Foreign Affairs", https://www.aspi.org.au/in-focus/election-policy-library/2013/2013-coalition-foreign-affairs-statement-the-coalitions-policy-for-foreign-affairs, September, 2013, p. 3.

[45] Song Wei: "On the Concept and Strategic Path Selection of the Indo-Pacific System in Australia", *Journal of Shanghai Jiaotong University (Philosophy and Social Sciences)*, Issue 2, 2016, p. 13.

[46] Song Wei, op. cit., p. 17.

"Indo-Pacific" is India and some members of the Indian Ocean Rim Association, which does not include most of the Indian Ocean countries on the east coast of Africa. A geostrategic concept that can tolerate African countries has deliberately avoided actually including African countries, reflecting Australia's attitude toward Africa.

Although the policy toward Africa during the period of the Abbott government does not seem to continue the previous "new contact policy," it returns to the track of Howard government. However, the existing results achieved during the Rudd and Gillard administrations and the path dependence on the formation of African policies also made the Liberal Party's three governments' depth and breadth of interaction with Africa higher than these same policies during the Howard government.

First, the governments of Abbott, Turnbull, and Morrison have lowered the amount of aid given to Africa. Fundamentally speaking, the ultimate purpose of Australia's official development assistance to Africa is to serve Australia's national interests, rather than simply focusing on humanitarian needs. In 2014, Australia's then-Foreign Minister Bishop proposed a new foreign aid paradigm—the "aid for trade" policy. In fact, "aid for trade" is not a new concept in the field of aid. "Aid for trade" refers to assistance given for the promotion of trade development. Aid is considered to be a catalyst for economic growth and poverty reduction; that is, the "'aid for trade' policy which uses foreign aid to connect businesses in developing countries to regional and global supply chains."[47] Replacing the foreign aid component of the "new engagement policy" with the "aid for trade" policy mainly reflects the Abbott government's four considerations: it does not want to link aid to the votes of non-permanent members of the Security Council; it reflects the pragmatism and utilitarianism policies of Australia and enhances Australia's political and economic interests; Australia still wants to streamline aid projects while the global economy is still sluggish; and it provides a new policy basis for future reductions in aid to Africa. Australia's development assistance budget for sub-Saharan Africa was A$186.8 million from 2014 to 2015, reduced to A$95.9 million from 2015 to 2016, and further reduced to 89.5 million from 2016 to 2017, but increased to A$119 million from 2019 to 2020. Although aid to Africa was reduced to a

[47] Julie Bishop, "A New Paradigm in Development Assistance – Harnessing the Private Sector", Speech at Queen's Hall, Parliament of Victoria, http://www.foreignminister.gov.au/speeches/Pages/2014/jb_mr_140429.aspx?ministerid=4, accessed on March 7, 2020.

greater extent than it had been during the Rudd and Gillard administrations, aid to Africa during the Abbott, Turnbull, and Morrison administrations was still higher than the Howard government's last year. Because of the sudden humanitarian relief, Australia actually has given more development assistance to sub-Saharan Africa than budgeted.[48]

Second, the political legacy left over from the "new contact" period continues to function. Although the Liberal government's expression of "looking west" is far less pronounced than that of the Labor government, the level of communication between Australia and Africa is still much higher than that of the Howard government. The mechanisms established or endowed with new connotations during the Rudd and Gillard period still have a certain degree of vitality and continue to provide the impetus for the development of further positive relations between Australia and Africa. The Abbott government continues to attach importance to the "Africa Down Under Conference" as a multilateral mechanism. Since 2007, the tradition of senior Australian officials attending the "Africa Down Under Conference" has not been changed. In 2014, Foreign Minister Bishop delivered a video speech at the "Africa Down Under Conference." In 2015, Bishop personally attended the conference to deliver a speech and announced the establishment of the "Advisory Group on Australia-Africa Relations."[49] After several years of operations, Australia's educational assistance to Africa, the African Scholarship Program, and the "Australian-African University Network" have also produced results. Based on development assistance in mining, Australia has trained a total of 450 masters and master's in mining in Africa since 2011.[50] In 2014, more than 900 African officials received management training in mining in Australia. Under the initiative of the "Australia–Africa Relations Advisory Group," the first "Australia-Africa Week" was held in 2006. "Australia-Africa Week" is itself built on the basis of the "Africa Down Under Conference." This is a diversified

[48] DFAT, "2015–2016 Development Assistance Budget Summary Mid-Year Economic and Fiscal Outlook Update", http://dfat.gov.au/about-us/corporate/portfolio-budget-statements/Documents/2015-16-development-assistance-budget-summary-feb16.pdf, accessed March 7, 2020; Africa region brief, https://www.dfat.gov.au/geo/africa-middle-east/Pages/africa-region-brief, accessed on March 7, 2020.

[49] Julie Bishop, "Address to Africa Down under Conference", http://foreignminister.gov.au/speeches/Pages/2015/jb_sp_150904.aspx, accessed on March 7, 2020.

[50] Julie Bishop, op. cit.

project that connects universities and business groups, diaspora communities, ministers, and senior officials in Australia and the African continent. The activities of "Australia–Africa Week" include the Australian-African University Network Forum, the Africa Down Under Conference, the Australian-African Trade, Investment and Cultural Expo, the Women's Leadership Forum, and the Africa–Australia Cup football match. The latest "Australia–Africa Week" was held in Perth, the capital of Western Australia, from September 1 to 7, 2019. Despite the acceleration of some existing cooperation mechanisms, there is no summit mechanism between Australia and African countries, and interaction and exchanges between national leaders are still limited.

Third, the increased interest of sub-national Australian actors in Africa has prompted the Australian government to continue to maintain its focus on Africa. During the period of the Abbott government, Australian subnational actors also began to interact with African regional organizations and countries at different levels, especially in the Western Australia region, which has important interests in Africa, and increasingly attaches great importance to "looking west." Western Australia Governor Colin Barnett is committed to promoting cooperation between Western Australia and Africa in the field of mineral resources and believes that Western Australia's advantages in the mining economy will help African countries build a complete mining industry chain-and-laws regulatory system. In February 2014, Barnett delivered the keynote speech as the first Australian politician invited to the Mining Indaba in South Africa. In 2014, Western Australia signed a memorandum of understanding with the Common Market for Eastern and Southern Africa to help members of the Common Market for Eastern and Southern Africa develop the mining economy.

Although the Liberal government's interest in the African continent is declining, the lasting results from the Rudd and Gillard administrations still promote the continued development of Australia's African policy. "Path dependence" makes this stage of the Australian government's policy toward Africa more active and proactive than that of the Howard government, but Australia's long-term interests and strategic vision for the African continent lack any forward-looking or clear insight. The African continent as a whole has not received attention in the "Foreign Policy White Paper" released by Australia in 2017 and the "Defense White Paper" released in 2016.

4 Conclusions

From 1885 to 2016, the form, depth, and breadth of the interactions between Australia and Africa have changed significantly. Before World War II, the internal logic of Australia's policy changes toward Africa was guided by their joint suzerain, Britain, and Australia's own policy changed along with the changing British policy toward Africa. Since the end of World War II, Australia has no longer been limited to British foreign policy. The internal logic of successive governments' policies on Africa is based on two points: one is the foreign policy choices made under the positioning of Australia's global role; the other is the domestic political and economic needs of Australia.

Based on their self-perceived national interests, Australian foreign policymakers rationally choose the role of the country and construct a foreign policy suitable for this constructed role. Australia's expectations and cognition of its own global positioning influence the direction and choice of its foreign policy. If Australia positions itself as a "responsible international citizen," for example, then Australia will accept and disseminate internationally prevailing norms and ethics. In the 1970s, the Whitlam government defined Australia as a "progressive international actor." Australia also showed firm support for anti-racism and anti-colonialism, thereby promoting the development of Australia–Africa relations. During the Rudd Government, Australia was defined as a "medium power with global interests" and considered Australia's diplomatic approach to be defined by "creative medium-power diplomacy." Therefore, during this period, Australia's geopolitical and economic key areas have been expanded, and Africa has become a key area of concern for Australia at the same time that the "new engagement policy" has also emerged. During the period of the Abbott and Turnbull governments, Australia's role has once again returned to being a "medium power with limited strength" or a "medium power based in the Asia–Pacific (Indo-Pacific)," so on the basis of maintaining full coordination and cooperation with the "Anglo World," Australian has paid more attention to developing relations with countries in the South Pacific, Southeast Asia, East Asia, and South Asia. During Abbott, Turnbull, and Morrison's administration, Australia's global ambitions decreased, again placing Africa at a relatively marginal position in its diplomacy.

Australia's policy toward Africa is also affected by Australia's own domestic political and economic factors. Australian mining companies have a strong lobbying capacity for the Australian government to be able to make decisions. When the world is in a period of resource booms, for example, Australian mining companies choose to invest in African iron ore, which is cheaper, to make more profits. Australia–Africa cooperation in ore resources has also promoted the in-depth development of the Australia–Africa economy in other directions. And when the resource boom period is coming to an end or there are uncertainties, Australia's investment in African ore resources will gradually decrease, which will affect the levels of cooperation between Australia and Africa in other fields. In addition, when the governments of Abbott and Turnbull had just come to power, they valued "domestic politics" over "international politics," which has also led to Australia's relative neglect of African affairs.

The evolution of Australia's African policy, especially Australia–Africa relations as they've existed since the twenty-first century, provides an inspiration and reference for the possibility of sustainable cooperation between other countries and African countries. The shift in interest caused by the change of government has affected the continuity of regional policies, but the creation and accumulation of institutional results helps to overcome the "decay period of interest." The "new engagement policy" of the Rudd and Gillard governments on Africa has given Australia and Africa a new impetus and institutional basis for the development of their cooperative relations. Although the Abbott government and Turnbull government have been cold on African affairs, the political legacy of the "new engagement policy" has pushed Australia–Africa relations to deepen. Australia–Africa relations have not demonstrated any significant setback. The more pragmatic Morrison government may gradually attach greater importance to the African continent on the basis of attaching importance to traditional geopolitical economic regions, and in doing so, continue to tap the value of Australia–Africa cooperation.

PART II

Africa and Its Relationships with Developing Countries

CHAPTER 10

Adjustments to India's African Policy Under the Modi Administration

Guoqing Xu

In order to enhance its status as a great power, increase its regional influence, and alleviate the geopolitical and economic challenges brought to India by the development of China's "Belt and Road" Initiative and existing China–Africa relations, the Modi administration—which took office in May 2014—has taken several steps. On the one hand, it has actively adjusted India's domestic development strategy and planned to promote the country's rapid economic development; on the other hand, under the guidance of the "great peripheral strategy" and the "Indo-Pacific strategy," India under Modi has not only stepped into Asia–Pacific affairs by deepening its relations with the US, Japan, and ASEAN and other countries and regions, but also accelerated the pace of cooperation with African countries. India–Africa relations have shown an all-round, deep-seated, and strategic development trend. The strengthening of India–Africa cooperation holds a certain degree of significance to

G. Xu (✉)
Institute of West Asian and African Studies, Chinese Academy of Social Sciences, Beijing, China
e-mail: wxy03@ssap.cn

X. Li and C. An (eds.), *The Changing World and Africa*,
https://doi.org/10.1007/978-981-16-4983-7_10

the enhancement of India's great power confidence and ability to increase its influence in Africa. India–Africa relations are expected to achieve a more sustainable level of development, but the developmental speed of bilateral relations will continue to be restricted by trade imbalance, a lack of mutual trust, and other factors. This paper attempts to analyze the new progress of India–Africa cooperation, the strategic considerations behind it, and to provide insights into the essence and trends of India's African policy, so as to provide a corresponding reference for China to formulate its own foreign security strategy and African policy.

1 The Main Contents of Modi Administration's Adjustment of African Policy

Since its independence, India's relations with Africa have experienced ups and downs, but they have generally gone through three stages. The first stage is from the beginning of the country's independence up through the 1990s. In order to safeguard national independence, safeguard the rights and interests of African Indians, and provide checks and balances to China and other factors, India actively supported the struggle against colonialism and racial discrimination in Africa. The second stage took place in the early 1990s. With the collapse of the bipolar structure, India's diplomacy shifted its focus to developed countries in Europe and the US in order to cope with domestic economic marketization and liberalization reforms. India–Africa relations were relatively weakened as a result. The third stage has been taking place since the beginning of the twenty-first century. Stimulated by the continuous development of China–Africa cooperation and other factors, India's National Congress Party-led government began to comprehensively adjust its African policy, and India–Africa cooperation has benefited, with India–Africa being fast-tracked and bilateral cooperation in the fields of diplomacy, economy, security, and the humanities having been continuously expanded and deepened.

The Enhancement of Africa's Diplomatic Strategic Position

For a long time, the development fields of India–Africa relations mainly focused on the humanities, the economy, and other fields, while political and diplomatic interactions with African countries were relatively underdeveloped—something that was considered to be a shortcoming of India's African policy. However, with Modi's coming to power, the Indian government has begun carrying out a "neighboring countries first" policy, regarded Africa as India's first foreign policy priority, and has stressed that with the global increase in its involvement in Africa, India needed to deepen its ties with African countries to ensure that Africa will not once again become a stage full of hostile ambitions.[1] All in all, the Modi administration's positive diplomatic measures toward Africa are mainly reflected in three aspects. First among these is the desire to expand the overall scale of diplomacy. In the previous first and second India–Africa Forum Summit, only 14 and 15 African countries, respectively, were invited. However, in October 2015, representatives of all 54 African countries were invited to attend the third India–Africa Forum Summit. It was the largest-scale Indian-African meeting since the Non-Aligned Movement Summit and the Commonwealth Summit held in New Delhi in 1983. During this period, Prime Minister Modi received 19 African heads of government within six hours. The second positive diplomatic measure toward Africa is the closing of high-level bilateral visits. As of November of the same year, the president, vice president, and prime minister of India have completed 29 visits to Seychelles, Nigeria, Kenya, Madagascar, and other countries, covering all regions of Africa. This is regarded by India's Vice President Venkaiah Naidu as an unprecedented diplomatic feat.[2] It is worth mentioning that in 2015, Indian Prime Minister Modi's first visit to Africa was to Seychelles and Mauritius. In October 2017, India's new President, Ram Nath Kovind, started his first overseas visit, visiting Djibouti and Ethiopia. Thirdly and finally, Africa's status in the diplomatic level has been enhanced by its African policies. In order to adapt to the expansion of the scale of the India–Africa Summit and the diversification of its cooperative project objectives, during the third India–Africa Forum Summit, India and Africa decided to upgrade their bilateral relations to a strategic level. Not only that, but India also made great efforts to establish more diplomatic agencies in Africa. Between the years of 2018–2021,

[1] "Prime Minister's Address at Parliament of Uganda During His State Visit to Uganda", July 25, 2018, http://www.mea.gov.in/Speeches-Statements.htm?dtl/30152/Prime_Ministers_address_at_Parliament_of_Uganda_during_his_State_Visit_to_Uganda, accessed on July 16, 2018.

[2] "Vice President Successfully Completes 3-Nation Africa Visit", November 7, 2018, http://orissadiary.com/vice-president-successfully-completes-3-nation-africa-visit/, accessed on January 19, 2019.

the number of India's permanent embassies in Africa will increase from 29 to 47.[3]

India–Africa Economic and Trade Cooperation

The Indian government is in the process of vigorously expanding the content of existing India–Africa economic and trade cooperation mechanisms. In 2014, the Indian Federation of Industry and Commerce held the 10th India–Africa economic and trade cooperation conference with the theme of a "new shared reality partnership." It was said that it would continue the "made in India" exhibition that had launched in Africa in 1995. The Ministry of Commerce and Industry of India stated that it had adjusted the tariff-free preferential scheme for the least developed countries launched in 2008, so that the proportion of India's tariff details covered by it was expanded from 94% to about 98%, so as to facilitate exports from underdeveloped African countries to India. During the 4th India–Africa Energy Conference in January 2016, India and other countries discussed the establishment of joint energy market coordination.

India hopes to strengthen cooperation with Africa in the fields of manufacturing, agriculture, and marine economy. The Indian government has promoted the "made in India" initiative in Ethiopia, South Africa, and other countries, and called on South Africa, Nigeria, and other countries to jointly produce military equipment with Indian enterprises. In August and September 2015, with the cooperation of Seychelles and other countries, India and Mauritius, respectively, held the first "IOR-ARC Dialogue Conference on the Blue Economy" and the first "IOR-ARC Ministerial Conference on the Blue Economy." During the third India–Africa Forum Summit, India and other participating African countries said that they would carry out closer cooperation in marine resource management, hydrographic surveying, and the sustainable development of fisheries. In 2017 and 2018, India successively held the first and second India–Africa ICT Expos in Lagos, Nigeria. In order to adapt to the expansion of the India–Africa Summit and the diversification of the objectives of the cooperation project, as well as to improve the efficiency of the project's implementation, India and Africa claimed that they would set up a formal

[3] "India to Open 18 New Embassies in Africa by 2021", http://www.africanews.com/2018/04/03/india-to-open-18-new-embassies-in-africa-by-2021/, accessed on March 1, 2019.

monitoring mechanism and adjust the cycle of the India–Africa Forum summit from every three years to every five years, raising bilateral relations to a more strategic level.

The Role of Overseas Indians

The Modi administration regards strengthening ties with African Indians overseas as an important agenda for cooperation with Africa. Firstly, India tries to maintain the historical heritage of overseas Indians in Africa. In 2016, Modi retraced the footstep of Gandhi's resistance to racial discrimination in South Africa and promised to provide US$1 million to fund the renovation of the Gandhi graduate library in the University of Nairobi in Kenya. The following year, the National Bank of India provided agricultural equipment to rebuild the Tolstoy farm established by Gandhi in South Africa. The Indian Minister of State for Foreign Affairs attended the opening ceremony of a museum reflecting Gandhi's life and international influence in the South African city of Durban.[4] In addition, there was originally a Gandhi statue in Jinja, Uganda, and in July 2018, Indian Prime Minister Modi said that a Gandhi Heritage Center would be set up in Jinja. It is worth noting that on June 21, 2015, India jointly held "International Yoga Day" activities in Nigeria, South Africa, and other countries. Secondly, India publicly reaffirmed the role of overseas Indians in India–Africa relations, stating that the 2.7 million overseas Indians living in Africa contribute to Africa's development and serve as a bridge for India–Africa cooperation. Former Vice President Ansari pointed out that the people of Indian origin in Rwanda were the most important factor in promoting the development of India–Rwanda relations, and said that he would provide financial support to Indian Rwandans.[5] The Delhi Declaration, which was the outcome of the 2015 India–Africa Summit, was the first to recognize the positive role of overseas Indians in India–Africa relations in the form of a document, pointing out that the overseas Indians in Africa have been instrumental to shaping the long-standing

[4] "Durban in South Africa Opens Doors to Gandhi Museum", *Financial Express*, October 30, 2017, http://www.financialexpress.com/lifestyle/travel-tourism/durban-in-south-africa-opens-doors-to-gandhi-museum/912203/, accessed on October 30, 2017.

[5] Edmund Kagire,"East Africa: India Woos Rwanda, Uganda to Expand Footprint", *The East Africa*, March 3, 2017, http://allafrica.com/stories/201703030681.html, accessed on April 25, 2017.

partnership between India and Africa. Thirdly, India expanded its contacts with overseas Indians in Africa. In 2015, India held the 13th Pravasi Bharatiya Divas. This differed greatly from previous ones, as about 20% of the participants were from French-speaking countries, and for the first time, the meeting discussed specific topics for cooperation with Benin and other French-speaking countries.[6] Fourthly, India made it more convenient to contact overseas Indians in Africa. In 2014, India extended an e-tourism visa to 46 countries, among which were Tanzania and South Africa, mainly to simplify the access procedures and to extend the stay time for those who go to India for business and medical treatment.[7] India approved the air service agreement signed with Rwanda and launched direct flights between the two countries. Fifthly, India ensured the safety of overseas Indians in Africa. During India's evacuation of tens of thousands of overseas Indians from Yemen in 2015, Djibouti, in the Horn of Africa, provided India with assistance from its sea and air facilities. Gurjit Singh, India's former ambassador to Ethiopia and Djibouti, said the move showed that the Indian government attaches great importance to the strategic position of Djibouti.[8]

Enrich the Connotation of India–Africa Security Cooperation

With its traditional influence in African countries, India has further strengthened its defense cooperation with Africa and expanded its strategic alliance and intelligence network in the Indian Ocean region.[9] Modi urged Seychelles to become a comprehensive cooperative partner of India, Sri Lanka, and the Maldives to strengthen maritime security.[10]

[6] Indian Ministry of External Affairs, "Official Spokesperson on the Ongoing Pravasi Bharatiya Divas-2015," January 9, 2015, http://pib.nic.in/newsite/PrintRelease.aspx?relid=114470, accessed on February 12, 2015.

[7] Indian Ministry of Tourism, "India Moves 12 Places Up from 52nd to 40th Position in Travel and Tourism Competitive Index of World Economic Forum," April 7, 2017, http://pib.nic.in/newsite/PrintRelease.aspx?relid=160715, accessed on May 6, 2017.

[8] Elizabeth Roche,"Ram Nath Kovind to Visit Africa This Week on First Foreign Trip as President," October 2, 2017, http://www.livemint.com/Politics/CQkQbBcmYUdTc1xkaWmgMN/Ram-Nath-Kovind-to-visit-Africa-this-week-on-first-foreign-t.html, accessed on October 12, 2017.

[9] *Indian Ocean strategic thinking: Committed to strategic leadership*, David Brewster, translated by Wu Juanjuan.

[10] Indian Ocean economies study, No. 1, 2016, P. 21Ladipu pakalatiWang Tingting.

He promised to support Seychelles in the construction of military infrastructure on Assumption Island, support the capacity building of their navy, hand over a missile speedboat and a second Dornier maritime reconnaissance plane to Seychelles, and launch a coastal radar surveillance plan off the coast of Mahe Island. In July 2016, Modi began an "Indian Ocean diplomacy" tour, during which he visited Mozambique, South Africa, Tanzania, and Kenya. India agreed to provide $500 million US dollars to support India's unconditional cooperation with Mauritius in maritime security and other fields.[11] In November of the following year, India hosted the Indian Ocean regional naval chief of staff meeting for the first time. During the meeting, India promised—also for the first time—to share real-time maritime intelligence with Malaysia, Mauritius, and ten other countries bordering the Indian Ocean in response to the threat posed by China. In addition, in June and July 2017, the Indian navy's most advanced stealth frigate, "Tarkash," successively visited Morocco, Nigeria, Angola, and South Africa to demonstrate India's military strength, conduct joint military exercises with the navies of these countries, and emphasize India's commitment to maritime security in Africa.[12]

India promotes deepening cooperation with African countries in non-traditional security fields such as ecology and counterterrorism. Shortly after taking office, the Modi administration held the first Indian Ocean dialogue meeting to explore cooperation opportunities in disaster management and other fields with countries around the Indian Ocean. Modi said that he would provide assistance to African countries in response to climate change and called on African countries to participate in the International Solar Alliance (ISA) launched by his government. The first and second India–Africa Forum Summit took nuclear disarmament as the main agenda of security and, in order to ensure national security, India and Africa called for nuclear weapons and other weapons of mass destruction to be eradicated completely. In 2016, Modi held consultations with Kenya, Mozambique, and other countries on anti-terrorism

[11] Indian Ministry of External Affairs, "India-Mauritius Sign Four Agreements," May 28, 2017, http://ddinews.gov.in/Home%20-%20Headlines/Pages/agreements.aspx, accessed on June 3, 2017.

[12] "Indian Navy Heading for W/Bay," June 8, 2017, https://www.newera.com.na/2017/06/08/indian-navy-heading-for-wbay/, accessed on June 9, 2017.

issues, stating that cooperation in cyber security and anti-terrorism should be strengthened.

Cooperation with the US, Japan, and Other Big Powers

Since Modi took office, India and the US, Japan, and other countries have had increasingly frequent consultations on the defense agenda as it relates to Africa. In January 2015, India and the US signed a joint strategic vision for the Asia–Pacific and Indian Ocean region, stating that in the expanded region from Africa to East Asia, the two countries would strengthen their partnership to support sustainable, inclusive growth, and regional connectivity.[13] In April of the same year, India and the US held the first consultation meeting on African affairs. In September, India and Australia held the first joint naval exercise. In December, India and Japan agreed to extend the special strategic global partnership between the two countries. Japan also joined the 2015 US and India "Malabar" joint military exercise as a permanent partner. In June of the next year, India and the US held the first "maritime security dialogue" meeting. India's trilateral cooperation with the US, Japan, and Australia was thereby deepened.[14]

Not only that, but Prime Minister Modi also expressed his intention to cooperate with Japan in pursuing further African affairs. This was after Japanese Prime Minister Abe called for the implementation of a free and open "Indo-Pacific" strategy at the 6th Tokyo International Conference on the development of African countries in 2016, so as to strengthen the close cooperation between Japan and India in Asia and Africa.[15] In May 2017, India and Japan launched the Asia-Africa Growth corridor concept document. The document proposes to give full play to these countries' respective advantages, build special economic zones in Africa,

[13] Indian Ministry of External Affairs,"US–India Joint Strategic Vision for the Asia–Pacific and Indian Ocean Region," January 25, 2015, http://mea.gov.in/bilateral-documents.htm?dtl/24728/USIndia_Joint_Strategic_Vision_for_the_AsiaPacific_and_Indian_Ocean_Region, accessed on December 6, 2016.

[14] Lou Chunhao, "New Situation Assessment of US–India Defense Cooperation", *China International Studies*, No. 1, 2017, p. 109.

[15] S. Majumder, "Trump's Isolationist Policy: Will It Nudge India—Japan Relations?" December 16, 2016, http://www.eurasiareview.com/16122016-trumps-isolationist-policy-will-it-nudge-india-japan-relations-oped/, accessed on December 18, 2017.

and strengthen cooperative exercises in African capacity building, infrastructure, health, agriculture, manufacturing, and personnel partnerships. Therefore, India plans to provide a $10 billion investment over the next three years, and Japan will provide $3 billion in the next three years.[16] In addition, on the occasion of his visit to Israel in July 2017, Modi said that he had discussed the issue of cooperation in third-party countries with Israel, stressing that both sides would deepen bilateral cooperation in Africa on the basis of existing cooperation with Africa, so as to boost Africa's future development.[17]

2 The Basic Motivation Behind the Modi Administration's Adjustment of Its African Policies

During the years of Modi's administration, India's national strength has continued to grow, and its desire to enhance its status as a great power has become increasingly strong. However, from India's point of view, China's rapid development and its "Belt and Road" Initiative have brought challenges to its geopolitical security environment. As far as India is concerned, drawing on African forces to strengthen India–Africa cooperation is not only aimed at maintaining sustainable economic development and maintaining domestic political situation, but also responds to the need to check and counterbalance China, enhance India's regional influence, and shape the international environment for India to become a world-class power.

India's Dream of Great Power

Modi's coming to power brings with it a new vision for India to realize its long-standing dream of being a powerful country. On the one hand, the Modi administration has already pushed India's economy out of the "growth dilemma." In 2010, India's GDP growth rate was 8.9%, while

[16] Lin Minwang, "Has the India–Japan Relationship Entered a 'Honey-Moon' Period?", *Contemporary World*, No. 10, 2017, p. 28.

[17] "In Israel, Modi Announces Direct Flights to India", *The Tower*, July 5, 2017, http://www.thetower.org/5167-in-israel-modi-announces-direct-tel-aviv-mumbai-flights/, accessed on July 12, 2017.

in 2011, 2012, and 2013, it fell to 6.7, 4.5, and 4.7%, respectively.[18] In 2015, India's economic growth exceeded that of China by 7.5%. According to the Modi administration's adjusted GDP statistical method, at the ratio of one dollar to 66.6 rupees, India's GDP reached about US$2.3 trillion in 2016, surpassing its former suzerain the UK (whose GDP for that same period was about US$2.29 trillion).[19] On the other hand, Modi himself has garnered the image of a political strongman in Indian politics. During the period in which he served as the chief minister of Gujarat, Modi promoted reform and created a miracle of rapid economic development. From 2001 to 2010, Gujarat had an average annual growth rate of 10.97%. When he won the general election in 2014, Modi vowed to make the twenty-first century be the Indian century and help 1.25 billion Indian people realize their dreams. Modi called for an assessment of the effectiveness of the United Nations over the past 70 years and advocated for the reform of the Security Council to make India a permanent member of it.

In 2014, Indian Foreign Minister Sushma Swaraj said that without African support, India, Germany, Japan, and Brazil would not be able to promote UN Security Council reform.[20] In 2015, India–Africa trade reached 61.963 billion US dollars, and India surpassed the US to become Africa's third largest trade partner, falling only behind the EU and China.[21] India's share of new investment in Africa increased from 3.3% in 2003–2008 to 6.1% in 2009–2015, while China dropped from 4.9 to 3.3% in the same period. In 2015, India added 45 new projects in Africa, making it the fourth largest investor in Africa, just after the US, the UK, and the United Arab Emirates.[22] India believes that the scale of India–Africa political exchanges does not match the increasingly close economic

[18] Lyu Zhaoyi, ed.: Annual Report on India's National Conditions (2016), Social Sciences Academic Press (China), 2017 edition, p. 3.

[19] Ibid.

[20] "Reform UN Security Council by 2015: Sushma Swaraj in NY Diplomatic Overdrive," *News Reality*, September 26, 2014, http://www.firstpost.com/politics/reform-un-security-council-by-2015-sushma-on-diplomatic-overdrive-in-ny-1730801.html, accessed on September 28, 2017.

[21] UNCTAD Handbook of Statistics online, October 28, 2016.

[22] "India Fourth Largest Greenfield Projects Investor in Africa: Arun Jaitley", *Energy Infra Post*, May 23, 2017, https://energyinfrapost.com/india-fourth-largest-greenfield-projects-investor-africa-arun-jaitley/, accessed June 12, 2017.

and trade ties already shared by the two political entities. In order to obtain substantial help from Africa in realizing its status as a great power, India emphasizes that it would provide Africa with more preferential policies and expand the level and scale of diplomatic cooperation. In addition, India believes that American interest in Africa has declined since Trump took office, and this provides an opportunity for India, which advocates increased development cooperation with Africa, to further deepen its ties.

The Adjustment of India's Development Plan

India's total economic growth is obvious, but its development tasks still hold many challenges. By 2014, India had one-third of the world's poor, with their average daily living cost hovering around less than \$1.25.[23] In 2013, India's export volume was only 1.7% of the world's, far lower than China's 11%. In 2014, India's dependence on foreign oil was as high as 77%, and coal accounted for 57% of primary energy consumption.[24] In 2014 and 2015, India's agricultural growth rate was—0.2 and 1.1%, respectively.[25] Affected by that year's drought, the output of beans in India decreased by 12% in 2014, and food became the main driving force to raise the consumer price index. According to the 2014–2015 global competitiveness report released by the World Economic Forum, India's infrastructure ranked only 87th among 148 countries with statistical data, down two places from 2013. According to the estimates of the Confederation of Indian industries, India needs at least \$1 trillion to meet its infrastructure development in 2014–2019.

To this end, India hopes to create a new development model and boost economic growth. In June 2014, the Modi administration released the "let India back on track - economic reform action agenda," emphasizing the strengthening of the construction of the manufacturing industry, grass-roots facilities, and other fields, so as to build India into an internationally competitive manufacturing center.[26] In September, India

[23] Du Youkang and Li Hongmei, "The Internal and External Environment of India's Development and the Strategic Support for Its Rise", *Research on Indian Ocean Economies*, No. 3, 2016, p. 86.

[24] Ibid.

[25] Xing Wanli, "Preliminary study on the impact of India's future energy demand on China's access to overseas energy", *Acta Geosciences Sinica*, No. 1, 2017, p. 39.

[26] India's Economic and Social Sciences Press, 2016, p. 92.

launched the "made in India" plan to relax the restrictions on foreign investment in the military industry, civil aviation, retail, and other fields, and to promote the proportion of manufacturing industry in GDP from 15 to 25% in 2022, achieving an average annual economic growth rate of 9–10%, so as to create employment opportunities for more than 12 million Indian youth entering the market every year.[27] As the world's fifth largest energy consumer, India is expected to double its energy consumption over the next 20 years. India plans to increase overseas energy assets acquisition. In order to expand its foreign trade, India announced a new foreign trade policy in March 2015 to promote the export volume of India to 900 billion US dollars by 2020.[28]

India attaches great importance to potential trade with Africa. Affected by the continuous slump of international commodity prices, the average annual growth rate of African GDP in 2015 was only 3.0%, lower than the 3.8% that was seen in 2014.[29] However, the business environment in Africa has improved significantly. In 2015, more than 30% of all global business regulatory reforms took place in Africa. In 2016, 14 out of 54 African countries grew by more than 5%. The African Development Bank predicts that Africa's economy will grow by 3.4% in 2017 and 3–5% in 2018.

In order to achieve sustainable economic development, the African Union Summit adopted agenda 2063 in June 2015, aimed at promoting peace and prosperity in Africa by declaring that it would focus on industrialization and the green economy, creating jobs, and achieving the goals of eliminating hunger in Africa by 2025 and eradicating war by 2020. The African Union declared 2015–2025 to be the decade of African oceans, officially regarding the blue economy as a new border for Africa's revival. Indian business circles believe that India–Africa trade is expected

[27] Wei Ling: Te "Indian Dream" of Modi Government, People's Forum, No. 15, 2016, p. 71.

[28] Et Bureau, "Government Unveils New Foreign Trade Policy Aims to Raise Exports to $900 billion by 2020," *The Economic Times*, April 2, 2015, http://auto.economictimes.indiatimes.com/news/policy/government-unveils-new-foreign-trade-policy-aims-to-raise-exports-to-900-billion-by-2020/46771769, accessed on October 5, 2016.

[29] Li Zhibiao, "Continued Commodity Slump Has Led to the Divergent Economic Development in Africa", Annual Report on Development in Africa (2015–2016), Social Sciences Academic Press (China), p. 188.

to increase to $117 billion by 2021.[30] India's revenue in Africa will rise to $160 billion in 2025 and is expected to have 2–5% of Africa's agriculture, 10% of electricity, and about 7% of information technology market share.[31]

Overseas Indians Have Increasingly Become an Important Component of India's Quest to Become a Great Power

In Modi's view, overseas Indians are the strength India can use to enhance its status as a great power. The reasons are as follows: first, Modi advocated relying on Hindu culture to consolidate the ruling position of the Bharatiya Janata Party (BJP) and expand India's national influence. The BJP has close ties with the National Volunteer Service Group, a Hindu religious organization. After the 2014 general election, Modi vigorously promoted the concept of a "Hindu identity" and gave it the label of "Indian culture," launching an "expansion movement" of members of the BJP. Most overseas Indians retain Indian cultural traditions and are responsible for expanding India's soft power. Modi stressed that every Indian is proud of their Indian heritage, and he was the first Indian Prime Minister to address the UN General Assembly in Hindi, the national language of India. As a Hindu Yoga enthusiast, Modi launched the "International Yoga Day" initiative in the United Nations. The following year, the Indian government not only established the Hindi Language Bureau under the Ministry of Foreign Affairs, but also established the Ministry of Yoga and Traditional Medicine, appointing the first "Yoga minister" in history.[32] Second, India regards overseas Indians as valuable diplomatic assets. Modi believes that India should not be limited to the South Asian continent, as he claims it exists in the hearts of every Indian in the world. India's Foreign Minister and Minister of Overseas Indian Affairs, Sushma Swaraj, also stressed that ensuring the security of overseas Indians was a

[30] "Bolstering Trade Links Between India & Africa, September 18", 2017, http://www.forbesindia.com/article/sponsored/bolstering-trade-links-between-india-africa/48157/1, accessed on September 23, 2017.

[31] Our Bureau, "India to Quadruple Revenues from Africa by 2025: McKinsey Report," *Business Line*, March 10, 2014.

[32] Long Xingchun, "The Resources, Practices and Enlightenment on the Public Diplomacy of India," *South Asian Studies Quarterly*, No. 1, 2016, p. 26.

top priority for the government. Modi's administration also issued regulations granting Indian card holders a lifetime Indian visa instead of the previous 15-year validity period.[33]

Third, connecting with overseas Indians is conducive to India's economic development. When developing the economy of Gujarat, Modi paid attention to the contact with overseas Indians and deepened foreign economic and trade cooperation. After assuming the post of Prime Minister, Modi issued his ten guiding principles of governance aimed at promoting economic and administrative reform, so as to promote the basic concept and direction of the "Gujarat model" throughout the country. Under the impact of international financial turmoil, Indian capital flight was serious. In the first six months of 2015, the net capital outflow of Indian institutions was as high as 8.8 billion US dollars. The Central Bank of India set up a deposit mechanism for overseas Indians to reverse capital outflow. In 2015, India received US$72 billion in remittances, making it the world's largest recipient of remittances.[34] As a result of this, Indian President Kovind said that the purpose of the Indian government maintaining sustainable and active contact with overseas Indians is to help them familiarize themselves with the changes in India and create the possibility for them to participate in the development of their home country.[35] Fourth, to contain China's "Belt and Road" initiative, the Modi administration put forward a "monsoon plan" and "blue economy strategy" that take advantage of the far-reaching influence of Indian culture in the region around the Indian Ocean, rely on the long history of trade exchanges among ancient Indian Ocean route countries and, under the guidance of India, jointly develop and utilize marine resources, promote economic and trade cooperation, and promote regional economic integration around the Indian Ocean.[36]

[33] Indian Ministry of External Affairs, "Official Spokesperson on the Ongoing Pravasi Bharatiya Divas-2015," January 8, 2015, http://www.mea.gov.in/media-briefings.htm?dtl/24661, accessed January 15, 2015.

[34] "World Bank, India Became Largest Remittances Receiving Country in 2015 Followed by China," December 19, 2015. http://newsonair.nic.in/news.asp?cat=Business&id=BN3276, accessed January 21, 2016.

[35] "India Keen to Strengthen ties with Diaspora: President Ram Nath Kovind," October 5, 2017, http://littleindia.com/india-keen-strengthen-ties-diaspora-president-ram-nath-kovind/, accessed October 7, 2017.

[36] Xu Changchun, "India's Economic Analysis and Outlook", *World Economic Analysis and Outlook (2015–2016)*, 2016, p. 144.

Although Africa's overseas Indians account for only 0.7% of India's total overseas Indians,[37] it has become an important link for the Modi administration to contact overseas Indians. Gandhi's idea of non-violence and non-cooperation originated from his practice of resisting racial discrimination imposed by white authorities in South Africa. So far, Gandhi enjoys a high reputation in Africa. It is worth emphasizing that most of the overseas Indians in Africa come from Gujarat, India.[38] South Africa, a large African country with nearly 1.2 million Indians, is the largest overseas Indian settlement. Six of the 16 cabinet members of the South African government during the Mandela period were of Indian origin. The proportion of Seychelles and Mauritian Indians was as high as 7 and 70.1%, respectively. Mauritius is the seat of the world Hindi Secretariat. The current President of Mauritius, Anerood Jugnauth, is also of Indian origin. Economically, there are fewer than 80,000 Indians in Kenya, but they make up 30–35% of the country's total economy.[39] As early as 1996–2003, the family income of Indian people in South Africa exceeded that of white people. In addition to southeast Africa and other traditional areas of India's foreign exchange, in recent years, the influence of Indian culture in western Africa has also spread. For example, the number of Hindus in Ghana has increased from 24 in the 1970s to about 3000 in 2012, and Hinduism has also spread from Ghana to Togo and other countries.[40]

The Stimulation of the "Belt and Road" Initiative Launched by China

The Modi administration believes that India's strategic environment is facing new challenges. In India's view, Russia's sale of advanced weapons such as Su-35 fighter planes and S-400 air defense missiles to China shows

[37] "Diaspora Networks, Cultural and Historical ties Bolster Diplomatic Drives," February 2, 2012, http://www.trademarksa.org/node/2435, accessed on January 6, 2017.

[38] Amrita Nayak Dutta, "PM Modi's Mann Ki Baat Creates Waves in Africa," June 5, 2017, http://www.dnaindia.com/india/report-pm-s-mann-ki-baat-creates-waves-in-africa-oz-2461674, accessed on June 8, 2017.

[39] Xu Guoqing, "Analysis of India's Cultural Diplomacy Towards Africa", *South Asian Studies*, No. 3, 2013, p. 131.

[40] "Hinduism Spreads in Ghana, Reaches Togo", *Africa Quarterly*, Vol. 51, No. 3–4, August 2011–January 2012, p. 10.

that Sino-Russian relations are warming up. Moreover, relations between Russia and Pakistan are gradually warming up as well. In 2016, Russia and Pakistan also held their first military exercise. India believes that although the main content of the "Belt and Road" Initiative is economic cooperation, it is of great significance for China to expand its diplomatic influence and strengthen its geopolitical status. It points out that China is accelerating its expansion into the Indian Ocean after declaring the South China Sea as its own maritime territory. The construction of the China–Pakistan Economic Corridor initiated by China and the first overseas military base set up in Djibouti, in the Horn of Africa, show that China is growing more expansionist in its global interests. It is an effort to strengthen the "pearl chain strategy" against India, which is bound to damage India's dominance in South Asia and the Indian Ocean. In order to counterbalance the "Belt and Road" Initiative, India hopes to build regional cooperation with India as the main focus. Prime Minister Modi even stated that "the Indian Ocean is the top priority of our diplomacy" and "we must bear the responsibility of shaping the future of the Indian Ocean."[41]

India has obviously accelerated the pace at which it adjusts its foreign and security strategies. Since taking office, Modi put forward the policy of "neighboring countries first" and adjusted the previous "Looking East" policy to the new "Act East" policy. In 2015, India launched the maritime strategy document, "Ensuring a Safe Ocean: India's Maritime Security Strategy." According to the document, the vision of a global and regional geostrategic environment has shifted from the Europe–Atlantic to the Indo-Pacific region, and this change has had a real impact on India's maritime security.[42] Compared with the Indian Ocean Strategy released in 2007, the new marine strategy in 2015 expands the geographical scope of Indian maritime strategic interests, taking the East and West Indian Ocean and Red Sea as the primary areas of Indian maritime focus, and taking the west coast of Africa, the Mediterranean, and other interest areas based on overseas Indian, overseas investment, and political reasons as the second most important area of maritime security interests.[43] In

[41] Wang Shida.

[42] India: from east to west, lookout, 22, 2016, p. 52.

[43] Lou Chunhao, "New Situation Assessment of US–India Defense Cooperation", *China International Studies*, No. 1, 2017, p. 112.

general, the new strategy document highlights India's main agenda in three aspects: seeking port construction in countries with friendly diplomacy and increasing its presence in strategic important places in the region; building a new profile of interest areas, expanding its east-oriented action, and the connecting the west policy.[44]

Moreover, India also pays attention to strengthening allied diplomacy with the US, Japan, and other countries. In January 2015, Prime Minister Modi suggested that the US, Japan, India, and Australia restore their loose security network. In June, India and the US signed a framework for defense relations, stressing that defense cooperation is an important pillar of bilateral relations. The following year, Modi affirmed that India was no longer hesitant and would "embrace" the India–US partnership. India and the US issued a joint statement, stressing that the two countries would strengthen their shared technology to a level equivalent to that of their closest allies and partners.[45]

Increasing Attention to Non-Traditional Security Issues

In recent years, the threat of terrorism has attracted more and more attention from India and Africa. According to the annual report of the global terrorism index released by the US Institute of Economics and Peace, in 2014 and 2015, eight and ten of the top 20 countries most affected by terrorism came from Africa, with India ranking sixth among them for the past two years.[46] After taking office, Modi publicly proposed that he would "take all necessary measures to safeguard India's national interests," and would never tolerate a recurrence of the terrorist attacks that were carried out on the Indian parliament building in 2001 and

[44] Luciane Noronha M. de Oliveira,"Reshaping India's Blue Economy Imperative – Analysis," May 6, 2017, http://www.eurasiareview.com/05062017-reshaping-indias-blue-economy-imperative-analysis/, accessed on May 20, 2017.

[45] Sadanand Dhume, "U.S.–India Relations: Balancing Progress and Managing Expectations," https://www.foreign.senate.gov/imo/media/doc/052416_Dhume_Testimony.pdf, accessed November 9, 2017.

[46] Institute for Economic and Peace, Global Terrorism Index 2015, p. 10. http://economicsandpeace.org/wp-content/uploads/2015/11/Global-Terrorism-Index-2015.pdf, accessed March 6, 2017.

Mumbai in 2008.[47] He called for further strengthening cooperation among countries to combat transnational terrorism.

The maritime security strategy document issued by India in 2015 was primarily concerned with non-traditional threats and state-sponsored terrorism, categorizing it as the navy's top priority.[48] The same year's Ministry of Defense of India's annual report claimed that terrorism and terrorist activities may be the most serious threat to peace and security, and stressed that further strengthening "regional cooperation" was an effective way to improve the ability to deal with non-traditional security challenges.[49] On the other hand, from 2010 to 2014, the prevalence of African terrorist organizations—represented by Boko Haram in Nigeria and Al-Shabaab in Somalia—grew. In 2014, the death toll caused by Boko Haram reached 6644, an increase of 31.7% over 2013, surpassing 6073 caused by Islamic State and ranking first among global terrorist organizations, The Peace and Security Council of the African Union urged all member states to combat terrorism and extremist ideas actively, continuously, and effectively. At the AU summits in January and June 2015, the establishment of regional anti-terrorism forces was put on the agenda, a Pan-African anti-terrorism strategy was planned, and an intelligence-sharing mechanism was established.

India and Africa also face heavy pressure from environmental degradation and climate change. India is the world's third largest emitter of greenhouse gases, with its per capita emissions ranking tenth in the world. In 2013, the health cost caused by environmental pollution reached 3.75 trillion rupees, which was about 3% of India's GDP. The Indian government has set a target of 175 gigawatts of renewable energy power by 2022, including 100 gigawatts of solar power and 60 gigawatts of wind power.[50] By 2030, India will increase the share of non-fossil fuels used

[47] Ding Hao, "Foreign Policy Adjustment and Its Influence of Modi Government of India", *China Military Science*, No. 1, 2017, p. 145.

[48] Ibid.

[49] Ministry of Defence, Government of India, Annual Report 2015–2016, New Delhi, 2016, p. 3.

[50] Xiong Yizhou, "Can India's New Energy Deal Last?" Social Sciences Weekly, 7th Edition, June 18, 2015, p. 1.

in energy consumption to 40%.[51] In 2015, Modi launched the International Solar Alliance (ISA) at the 21st United Nations Climate Change Conference, trying to raise $1 trillion in investment by 2030 by focusing on the development of solar energy in equatorial countries with abundant sunshine but a lack of funds. At present, more than 600 million Africans in sub-Saharan Africa have no access to electricity, and 600,000 people die every year due to smoke from cooking fires.[52] According to the IEA, solar panel prices fell by 75% from 2009 to 2014, and the proportion of renewable energy in Africa's total energy production capacity will increase to 40% by 2040.[53]

3 Three Comments on Modi's New African Policy

As a latecomer to African international relations, India has always emphasized the uniqueness of its cooperation with Africa in terms of foundation, subject, and agenda, to which the Modi administration is no exception. Moreover, in order to highlight its own unique characteristics and adapt to India's dream of becoming a powerful country in the new era, Prime Minister Modi has adjusted the ethical factors to which India has long attached great importance and injected more strength into India's diplomatic strategy. His new policy toward Africa will play a certain role in enhancing India's confidence in the great power and deepening fields of India–Africa cooperation. However, this approach needs to be more effectively observed in the future.

[51] Xiong Yizhou, "Can India's New Energy Deal Last?" Social Sciences Weekly, 7th Edition, June 18, 2015, p. 1.

[52] Chen Lijun, "He Ruifang: India's Role in Promoting BRICs Cooperation Mechanism Construction, BRICs Development Report (2017), Social Science Literature Press, p. 62, *China Electric Power News*, October 19, 2015.

[53] Anna Leach, "Africa Could Lead World on Green Energy", *Guardian*, November 11, 2015. https://www.theguardian.com/global-development-professionals-network/2015/nov/11/africa-could-lead-world-on-green-energy-says-iea-head, accessed on November 16, 2016.

Reflect India's Overall Coordination of African Diplomacy and Internal Affairs

Diplomacy is the extension of internal affairs, and all diplomatic activities are based on internal affairs. To a certain extent, the Modi administration's policy on Africa is not only intended to consolidate the ruling position of the BJP in India, but also to enhance India's international influence. For example, in terms of culture, the Modi administration contacts overseas Indians in Africa and other places to promote yoga based on Hindu practices, which is not only conducive to expanding India's cultural soft power and attracting overseas Indians to participate in India's economic construction, but is also in line with the BJP's promotion of Hindu culture in India and expanding the demand of members of the BJP. In terms of values, Modi believes that democracy is a unique advantage held by India. In his election as Prime Minister, the voter turnout hit a new record high. Modi has used his journey "from grocery clerk to prime minister" as an example of this democratic movement. He stressed that the world should respect Indian democracy and affirmed that India's democratic experience would be publicized in mechanisms such as the India–Brazil–South Africa Dialogue Forum. India also used democratic politics and other values to narrow its relations with Western countries and jointly set foot in African affairs. Modi stressed the multi-dimensional development partnership between India and Africa, saying India has been helping African capacity building.[54] In order to meet the needs of the manufacturing industry and strengthen scientific and technological innovation, the Modi administration set up a Ministry of Skills Development (MSD). In 2017, India also set up an India–Africa knowledge exchange mission to strengthen cooperation with African countries in the fields of energy and skills development.[55]

[54] Luciane Noronha M. de Oliveira, "Reshaping India's Blue Economy Imperative – Analysis," May 6, 2017, http://www.eurasiareview.com/05062017-reshaping-indias-blue-economy-imperative-analysis/, accessed on May 10, 2017.

[55] "Union Minister Dharmendra Pradhan Welcomes the African Nations at India Africa Knowledge Exchange Mission," October 30, 2017, https://www.easterneye.eu/union-minister-dharmendra-pradhan-welcomes-african-nations-india-africa-knowledge-exchange-mission/, accessed on November 5, 2017.

Show the Strength of India's African Policy

In order to showcase India's greatness and its unique characteristics, the Modi administration has incorporated more strength factors into its African policy. First among these was to downplay the political heritage of the Indian National Congress's relations with Africa. Modi did not participate in the 60th anniversary celebration of the Asian-African Conference in April 2015, and the Indian representatives did not mention former Indian Prime Minister Nehru, who was the initiator of the Asian-African Conference. The leaders of Namibia, Egypt, and other African countries expressed their gratitude to Nehru and the Indian National Congress for their support of Africa's liberation. However, Prime Minister Modi and Foreign Minister Sushma Swaraj did not mention Nehru or Indira Gandhi in their speeches at the conference.[56] In September of the following year, Prime Minister Modi was absent from the Non-Aligned Movement Summit. The second way in which the Modi administration attempts to show the strength of India's African policy is to downplay ideological factors and add realism. During the Cold War, the moral elements of India's diplomacy toward Africa were very obvious. However, in recent years, stimulated by the enhancement of national strength and the rapid development of China, India's value-oriented diplomacy has been weakened. From 1998 to 2004, when BJP was in power, the Vajpayee administration strengthened economic and trade cooperation with Africa while pursuing realism, abandoning the moderate foreign policy pursued by the Indian National Congress-led government that insisted on strength and great power diplomacy. After taking office, Prime Minister Modi advocated "strength against strength" and further carried forward the strength policy of the BJP. Compared with India's absence at the Asian-African Conference and the Non-Aligned Movement Summit, Prime Minister Modi changed his previous attitude of irreconcilability with the US, giving top priority to relations with the US and visiting the US four times in two years to enhance the level of cooperation among India, the US, and Japan in various fields. This was undertaken using the Indo-Pacific framework and in order to become the de facto quasi-allies of the US. Out of concern for China's increasing influence in Africa under

[56] Arati R. Jerath, "When Africans Differed with Modi," November 4, 2015. http://www.thestatesman.com/news/opinion/when-africans-differed-with-modi/101461.html, accessed on December 6, 2016.

the Belt and Road Initiative at the same time that the U.S.'s influence is declining, India is deepening its partnership with African countries. At the same time, it has begun to engage in military and economic relations with the US, Japan, and other countries, and has started substantive cooperation on military and economic issues in Africa.

Help India to Enhance Its Confidence as a Great Power

The India–Africa Cooperation Initiative enhances India's position among African countries. Forty-one African heads of government attended the India–Africa Forum Summit in October 2015. On May 25, 2017, the 52nd organizational meeting of the African Development Bank in Ahmedabad was held in India. For the first time since the establishment of the African Development Bank in 1964, India became the fourth non-African Member State to host this meeting. The participation of the heads of three West African countries—Senegal, Benin, and Cote d'Ivoire—was a milestone in India–Africa relations. During this meeting, Cote d'Ivoire, Somalia, and Ghana signed the International Solar Alliance (ISA) agreement, bringing the number of signing countries to 31.[57] India also signed a series of partnership agreements with 13 African countries, including Zambia, to strengthen cooperation in agricultural machinery, farmer credit, and science and technology.[58] South Africa appreciated India's preferential investment policies in defense and other fields, which created opportunities for African countries' private sectors to participate in economic and trade cooperation with India. Kenyatta, President of Kenya believes that Indian people are part of the country's society and have made indelible contributions in the country's fight for freedom and resisting colonial rule and, further, are a major contributor to national

[57] Bytesapp, "India to Invest $2 bn for Solar Projects in Africa," May 23, 2017, https://energyinfrapost.com/india-invest-2-bn-solar-projects-africa/, accessed on June 8, 2017.

[58] David Ibemere, "13 African Countries Sign Deal with India On Supply of Farm Machinery," May 31, 2017, http://thenerveafrica.com/11775/13-african-countries-sign-deal-with-india-on-supply-of-farm-machinery/, accessed on June 22, 2017.

tax revenue.[59] At the end of 2016, Tanzania, Kenya, and other countries held Indian cultural festivals.

The development of India–Africa relations is conducive to enhancing India's regional and international influence. Mauritius, Seychelles, and other countries have participated in the "blue economy" strategy advocated by India. South Africa expressed support of India's assumption of the rotating presidency of the Indian Ocean Rim Association for Regional Cooperation from 2017 to 2019. In 2017, India and the European Union declared the importance of maritime security, stability, and connectivity in the development of the blue economy, announcing that the two sides would strengthen maritime security cooperation in the Indian Ocean.[60] After that, Italian Prime Minister Paolo Gentiloni promised to coordinate with India on the cooperative initiatives with Africa and to support African peace and development, stating that Italy expected India to become an observer state of the EU-African Union Summit in the future.[61] It is worth mentioning that, to a certain extent, strengthening cooperation with Africa is conducive to cooperating with India's "Act East" policy and enhancing India's voice in Asia–Pacific region. The 2014 joint statement of India and the US indicated that these two countries promised to cooperate more closely with other Asia–Pacific countries and stressed the importance of trilateral dialogue with Japan, believing that it was necessary to ensure maritime safety and the freedom of navigation and overflight (especially in the South China Sea). The joint strategic vision for the Asia–Pacific and Indian Ocean region signed by India and the US emphasizes the fact that a closer India–US partnership is indispensable for promoting peace, prosperity, and stability in the Asia–Pacific and Indian Ocean regions.[62] In addition, Modi's proposal for "International

[59] FICCI, "Museveni Asks Indian Automobile Manufacturers to 'Make in Uganda'", February 23, 2017, http://ficci.in/ficci-in-news-page.asp?nid=12268, accessed on March 26, 2017.

[60] "Somalia: India and EU to Enhance Maritime Security Off Somalia Coast", *Shebella Media Network*, October 7, 2017, http://allafrica.com/stories/201710090182.html, accessed October 5, 2017.

[61] Indian Ministry of External Affairs, "Narendra Modi, Paolo Gentiloni Emphasise Importance of Regular Meetings", October 31, 2017, http://www.firstpost.com/india/india-italy-joint-statement-full-text-narendra-modi-paolo-gentiloni-emphasise-importance-of-regular-high-level-meetings-4184105.html, accessed November 3, 2017.

[62] Lou Chunhao, "New Situation Assessment of US–India Defense Cooperation", *China International Studies*, No. 1, 2017, p. 113.

Yoga Day" at the UN General Assembly was supported by 177 countries, including South Africa. India believes that this is the UN General Assembly initiative with the most support from the largest number of countries. Although the third India–Africa Summit document does not explicitly support India as a member of the UN Security Council, India believes that it has become one of the few countries that can hold a summit with all African countries. Nigeria and other major African powers and the AU have reached consensus with India on Security Council reform.

India–Africa Relations Face Many Difficulties

In the future, India's relations with Africa will need to overcome many difficulties. At the political level, as a supporter of the African liberation movement and the founder of the Non-Aligned Movement, India has long regarded "strategic autonomy" as the standard by which to act. However, the current diplomatic action of the Modi administration in Africa reflects, to a certain extent, India's policy of "strategic autonomy" and "non-alignment." On the other hand, compared with the "Belt and Road" Initiative, which is advocated by China for mutual benefit and win–win results and does not seek a sphere of influence, India's many initiatives toward Africa have strategic demands such as seeking regional leadership. This runs counter to the "united and self-improvement" advocated by Africa. In terms of economy and trade, although India publicizes its demographic dividends and market potential, its business environment is still unsatisfactory, which not only restricts the implementation of India's manufacturing strategy, but also reduces foreign investment confidence in India. According to the World Bank's 2016 business environment report, India ranked 130th in the Business Facilitation Index of 189 countries and regions in the world. Although India rose twelve places, up from 142nd in 2015, this is still far less than China's 84th place. Under the resistance of trade unions and farmers, it is difficult for Modi administration to make substantive breakthroughs in the core areas of reform such as labor and land acquisition. In addition, India's imports from Africa are dominated by energy, and its economic and trade deficit with Africa is difficult to change. It is not only contrary to Africa's desire to change the export market of raw materials, but also limits India's trading partners in Africa to a small number of energy rich African countries. According to the statistics of the United Nations Conference on Trade and Development, in 2018, India imported 20.048 billion US dollars of oil from Africa, accounting for 47.5% of India's total import trade from Africa, and India's trade deficit with Africa reached 15.207 billion US dollars.[63] It is estimated that India's cumulative oil imports will reach 3.777 billion tons from 2015 to 2030, becoming the main driver of global oil consumption.[64] As far as the field of humanities is concerned, India's increasingly frequent incidents of discrimination against Africans cast a shadow on India's image in Africa and cultural exchanges between India and Africa. In May 2016, a Congolese (DRC) teacher who was teaching in India was killed and 42 African countries accused India of not seriously investigating the case. They also called on African countries to refuse to participate in India's "Africa Day" activities and stop sending students to India. Five Nigerians were detained by Indian police for the death of an Indian boy on charges of providing drugs. Many Indians also attacked African students. After that, although the suspected Nigerians were released for lack of evidence, African diplomatic envoys in India held a special meeting to express concern about the safety of Africans in India. This involved condemning the Indian government's failure to enforce strong legal action against criminals, calling on human rights bodies such as the UN Human Rights Council to carry out independent investigations and provide a comprehensive investigative report to the African Union Commission.[65]

4 Conclusion

The change in the Modi administration's policy toward Africa and the new measures taken by it are inseparable from the growth of India's national strength and the concurrent strategic competition currently intensifying between China and India. It is an important part of India's commitment to the dream of becoming a big country and speeding up its overseas strategic layout. It is also the epitome of the deepening geopolitical game between China and India. India has always regarded itself

[63] UN Comtrade Database, December 20, 2019.

[64] Xing Wanli, "Preliminary Study on the Impact of India's Future Energy Demand on China's Access to Overseas Energy," *Acta Geosciences Sinica*, No. 1, 2017, p. 43.

[65] Kallol Bhattacherjee, "African Envoys Seek U.N. Probe into Noida Attacks", *The Hindu*, April 3, 2017, http://www.thehindu.com/news/national/hrc-should-investigate-anti-african-attacks-african-envoys/article17768164.ece, accessed on May 16, 2017.

as the "leader" of the third world. With the development of China and the advancement of the "Belt and Road" Initiative, India, whose overall strength is far weaker, is strengthening its relations with the US and Japan, and at the same time, also counting on Africa's support in seeking a permanent member of the UN Security Council. It also relies on African support for competing with China in the fight for leadership in Asia, in developing countries, and the world stage, and of consolidating their influence in the Indian Ocean region.

China and India have border disputes, but they are both big developing countries. Facing the deepening of India–Africa relations, China must, on the one hand, maintain strategic confidence. China's GDP is five times that of India's, so it does not need to fall into vicious competition with India in Africa-related affairs, thereby forcing African countries to choose sides between China and India. Such a situation would be unfavorable to South–South cooperation and the construction of the "Belt and Road" Initiative. China can play to its advantages in infrastructure, manufacturing, and other fields. China should participate in the process of African integration and encourage the transfer of labor-intensive industries with technological advantages to Africa in order to meet the strategic needs of upgrading domestic industrial structures. China can also actively participate in the UN 2030 agenda for sustainable development and strengthening the level of cooperation between China and Africa in the fields of new energy, ecological agriculture, and the marine economy. On the other hand, it should also be noted that given the lack of strategic mutual trust between China and India, India's advancement of China-oriented interests in Africa is bound to worsen the strategic environment for the implementation of the "Belt and Road" Initiative in the region. In this regard, China should not only attach greater importance to the development of a multilevel dialogue mechanism with India and enhance mutual trust, but also take certain preventive measures. For example, in terms of the economy, India has certain technical advantages in health, agriculture, information technology, and other fields. In 2015, India and the US held the first health dialogue meeting, which prompted Chinese enterprises to avoid fierce competition with Indian companies in health, agriculture, and other fields when necessary, and also reduce unnecessary costs. Culturally, China should pay attention to the construction of soft power in Africa, which can not only make use of its own development experience, but also actively participate in the economic and social development process of Africa and support the integration of Chinese people into local society. In terms of security, compared with India, China, as Africa's largest trading partner, has lagged behind in security cooperation with Africa. This should remind China that while strengthening its security cooperation with Pakistan and other countries, China should speed up the pace of China–Africa defense cooperation. China should actively build a variety of security cooperation mechanisms with African countries, promote the stability and balance of defense forces in the region, provide guarantees for further China–Africa cooperation, and alleviate China's defense pressure in the Asia–Pacific region to a certain extent.

CHAPTER 11

The Features and Limitations of Turkey's Foreign Policy Toward Africa

Min Wei

Turkey's foreign policy toward Africa has made great strides in recent years, and has created new markets and investment potential that will enhance Turkey's role as a major player in the politics of resource allocation in the international economic system.[1] As an emerging economy, Turkey has passed through the three levels of soft power, economic diplomacy, and security systems in only two decades. Turkey's foreign policy toward Africa not only strengthens its political and economic ties with African countries and maximizes its national interests, but also greatly

[1]Chigozie Enwere and Mesut Yilmaz, "Turkey's Strategic Economic Relations with Africa: Trends and Challenges", *Journal of Economics and Political Economy,* Flight. 1, Issue 2, 2014.

The main research field is the world economy, the industrialization and development in the Middle East countries.

M. Wei (✉)
Institute of West Asia and African Studies, Chinese Academy of Social Sciences, Beijing, China
e-mail: wxy03@ssap.cn

X. Li and C. An (eds.), *The Changing World and Africa*,
https://doi.org/10.1007/978-981-16-4983-7_11

enhances Turkey's influence and discursive power in regional affairs. The African continent, both as individual states or as a regional organization, has responded positively to Turkey's participation. The African Union (AU) granted Turkey observer status in 2005 and, in 2008, also declared Turkey to be a "strategic partner" of the African continent. In January 2013, Turkey became a non-regional member of the African Development Bank and the African Fund. In 2017, Turkey's first overseas military base was established in Mogadishu, and in December of the same year, Turkey reached a lease agreement with the Sudanese government to rebuild Swakin, Africa's oldest island on the Red Sea coast. In December 2019, Turkey sent a high-profile military presence to Libya. Turkey's foreign policy orientation and behaviour toward Africa has aroused widespread concern in the international community.

After coming to power, the Justice and Development Party proposed shaping the image of a "New Turkey," and its subsequent national strategy of "New Ottomanism" became the basis of Turkey's policy toward Africa.[2] "New Ottomanism" defines Turkey as a country that can live in peace with its own Muslim identity and which "has a sense of greatness and self-confidence"; its policy objective is to create a new Turkey based on generations of devout Muslims.[3] In order to achieve this goal, the Justice and Development Party has adopted a continuous multilevel foreign policy. On the one hand, while firm secularism remains in the mainstream of modern Turkish ideology, successive governments have been unable to resist the temptation to exert influence over Turkish society by controlling religious institutions. On the other hand, the government also intends to establish a modern model of the Islamic State and expand its voice and influence in the Islamic world.[4] In the process of forming this new state orientation, Turkey has emphasized its concept, construction, and cultural factors, gives full play to its advantages in education, culture, and humanitarian assistance and, taking soft power

[2] Asya Akca, "Neo-Ottomanism: Turkey's Foreign Policy Approach to Africa", https://www.csis.org/neo-ottomanism-turkeys-foreign-policy-approach-africa, achieved on February 2, 2020.

[3] Zan Tao, "An Analysis of Contemporary Turkey's Foreign Strategy", *China International Strategy Review*, 2011, pp. 279–289.

[4] Jenny White, *Muslim Nationalism and the New Turks*. Princeton, NJ: Princeton University Press, 2013, p. 5.

as its starting point, promotes the furthering of cooperation with non-state actors. Turkey has long followed two principles in its policy-based practice; one of these is the pursuit of "New Ottomanism" to enhance Turkey's status in the ideological sphere, to expand Turkey's influence in the outside world (especially in the Islamic world), and to maximize the economy and security as priorities in its external relations in accordance with the principle of realism. Some scholars argue that in Turkish diplomatic practice, Turkey has often conducted a Realpolitik foreign policy wherein Turkey's security and economic priorities have played a greater role in shaping Turkey's foreign policy in an international context. Moreover, Turkish foreign policy has been most successful when it has acted in a manner consistent with the principles of Realpolitik, and conversely unsuccessful when it pursued ideological policies.[5] A foreign policy rooted in ideology is less flexible and less capable of adjusting to current international dynamics.[6] In this paper, the Justice and Development Party's foreign policy toward Africa is a rational choice based on realism and constructivism, and its core is still Turkey's national interest. In the course of implementation, using the influence of the modern Turkish model and the cultural heritage of the Ottoman Empire to maximize the national interest by demonstrating its advantages in enterprising and humanitarian foreign policy on the one hand,[7] and on the other hand, using Somalia in the Horn of Africa as a breakthrough point to enhance its influence and voice in regional affairs, Turkey has finally realized the transformation from soft power to hard power and achieved a dual policy effect. Based on the evolution of Turkey's Africa policy, this paper analyzes the features and limitations of Turkey's Africa policy, and finally evaluates the impact of the policy.

[5] Seyfi Taşhan, "The Factors Impacting Turkish Foreign Policy", *External Policy-Foreign Policy*, Vol. XXXXIII, No. 1, 2016, http://foreignpolicy.org.tr/dis-politika-foreign-policy-vol-xxxxiii-no-1-2016-an-essay-on-the-factors-impacting-turkish-foreign-policy-seyfi-tashan/, achieved on March 10, 2020.

[6] Szymon Ananic, "Alone in Virtue the "New Turkish" Ideology in Turkey's Foreign Policy", *Point of View*, No. 49, April 2015, p. 6.

[7] Official website of the Ministry of Foreign Affairs of Turkey, Turkey's Enterprising and Humanitarian Foreign Policy/Rep. of Turkey Ministry of Foreign Affairs, http://www.mfa.gov.tr/synopsis-of-the-turkish-foreign-policy.en.mfa, achieved on February 20, 2020.

1 The Evolution of Turkey's Soft Power and Africa Policy

Soft power is a staple of daily democratic politics. The soft power of a country rests primarily on three resources: its culture (in places where it is attractive to others), its political values (when it lives up to them at home and abroad), and its foreign policies (when they are seen as legitimate and having moral authority).[8] At its core is attraction rather than force deterrence or economic transactions, so soft power depends on how we shape our goals. Policies based on broad and far-sighted national interests are more attractive than narrow and short-sighted policies. The history, cultural heritage, and political values of the new Turkish model and the Ottoman Empire have become the cornerstone of Turkish diplomacy. In the early days of the formation of foreign policy toward Africa, Turkey highlighted its institutional and cultural advantages by assisting African countries in establishing peace and development plans, while soft power-based public diplomacy became an important means for Turkey to develop relations with Africa.

Turkey's Soft Power

A well-known Turkish politician, Ahmet Davutoglu, believes that a country's strength can be measured by the economic, cultural, and diplomatic activism of its foreign policy. He put forward the concept of "Strategic Depth,"[9] arguing that Turkey has a clear diplomatic "strategic depth" and "geographical depth" and thus has the natural advantage of becoming a "central country" and "positive contributor" to international politics.[10] This strategic thinking is to be the guiding principle of Turkey's policy toward Africa.

First of all, to this end, deep historical and cultural roots have become important components of soft power in Turkey. As the successor of the Ottoman Empire, Turkey has abundant resources in culture, education,

[8] Joseph S. Nye, Jr., Public Diplomacy and Soft Power, http://sage.cnpereading.com/paragraph/download/10.1177/0002716207311699, achieved on March 7, 2020.

[9] K. Kirişçi, "The Transformation of Turkish Foreign Policy: The Rise of the Trading State" (2009). *New Perspectives on Turkey*, 40: 29–57.

[10] Loannis N. Grigoriadis, "The Davutoglu Doctrine and Turkish Foreign Policy", Hellenic Foundation for European and Foreign Policy, Working Paper Nr 8/2010, p. 5.

art, poetry, literature, architecture, and more. As early as the ninth and tenth centuries, Turks established Tulunids, which ruled today's Egypt, while at the height of the Ottoman Empire, its influence spread over Egypt, Tunisia, Algeria, Libya, Morocco, and Ethiopia and Somalia in the Horn of Africa. Not only did Turkey experience no colonization on the African continent, but the Ottoman Empire actively prevented colonial expansion in North Africa. Similarly, Ottoman Empire successfully fended off Western colonial invasions of North and East Africa.[11]

Second, the political system and its associated values have become an important component of Turkey's soft power. In the field of international relations, the recognition and acceptance of a country's system and policy is the ultimate embodiment of a country's soft power. Turkey was once seen as an example of the successful establishment of a secular constitutional system in the Islamic State, and the modernization of secularism—or Kemalism—is an important Turkish legacy. Multiple actions have struck a blow against Turkish–African relations, including the practice of pursuing a pro-Western foreign policy during the Cold War, Turkey's pro-Western stance during the Suez Canal crisis, diplomatic tensions between Turkey and the African states of the Non-Aligned Movement, Turkey's pro-Western attitude in the 1955 Bandung Conference, and Turkey's abstention in the United Nations General Assembly 1956 vote for Algeria's independence were dealt.[12] Turkey only established diplomatic relations with a few African countries, such as Morocco and Tunisia, as well as Ethiopia, Ghana, and Nigeria in sub-Saharan Africa. In the late 1960s, with the rise of Islamist parties and conflicts with the West in Cyprus, Turkey began to seek a more independent foreign policy and sought to exercise power or influence beyond its borders in order to change the geopolitical situation.[13] By the end of the Cold War, Turkey had tried to change its traditionally pro-Western strategic position, increasingly displaying the image of a regional "central state" and making

[11] Ibid.

[12] Volkan İpek, and Gonca Biltekin, "Turkey's Foreign Policy Implementation in Sub-Saharan Africa: A Post-international Approach". *New Perspectives* on Turkey, 49, 2013, pp. 121–156. https://doi.org/10.1017/S0896634600002065.

[13] Zbigniew Brzezinski, *The Grand Chessboard: American Primacy and its Geostrategic Imperatives*, New York: Basic Books, 1997, pp. 40–41, 47 and 149–150.

no secret of its attempts to become a global power. Thanks to the new geopolitical vision and the rising reality of new Turkey, the secularization of the modern Islamic State is among the factors that laid the foundation for Turkey's soft power and institutional appeal.

Introduction to the Concept of a "New Africa"

During the government of Turgut Ozal (1983–1993), Turkey initiated economic reforms and structural adjustments. The rejection of Turkey's application to join the European Union in 1989 and subsequent changes in the post-Cold War pattern prompted Turkey to turn its diplomatic thinking away from pro-Western countries and actively seek to develop cooperation with the Middle East, Central Asia, the South Caucasus, and the Balkans. After the collapse of the Soviet Union, Turkey targeted schools, mosques, students, and imams in the newly established countries in Central Asia and the Balkans to begin its cultural infiltration. In the late 1990s, with the rise of economic integration, Turkey's structural adjustment was beginning to bear fruit and its economic power increased significantly under the guidance of neo-liberalism. The emerging Turkish middle class desperately needed to expand into the new global markets, and Turkey's social elite turned its attention from Europe and the West to Turkey's immediate neighbors and neighboring countries. Turkey adopted the Africa Action Plan in 1998 with the aim of opening up to Africa and creating a new image of Africa in Turkish society.[14] The plan sets out the concept of "New Africa," integrates North Africa and sub-Saharan Africa, two geographical concepts, into one, removed the negative impression of sub-Saharan Africa as a byword for ethnic conflict, disease, and natural disaster in Turkish society,[15] diluted Turkey's image as a traditional ally of the North African countries, and created a new image of Turkey as a constructive participant in the Africa continent. The introduction to the concept of "New Africa" expanded Turkey's diplomatic orientation and marked Africa's formal entry into Turkey's diplomatic horizon; however, at this time, the Turkish government had not yet had

[14] Official website of the Ministry of Foreign Affairs of Turkey-Africa Relations, http://www.mfa.gov.tr/turkey-africa-relations.en.mfa, achieved on February 2, 2020.

[15] Mehmet Özkan, "Turkey's Rising Role in Africa", *Turkish Policy Quarterly*, Vol. 9, No. 4, 2015, p. 94.

time to consider the formulation of a complete foreign policy toward Africa.[16]

The plan established a road map for the future improvement of Turkey's foreign relations with African countries, including the following aspects: politically, inviting high-level officials from African countries (presidents, prime ministers, and ministers) to visit Turkey, increasing the number of embassies in African countries, increasing political consultations on African affairs with international organizations such as the United Nations and Islamic Conference Organization[17]; and adopting the Africa Action Plan as the fundamental document underlying of Turkey's policy toward Africa. However, the 1999 earthquake and the 2000–2001 financial crisis delayed the implementation of the plan.

In 2002, the AKP won the election, took to the Turkish political scene, and issued six basic principles of foreign policy, stressing that future Turkish foreign policy would focus on multidimensional and active regional diplomacy.[18] The Ministry of Trade formulated the Strategy for the Development of Economic Relations with African Countries in 2003, proposing 16 initiatives to promote economic development, including the need to strengthen trade with African countries, to increase investment in Africa, to encourage small and medium-sized enterprises to enter the continent, etc. The Turkish Office for International Cooperation and Coordination (TiKA) is responsible for promoting Turkey's economic and trade cooperation with African countries. In the area of education and culture, the Turkish Religious Affairs Bureau (Diyanet) is still responsible for the promotion of cultural education and imam exchange projects in Africa.

[16] Chigozie Enwere, and Mesut Yilmaz, "Turkey's Strategic Economic Relations with Africa: Trends and Challenges" *Journal of Economics and Political Economy*, Vol. 1, Issue 2, 2014.

[17] Official website of the Ministry of Foreign Affairs of TurkeyTurkey–Africa Relations, http://www.mfa.gov.tr/turkey-africa-relations.en.mfa, achieved on February 2, 2020.

[18] Note: The six core principles of Turkey's foreign policy are: the balance of security and freedom; See Ahmet Davutoglu, "Zero Problems in a New Era", https://foreignpolicy.com/2013/03/21/zero-problems-in-a-new-era/, achieved on April 26, 2018.

"Opening Up to Africa"—The Initial Formation of the Foreign Policy Framework

In order to build an overall diplomatic approach toward the Africa continent, expand Turkey's diplomatic and commercial presence across the continent, and increase Turkey's influence on the continent, the government declared 2005 as the "Year of Africa" and launched a comprehensive "Opening up to Africa" policy. Then-Prime Minister Recep Tayyip Erdogan visited Ethiopia and South Africa, becoming the first Turkish Prime Minister to make an official visit to the south equatorial region in Africa. In the same year, the African Union granted observer status to Turkey. In 2007, Turkey hosted the first LDC Cooperation Summit. At the Tenth Summit of the African Union in January 2008, Turkey became a "strategic partner" of the Africa continent. In August, the first Africa–Turkey Partnership Summit was held in Istanbul, opening up the process to initiate stable and sustainable cooperation among Turkey and African countries. Two documents, the Istanbul Declaration on Turkey–Africa Partnership: Cooperation and Solidarity for a Common Future and the Cooperation Framework for Turkey–Africa Partnership were adopted. The First Turkey–Africa Cooperation Summit of 2008 established a follow-up mechanism, that of setting out the areas and framework for cooperation between the two sides and jointly developing the Implementation Plan 2010–2014. The objective of this is to consolidate and develop Turkey–Africa cooperation at all levels and in all fields, to establish stable and long-term partnerships based on common interests and equality, and to strengthen cooperation in the field of special interests. As Turkey's foreign policy toward Africa continued to become more refined and detailed, the African Development Bank accepted Turkey's application to join, and in the same year Turkey became a member of the African Intergovernmental Development Partners Forum (IGAD). In 2009, 50 of the 53 African countries supported Turkey as a non-permanent member of the United Nations Security Council. In January 2013, Turkey became a non-regional member of the African Development Bank and African Development Fund, directly involved in investment and financing activities in various African countries.

Building a "Partnership" Policy: From Soft Power to Hard Power

The 2011 upheavals and the subsequent dramatic changes in the Middle East and international environment caused by NATO troops in Libya have had an important impact on Turkey's foreign policy. At this time, the situation in Turkey has also changed, with the Justice and Development Party and the "Gülen Movement"[19] having split in 2013. The Justice and Development Party (AKP) entered its third term in 2014, when Erdogan was elected President, and began to focus on promoting culture, the economy, and security, in its pursuit of a systematic partnership with Africa. Simultaneously, Africa became a priority area for Turkey's foreign policy.

In November 2014, the second Turkey–Africa Partnership Cooperation Summit was held in Malabo, Equatorial Guinea. The theme of the meeting was "a new partnership model for strengthening sustainable development and integration." At this summit, the Declaration and Joint Implementation Plan 2015–2019 were adopted and the seven principles for foreign policy toward Africa were first proposed. These include: firstly, establishing closer political relations by intensifying bilateral high-level visits and, by acting as the voice of Africa, defending the legitimate rights and interests of African countries in bilateral and multilateral talks. Secondly, supporting African countries in the economic sphere to overcome their difficulties through trade, investment, and humanitarian assistance. Thirdly, playing a role through diplomacy in the peaceful settlement of disputes on the continent. Fourthly, providing assistance to the continent to make further progress in the areas of democracy and good governance. Fifthly, giving support to the international and regional organizations of the continent in order to increase dialogue, understanding, and peace in the region. Sixthly, actively participate in African peacekeeping missions in Africa. Seventhly, in accordance with the policy of the

[19] Gülen Movement originated in the late 1990s, it was originally a youth social network and later developed into a transnational educational, media, and business organization, recognized by Turkey and the Muslim world, headed by Turkish religious figure Fethullah Gülen, the organization's branch has spread over more than 120 countries around the world. After July 15, 2016 Coup Attempted, which caused 251 people death, and near 2200 people injured. The Turkish government declared that The Gülen organization staged the coup and identified the Gülen Movement as a terrorist organization.

African Union, adhere to the principle of an African solution for African problems.[20]

In May 2017, the Turkish Foreign Minister announced that the "opening up to Africa" policy had been successfully transformed into more of a "partner" policy, which is intended to usher in a new era of integration of soft power and hard power. In September, Turkey became the 55th country to open a military base in Africa when a $50 million military base was built in Somalia. In December, Turkey reached an agreement with Sudan to jointly combat terrorism in East Africa and the Horn of Africa, strengthen military cooperation in the Red Sea, and gain influence on the continent.

2 The Features of Turkey's African Policy

Turkey's foreign policy toward Africa began with soft power and public diplomacy, and the scope and influence of cultural penetration on the continent has expanded, culminating in an increase in the range from soft power to hard power. At present, Turkey's position on the African continent has been established, and the foreign policy toward Africa not only explains the principles of Turkey's enterprising humanitarian foreign policy after the AKP came to power, but also reflects its unique perspective and approach at the regional diplomatic level, which to some extent is undoubtedly successful.

Turkey's Soft Power—An Entry Point for African Policy

From the end of the Cold War to the first decade of the Justice and Development Party's rule, Turkey pursued a more independent foreign policy and sought to build further relations with its neighbors. In 2010, the Turkish Prime Minister's Office set up the Office of Public Diplomacy as a platform for the implementation of the country's soft power, as well Turkey's soft power diplomacy with African countries. This was intended to promote educational and cultural cooperation and promote the construction of schools and the Yunus Emre Turkish Cultural Centre. Turkey is committed to assisting the African countries in achieving their national education goals. Since 1991, Turkey has established schools,

[20] TURKEY-AFRICA RELATIONS / Rep. of Turkey Ministry of Foreign Affairs, http://www.mfa.gov.tr/turkey-africa-relations.en.mfa, achieved on February 20, 2020.

libraries, and laboratories in nearly 20 countries in Africa and provided technical equipment to numerous universities. A large number of scholarship programs have been offered to African countries, providing 1239 scholarships per year for students from Africa. There are currently 5437 higher education students and 116 visiting professors or research assistants from African countries studying or working in Turkey.[21]

Turkey also uses mosque diplomacy to increase its influence in the Islamic world. Since its inception, the Turkish Cooperation and Coordination Agency (TİKA) has carried out nearly 30,000 projects and activities across five continents through its 62 Program Coordination Offices located in 60 countries, and has implemented more than 140 mosque restoration projects to protect the common history and cultural heritage of the Ottoman Empire. The restoration of the Ketchaoua Mosque in Algeria, a UNESCO World Heritage Site, and the restoration of the Tombs of King Najashi and His Companions in Ethiopia are the most well known. Over the past five years, Turkey has helped African countries restore more than 50 mosques, 10 tombs, and various complexes (mosques, mausoleums, castles, bridges, clock towers, markets, schools, and various administrative buildings) and made them available to visitors.[22]

These policies have brought great rewards to Turkey, not only by establishing a good national image in Africa, but also by expanding its political, cultural, and religious influence. More importantly, by using soft power diplomacy, Turkey has cultivated a large number of elites in various fields in Africa and has become a strong guarantor for the future development of Turkey–Africa relations.

Somalia as a Breakthrough to Enhance Turkey's International Influence in the Horn of Africa

Turkey's foreign policy toward Somalia has not only gained its soft power status, which had long been neglected by the West in resolving the Somali problem, but Turkey has also fully supported Somalia's reconstruction. With Somalia as a breakthrough, Turkey has completed its

[21] Ibid.

[22] Official website of the Turkish Cooperation and Coordination Agency, https://www.tika.gov.tr/en/page/about_us-14650, achieved on March 10, 2020.

overall diplomatic approach to the Horn of Africa, with the "Somalia model" becoming Turkey's diplomatic success story. The Horn of Africa has long been a major focus of international attention because of war, famine, piracy, and the war on terror. Turkey, on the other hand, has taken a step-by-step approach, drawing on the legacy of the Ottoman Empire, and has become an important stakeholder in Somalia.[23] The primary aspects characterizing Turkey's foreign policy in Somalia are: first, to actively organize or participate in bilateral and multilateral consultation mechanisms. Turkey hosted the Istanbul Somalia Conference on two occasions within the framework of the United Nations, from 21 to23 and 31 May 2010. Since 2013, Turkey has been active in promoting the talks between the federal government of Somalia and the Somaliland administration. In June 2014, the Turkish consulate general in Hargeisa became operational. The High-level Partnership Forum for Somalia was hosted in Istanbul from 23 to 24 February 2016. The second aspect is characterized by frequent high-level visits. Since 2011, Erdogan has visited Somalia three times as Prime Minister or President, and he has launched a humanitarian initiative for Somalia that includes not only economic relations, humanitarian assistance, and development assistance for Somalia, but has also included a focus on Turkey's willingness to help Somalia restore political and public order and security. Third, humanitarian assistance and direct investment go together. Since the outset of President Erdogan's humanitarian initiatives, the Turkish Agency for International Cooperation and Coordination, the Turkish Red Crescent Society, and various Turkish non-governmental organizations have carried out Turkey's largest international assistance operation to date, providing humanitarian and technical development assistance to Somalia amounting to more than $1 billion. Projects in health, education, municipal services, and infrastructure, including the country's largest hospital (with 200 beds) and the largest Turkish embassy council in the world were officially inaugurated in Mogadishu.[24] At the same time, Mogadishu International Airport and Mogadishu Sea Port are run by Turkish companies in Somalia. During this period, Turkey participated in the reconciliation of relations between

[23] Mehmet Ozkan, "Turkey's Political-Economic Engagement With Africa", In Justin van der Merwe, Ian Taylor, and Alexandra Arkhangelskaya (eds.), *Emerging Powers in Africa*, 2016, pp. 217–220.

[24] Ibid.

Ethiopia and Eritrea, and actively operated the construction of a new military base in Djibouti. Turkey's success in Somalia has greatly enhanced Turkey's image and status on the African continent and has served as a good model for the full implementation of Turkey's policy toward Africa.

Economic Factors are One of the Main Objectives of Turkey's Africa Policy

Turkey's foundational documents on Africa—the African Action Plan (1998) and the Strategy for the Development of Economic Relations with African Countries (2003)—were shaped by dual political and economic factors. Politically, Turkey's EU membership has historically been frustrated and it has subsequently sought foreign policy diversification. At the same time, there is an urgent need for the resources necessary for economic development and new commodity markets. Close economic relations with African countries have become one of the main objectives of the policy toward Africa. The main measures taken by Turkey in carrying out economic diplomacy toward Africa include signing the Bilateral Investment Agreement and the establishment of a joint mechanism between the Business Committee and the Economic Commission. Turkey has opened commercial consulates in 26 African capitals. The Turkish Council for Foreign Economic Relations has established business councils in 19 sub-Saharan African countries. Since 2003, Turkey has signed bilateral investment agreements with 28 African countries and signed trade and economic cooperation agreements with 38 African countries, as well as economic and technical cooperation agreements and implementations of technical assistance programs for Africa.[25]

Following this, Turkey intends to expand the goods and services trade with African countries. In the area of international trade, Turkey's current account has been under a deficit for a long time because of its inability to supply its own oil and gas resources. In recent years, the African continent has become a rare area of trade surplus for Turkey. Turkey's bilateral trade with Africa was only $5.47 billion in 2003 and

[25] http://www.balkanalysis.com/turkey/2011/01/01/turkey%E2%80%99s-developing-role-in-africa-interview-with-mehmet-ozkan-and-birol-akgun/, achieved on March 10, 2020.

gained $20.6 billion in 2018, an increase of about fourfold.[26] In 2018, Turkey's foreign trade amounted to US$374 billion and trade with Africa accounted for about 4.89% of that. Exports to Africa accounted for about 7.6% of total exports and imports accounted for 2.8% of total imports.[27] Turkey's main trade partners are still concentrated in North Africa, with Egypt, Algeria, Morocco, and Libya in the top 25 countries. The largest annual trade between sub-Saharan African countries in 2018 was South Africa, with $1.91 billion in trade. This was followed by Senegal, Nigeria, and Ethiopia, with total trade exceeding $300 million.[28] In recent years, Turkey's share of Africa's total trade has increased every year, and Turkish-branded products have gained some influence on the continent.

In the area of trade in services, Turkish contractors are one of the main contractors on the African continent. African countries account for about 21% of the total international operations of Turkish contractors, of which North Africa accounts for 19%. To date, Turkish contractors have carried out more than 1150 projects in Africa, valued at $55 billion.[29] At the same time, Turkey's foreign direct investment on the African continent is also generating positive returns for the Turkish economy, with Turkey's investment stock in Africa estimated to have been about $6.5 billion from 2003 to 2018, and Turkish investment creating significant jobs for African countries.[30] Priority areas for Turkish enterprise investment are agribusiness, rural development, civil defense, water resources management, development, security, health, and transportation for small and micro-enterprises.

Subsequently, official development aid and humanitarian assistance have always been important parts of Turkey's Africa policy. Turkish government agencies, non-governmental organizations, and enterprises

[26] Turkey's Trade Volume with Africa Reaches US$20.6Bn, http://www.businessworldghana.com/turkeys-trade-volume-with-africa-reaches-us20-6bn/, achieved on March 17, 2020.

[27] https://tradingeconomics.com/turkey/exports-by-country, achieved on March 17, 2020.

[28] Ibid.

[29] Turkey's Trade Volume with Africa Reaches US$20.6Bn, http://www.businessworldghana.com/turkeys-trade-volume-with-africa-reaches-us20-6bn/, achieved on March 17, 2020.

[30] Ibid.

are all actively involved in official aid and humanitarian assistance projects initiated by the World Health Organization, the Food and Agriculture Organization of the United Nations, and the Red Crescent Society. Health sector cooperation is Turkey's main channel for cooperation with Africa. Turkey has signed health cooperation agreements with some 20 African countries and has also implemented different healthcare projects in different countries. Within the framework of the Cataract Project in Africa (implemented in four countries: Niger, Somalia, Ethiopia, and Sudan), 53,000 cataract patients were provided with surgery. Between 2007 and 2014, Turkey provided health checks to some 20 countries in Africa. More than 280,000 Africans have undergone medical examinations and are being treated even while more than 1000 African patients are unable to receive treatment in their home countries.[31] The Ministry of Health of Turkey also provides vocational training in the medical sector to African countries. The Recep Tayyip Erdogan Hospital in Somalia, the Turkish-Sultana Research and Training Hospital, the Juba Education Hospital in South Sudan, and the Black Lion Hospital in Addis Ababa, Ethiopia, are among the key cooperation projects of the Ministry of Health of Turkey.

Turkey became the world's third largest humanitarian donor in 2013 and 2014. According to the World Humanitarian Aid Report, including the contribution of Turkish non-governmental organizations in the areas of health, education, and capacity-building, Turkey's development aid and humanitarian assistance amounted to $3.3 billion in 2014 and made up 0.42% of its gross national income. Sub-Saharan Africa's share was $1536. In recognition of Turkey's work on humanitarian assistance and development assistance, the first United Nations Humanitarian Summit of 2016 was held in Istanbul, with Turkey becoming a leading country for humanitarian assistance to Africa. In 2017, Turkey will become Africa's second largest humanitarian aid country after the US.

Counter Terrorism Security Cooperation and Military Presence

Since 2011, as relations between Turkey and Africa have deepened, Turkey has sought a foreign policy integrating soft and hard power by actively presenting African countries with a "strong and secure" image

[31] TURKEY–AFRICA RELATIONS/Rep. of Turkey Ministry of Foreign Affairs, http://www.mfa.gov.tr/turkey-africa-relations.en.mfa, accessed on February 20, 2020.

of the new Turkey. Military and security cooperation has become an important part of Turkey's foreign policy toward Africa. Military and security cooperation in Turkey's relations with Africa mainly occurs in four areas. First, Turkey attaches great importance to peace and stability in Africa and actively participates in United Nations peacekeeping operations, including in the MONUSCO/Democratic Republic of the Congo, MINUSMA/Mali, MINUSMA/CAR, UNAMID/Darfur, UNMISS/South Sudan, UNOCI/Cote d'Ivoire, and UNMIL/Liberia existing nine peacekeeping missions in Africa. Second, Turkey attaches great importance to cooperation with African countries in the military and security fields. Turkey has signed military and defense industry cooperation agreements with Ethiopia, Nigeria, and South Africa covering arms sales, personnel training, and joint military exercises, and has steadily promoted military cooperation with Kenya, Tanzania, Uganda, Cameroon, and other countries. In 2014, the Turkish naval fleet made a voyage along the African coastline, arriving in 24 African countries and stopping at 25 ports, 19 of which for the first time.[32] Third, Turkey established military bases as a key point in permanent military relations. In September 2017, Turkey completed a $50 million military base in Somalia, mainly to train Somali soldiers and fight the extremist group al-Shabab. In December, Turkey reached an agreement with Sudan to jointly combat terrorism in East Africa and the Horn of Africa and to strengthen military cooperation in the Red Sea. Fourth, Turkey emphasized cooperation in education and training for military personnel and police. By the end of 2018, Turkey had provided military training to more than 3000 military personnel from more than 20 African countries. In recent years, there has been a significant increase in the number of training programs and trainees.

Focus on Improving Bilateral and Multilateral Relations to Build Comprehensive and MultiDimensional Cooperation Mechanisms

Following the introduction of its "opening up to Africa" policy, Turkey began to attach greater importance to the development of bilateral relations with African countries and to the establishment of multidimensional cooperation mechanisms with regional and sub-regional organizations

[32] TURKEY–AFRICA RELATIONS/Rep. of Turkey Ministry of Foreign Affairs, http://www.mfa.gov.tr/turkey-africa-relations.en.mfa, achieved on February 20, 2020.

and extra- or regional organizations. Since 2009, Turkey has contributed $1 million annually to the African Union, as well as to other regional organizations. Turkey's cooperation mechanism with Africa includes four levels, one of which is to actively build bilateral cooperation mechanisms with African countries and expand Turkey's political representation. Since 2003, Erdogan has visited Africa more than 30 times as Prime Minister and President, visiting 23 countries. The number of Turkish embassies in Africa also increased, from twelve in 1996 to 44 in 2018.[33] Second among these cooperation mechanisms is to actively organize, host, or participate in the activities of regional and sub-regional organizations. Turkey is a strategic partner of the African Union and an extraterritorial member of the African Development Bank. Turkey has been endorsed by sub-regional organizations and has been actively involved in the meetings and events held by sub-regional organizations. In 2005, the Turkish Embassy in Abuja was endorsed by the Economic Community of West African States (ECOWAS), and in 2010, the Turkish Embassy in Dar es Salaam was accredited by the East African Community (EAC). In 2012, the Turkish Embassy in Addis Ababa was endorsed by the Intergovernmental Development Partners Forum Authority for Africa (IGAD). In the same year, the Turkish Embassy in Lusaka was endorsed by the Common Market for Eastern and Southern Africa (COMESA). In 2013, the Turkish Embassy in Libreville was accredited by the Economic Community of Central African States (ECCAS).

Third among these involves actively participating in Africa-related activities and projects initiated or organized by various United Nations agencies. In March 2011, Turkey and the United Nations Development Programme (UNDP) signed the Framework Agreement on Partnership, which elevated the partnership to the global and regional levels and established a dynamic global partnership that greatly facilitated Turkey's cooperation with African, Arab, and Asian countries. Turkey, which chaired the G-20 in 2015, organized events under Africa's Agenda 2063. Regarding this, a high-level meeting on energy access in sub-Saharan Africa was held for the first time in the framework of the G20 in October 2015. Turkey also organizes, hosts, and participates in bilateral and multilateral meetings with African countries. To date, the conference

[33] Ibid.

mechanism has been an important way for Turkey to achieve its strategy for Africa.

Fourth, Turkey wants to actively explore and build tripartite cooperation mechanisms with other regional organizations or extraterritorial countries to promote cooperation with Africa. Within the framework of the G-20 presidency, Turkey organized a high-level meeting on energy access in sub-Saharan Africa. The African-Middle East-Gulf Region Commission was established to promote tripartite cooperation through contacts and ongoing exchanges between Turkish entrepreneurs and professionals and institutions from African and Middle Eastern countries. In August 2019, on the occasion of the 7th Tokyo International Conference on African Development (TICAD), the "Turkey–Japan–Africa Partnership" conference was held to explore cooperation with Japan to promote sustainable development in African countries through trade, capacity building, skills development, health cooperation, energy, and infrastructure development.

The Turkish Bureau of International Cooperation and Coordination

Promoting the in-depth development of African policy on behalf of the Turkish Bureau for International Cooperation and Coordination, Turkey's foreign policy toward Africa has been gradually refined in practice, and this successful experience is uniquely valuable to Turkey's overall foreign policy system.[34] During this period, the role of institutions such as the Turkish Bureau for International Cooperation and Coordination (TİKA) has been instrumental. The Turkish Office for International Cooperation and Coordination is the coordinating body for Turkish government agencies, universities, non-profit organizations, and the private sector. Founded in February 1992, its original purpose was to strengthen relations with the states of Central Asia, the Caucasus, and the Balkans, which were newly established after the collapse of the Soviet Union. Africa has been a priority area of work for the Agency since the

[34] Turkey's Developing Role in Africa Interview with Mehmet Ozkan and Birol Akgun, http://www.balkanalysis.com/turkey/2011/01/01/turkey%E2%80%99s-developing-role-in-africa-interview-with-mehmet-ozkan-and-birol-akgun/, achieved on March 7, 2020.

Justice and Development Party came to power, and the Turkish Government has indicated its willingness to support agenda 2063, proposed by the AU, through TİKA.[35]

As Turkey's main implementing agency for Africa, TİKA's primary goal is to assist African countries to develop their healthcare, education, agriculture, environmental protection, and infrastructure. This includes the organization of trade, construction, and procurement delegations, participation in and the organization of trade fairs, increased banking operations, and support for the accession of African countries to the World Trade Organization. This also includes regular dialogues with African leaders, the opening of new business offices, legal framework agreements, and free trade agreements with different countries.

TİKA has strengthened and increased humanitarian, technical, and development assistance to Africa in the areas of education, health, agriculture, infrastructure, and capacity building. On the basis of long-term non-cooperation, the four main areas of this work have been further defined, namely, the development of social infrastructure (education, healthcare, water, and sanitation), the development of economic infrastructure (infrastructure related to transport and commerce), the promotion of the development of the productive sector in African countries (agriculture, forestry, animal husbandry, industry, and commerce), and the provision of emergency and humanitarian assistance to African countries. Turkey also provides assistance to Africa through international organizations. From 2002 to 2017, Turkey's official development aid increased from $85 million to $8.12 billion. In 2017, Turkey became the world's largest recipient of humanitarian assistance. With 45 planning coordination offices on the African continent, it is one of Africa's largest development partners, operating throughout the continent, most notably in Somalia and Darfur.[36]

In its cooperative efforts with Africa, the diplomatic approach of promoting the comprehensive development of Turkey–Africa relations, along with the combination of officials and people and the comprehensive mobilization of soft power,[37] has become an important reason for

[35] Official website of the Turkish Cooperation and Coordination Agency, https://www.tika.gov.tr/en/page/about_us-14650, achieved on March 10, 2020.

[36] Ibid.

[37] Zhang Chun: Turkey's Africa Strategy and Policy Review, West Asia Africa, No. 9, 2011, pp. 53–67.

the success of Turkey's policy toward Africa. In the eight years between 2006 and 2015, the number of African tourists visiting Turkey increased about fourfold, from 210,000 to 885,000. Turkish Airlines (THY) has strengthened Turkey's presence on the continent and has become one of the major international airlines connecting the continent with the world, with 48 destinations in 31 countries and regions in Africa.

3 The Limitations of Turkey's African Policy

Turkey's diplomatic involvement in the African continent transcends traditional inter-state interactions, and education, culture, humanitarian activities, and private sector investment not only provide Turkey with a deep cultural and commercial presence on the continent, but also enhance Turkey's national status and image in Africa. However, since 2017, Turkey has frequently intervened in Africa by radical means based on hard power, and these actions have attracted the attention of the entire international community and exposed the limitations of Turkey's policy toward Africa, which are now at stake.

The Constraints of Turkey's Economic Development

The development of Turkey's domestic economy has become a hard constraint on Turkey's Africa policy in recent years. After the rapid growth of the two "golden decades" of the 1980s and the first decade of the twenty-first century, especially the first decade of the Justice and Development Party, the Turkish economy created its own self-serving myth, and now Turkey, once only the world's 16th largest economy, is regarded as one of the most promising emerging economies after the BRICS. Subsequently, the short-term impact of the domestic political situation and the external environment, coupled with the impact of the entire economic development cycle, has caused Turkey's economic growth to fluctuate considerably, resulting in its rating as the world's 18th largest economy. The lira crisis in August 2018 dealt a major blow to the Turkish economy, and in October 2018, during the second Turkey–Africa Economic and Business Forum, Turkey signed three cooperation agreements with Senegal, Zimbabwe, and the African Union Commission, but the lira crisis severely limited the trade. President Erdogan has called on African countries to use local currencies to deal with Turkey in trade and to mitigate currency risks for industry. At present, Turkey is still in the post-crisis recovery period, economic development still has weak growth, high inflation, high unemployment, and is vulnerable to external shocks and other structural problems. As a result, Turkey's own economic strength and growth efficiency directly affect its investment and cooperation in Africa.

The Impact of the Political Situation in Turkey

The Justice and Development Party (AKP), which split from the Gülen Movement in December 2013, has also had an impact on African countries, particularly in civil society groups such as chambers of commerce and educational associations that were active in early years in Africa. These civil society groups are mostly Gülen followers and have been since the launch of the Action Plan for Africa in 1998, which resulted in Turkish schools in Africa being sponsored or operated directly by the Gülen Movement. The Gülen Movement not only affected a large number of social elites and nurtured the children of many government officials, but also created a rapport with the local population in sub-Saharan Africa, particularly Tanzania, Kenya, Mozambique, Uganda, and South Africa, operating as a source of Turkish soft power in the region. The Gülen movement has been discredited by the Turkish government and has since strengthened ties with local African communities to compensate for losses at home. After the failed coup attempt in 2016, Erdogan characterized the Gülen movement as a terrorist organization, demanding that African countries cooperate with the closure of Gülen schools and the expulsion or extradition of Gülenists. Although six African governments have agreed to Turkey's demands, Turkey's desire to remove the influence of the Gülen movement in Africa will not be easy to accomplish, and Turkey may still face the Gülen problem for a long time to come. The closure of schools would not only cost African countries with educational resources, but would also have a negative impact on Turkey's long-standing good image in Africa.

The Transition Toward Hard Power in Africa Has Intensified Competition with Middle Eastern Countries in the Horn of Africa

After the Arab Spring, Turkey's competition and confrontation with Middle Eastern countries have been spreading into Africa, and thus Turkey is entering a new round of geo-competition and sectarian struggle.

Since 2013, in addition to Turkey, Saudi Arabia and the United Arab Emirates in the Middle East have stepped up their military presence in Africa in order to counter Iran's growing influence in the region. Saudi Arabia has signed a series of agreements with Djibouti, reached consensus with Sudan, and conducted cooperative military exercises. The UAE has a military base in Eritrea and has signed similar agreements with semi-autonomous regimes such as Somaliland and Puntland in the north-eastern corner of Somalia. While the two countries' immediate goals are not directed at Turkey, given that the Gulf Cooperation Committee split and Turkey allied with Qatar, Turkey is increasingly on the opposite side of Saudi Arabia and the United Arab Emirates, which are themselves under strategic pressure from Turkey in North Africa.

At the same time, while Turkey is very concerned about sub-Saharan Africa, Egypt, a key AU member that is increasingly cooperating with Saudi Arabia and the United Arab Emirates, will largely offset Turkey's efforts across the African continent. Turkey's growing military cooperation with Sudan and Somalia, as well as its recent military presence in Libya, has also intensified tensions between Turkey and regional countries such as Saudi Arabia and the United Arab Emirates. In the future, Turkey is unlikely to become embroiled in its familiar confrontations and conflicts in Africa.

Radical Diplomacy in Libya Created Multidimensional Tensions

On November 27, 2019, Turkey and the Libyan National Unity Government signed a memorandum of understanding on military cooperation and a maritime delimitation agreement. On December 17th, Turkey sent troops to help Libya's national unity government to counter Saudi and UAE-backed Haftar forces. This was a big move for Turkey in Africa, after everything that had happened in Somalia. North Africa has gone from what Turkey envisaged as a "strategic depth belt" to a battlefield after Syria, and Turkey's diplomatic environment is now becoming increasingly complex. At the big power level, Turkey has been at loggerheads with the US over the Syrian Kurdish issue and the purchase of the Russian S-400. Although there is still much cooperation with Russia and Iran on Syria, with the "Islamic State" dispersed and Syria's post-war arrangements up for debate, the conflicts between the three players have also gradually surfaced. At the regional level, Turkey has been isolated by Saudi Arabia, the United Arab Emirates, and other countries for its support of

Libya's unity government. At the same time, the Mediterranean demarcation agreement with Libya touches on the interests of Egypt, Greece, Cyprus, Israel, and Italy, and they are unlikely to reach an agreement in the short term. At present, African countries, including Egypt, strongly oppose Turkey's troops to Libya, Tunisia, and Algeria, claiming that the situation remains neutral. Whether Turkey can achieve its goal remains to be seen.

4 Conclusion

Given the transformation of its overall diplomatic thought from "pro-Western" to one more based on "multi-balanced diplomacy" after the Cold War, Turkey's foreign policy toward Africa not only achieves realistic political and economic goals, but also is helping Turkey to achieve its strategic goal of becoming an influential global power. This is a successful example of the country's "new Ottomanism" foreign policy. Turkey's foreign policy model for African countries usually begins with soft power diplomacy followed by economic activities, including infrastructure development, food security, health care, and human capital development, and only then followed by involvement in military and security areas. Turkey provides African countries with experience in local and national governance which, in turn, gradually influences African policies and decisions and forges closer partnerships. With the establishment of the Partnership policy, Turkey is now actively involved in the reconstruction of the Somali state and the establishment of a military base in Mogadishu, and if Turkey takes its foreign policy toward Somalia as an opportunity to transform Turkey's hard power in Somalia into a clear effort to promote peace and economic prosperity and to ensure that the African continent remains a priority on Turkey's foreign policy agenda while sharing its development experience with African countries for decades to come, then Turkey will undoubtedly effectively balance its economic and political interests for the continued furtherance of Turkey's relations with Africa. In the future, Turkey's long-term interests in Africa will depend not only on the provision of assistance or commercial relations to Africa, but also on the long-term stability of institutional mechanisms based on Turkish–African partnerships.

CHAPTER 12

Saudi Arabia's Aid to Africa: Motivations, Contributions, and Trends

Mo Chen

In the 1970s, while the General Assembly of the United Nations for the First Development Decade's aid program ended up in frustration and the aid program of the Second Development Decade also foresaw a bleak outlook due to the economic stagnation and inflation being experienced by developed countries, Saudi Arabia and other OPEC countries in the Middle East broke away from this trend to join the group of major donor countries. Using their huge share of petrodollars, they thus became a major force in South–South cooperation, offering fresh support for solving Africa's economic hardships.

Mo Chen's main research interests are energy issues and Middle East economics. This paper is translated by Tian Chunlin, Kang Lei, and Chen Weiguo.

M. Chen (✉)
Institute of West Asia and African Studies, Chinese Academy of Social Sciences, Beijing, China
e-mail: wxy03@ssap.cn

X. Li and C. An (eds.), *The Changing World and Africa*,
https://doi.org/10.1007/978-981-16-4983-7_12

1 The Formation of the Saudi Aid System and its International Status

As early as the 1950s, Saudi Arabia had already started to provide a small number of aids to other countries, but this aid was discreet, sporadic, and on a small scale, mainly because Saudi Arabia's petroleum was completely controlled by the Western oil consortium, which adopted a concession system to maximally obtain the profits from petroleum and to protect the interests of the petroleum-consuming countries in the West. These countries desperately expanded the Middle East's crude oil production while lowering crude oil prices. Therefore, the Saudi royal family and its government could only get a small amount of oil field royalties from their original exports, and their foreign aid capabilities were severely constrained as a result. Since that time, Saudi Arabia has been struggling to recover its oil rights and interests from the consortium, achieving equal sales profits in the 1950s, and raising oil prices with Organization of Arab Petroleum Exporting Countries (OAPEC) during the Yom Kippur War (the Fourth Middle East War) in the 1970s. In addition, Saudi Arabia also completed the nationalization of its oil industry in 1980. These major changes greatly increased Saudi Arabia's oil revenue and improved its foreign aid capabilities. Since the 1970s, Saudi Arabia has gradually established a relatively complete foreign aid system, becoming one of the world's foremost countries offering aid. The development of Saudi foreign aid could be illustrated from the following aspects.

Firstly, a government-led aid organization was established by the country to develop its foreign aid. In 1974, Saudi Arabia established a government organization specializing in development projects aid, the Saudi Fund for Development (SFD), which officially started running in 1975. With an initial capital of 10 billion Saudi Riyal (approximately 3.7 billion US dollars), the fund increased its capital to 31 billion Saudi Riyal (approximately 13.7 billion US dollars) in 1991. In 2015, Saudi Arabia established the King Salman Humanitarian Aid and Relief Center (KSrelief), specializing in humanitarian and relief assistance. As a result, a Saudi foreign aid organization system centering around these two organizations was formed. In addition to these two official aid organizations, there are also several semi-official ones. For example, the Saudi Red Crescent Authority (SRCA) is a humanitarian aid distribution institution and essentially a government foreign aid organization. A group of "public companies" are also funded by the government and run by respective relief committees, aiming to raise private funds to implement assistance for improving people's livelihoods. Saudi royal family members have also established private charities, such as the Alwaleed Bin Talal Foundation, the Sultan Bin Abdulaziz Al Saud Foundation, etc. Coverage of the assistance provided by these semi-official and royal foundations, however, is incomparable with that of the two major aid organizations.

Secondly, Saudi Arabia has established a decision-making and implementation system for aid. In Saudi Arabia, all the major foreign aid projects are decided or approved by the royal family with whatever necessary cooperation is needed to be offered by the royal office and the government departments, including the Ministry of Foreign Affairs, the Ministry of the Interior, and the Ministry of Finance. The Ministry of Foreign Affairs mainly collects necessary information about recipient countries from their embassies. The Ministry of Finance is responsible for allocating government funds. The Ministry of the Interior is accountable for supervising and implementing the aid itself. The royal family's aid decisions are primarily implemented by official institutions such as the King Salman Humanitarian Aid and Relief Center, the Saudi Fund for Development, the Ministry of Foreign Affairs, the Ministry of Education, and the Ministry of Health, as well as semi-official organizations such as the Saudi Red Crescent Authority, other related "public companies," and the royal family foundations. Last but not least, royal family member foundations also contribute. The Saudi Fund for Development generally implements developmental project assistance, while other aid organizations mainly implement humanitarian and emergency relief assistance. Among other aid organizations, the king's palace plays a pivotal role in mediating between the royal families and the relevant decision-making consultation departments, as well as the relevant implementation departments.

Thirdly, bilateral and multilateral aid channels have been formed. Saudi Arabia's aid distribution channels are mainly bilateral, but regional and international multilateral aids have also grown to become important channels for Saudi Arabia's foreign aid. Saudi Arabia has also contributed a large amount of funding to multilateral aid agencies such as the Arab Fund for Economic and Social Development, the Islamic Development Bank, Arab Bank for Economic Development in Africa (BADEA), Arab Monetary Fund, African Development Bank, International Development Association, the United Nations Development Programme, the United Nations Educational, Scientific and Cultural Organization (UNESCO),

United Nations Conference on Trade and Development (UNCTAD), OPEC Fund for International Development, International Fund for Agricultural Development, the Global Fund to Fight AIDS, Tuberculosis and Malaria, the United Nations Children's Fund, the World Health Organization (WHO), United Nations High Commissioner for Refugees (UNHCR), and International Committee of the Red Cross. Since 1992, Saudi has strengthened its coordination with recipient countries and discussed how to improve both the effectiveness of aid and how best to correspond with the recipient countries' priority development plans through frequent consultation meetings involving both parties.

Fourthly, a multilevel coordination mechanism for aid has been established. While aiming at improving aid efficiency and exchanging aid experiences, Saudi Arabia participated in a variety of aid-related international coordination mechanisms. The first of these was strategic coordination at the summit level, which often determines the basic guidelines for aid and which has outlined many strategic aid policies for Saudi Arabia. For example, Saudi Arabia increased its aid to African countries in the 1970s and participated in the construction of the Arab Bank for Economic Development. The decision to give aid was based on a political decision made by the Arab Summit in 1973, prompting African countries to ally with Arab countries in the Yom Kippur War (the Fourth Middle East War) and to break off diplomatic relations with Israel. Saudi Arabia participated in the construction of the OPEC Fund in 1976, which was, itself, a political decision made by the OPEC summit to help underdeveloped oil-importing countries to cope with rising oil prices. The second level of coordination occurs at the level of aid agencies and is aimed at improving work efficiency. In 1975, Saudi Arabia and the other nine Arab Islamic aid agencies jointly formed the Arab Coordination Group (ACG), a coalition of financial institutions for national and regional development to promote aid standardization, exchange aid experience, and release relevant statistics. In 2018, Saudi Arabia's cooperation with the Development Assistance Committee (DAC) of the OECD also saw substantial progress due to the fact that Saudi Arabia became "a participating country of the Development Assistance Committee" and contributed to aid-based statistical analysis. The third level of coordination takes place between Saudi Arabia and its aid recipient countries, aiming at better understanding their needs and further improving aid efficiency. A coordinating mechanism for regular exchanges between the Arab Coordination Group (ACG)—made up of financial institutions for national and regional development—and the recipient countries was established in 1992.

Fifthly, Saudi Arabia has become one of the world's most significant donor countries. Following more than 40 years of construction and development, the country has already become one of the world's major donor countries, with the scale of its aid ranking tenth in the world. In 2018, Saudi provided a total of 4.34 billion US dollars in the form of various aids,[1] ranking 10th worldwide, right after the US, Germany, the UK, the European Union, Japan, France, Sweden, the Netherlands, and Canada. It has grown to be one of the world's major countries providing foreign aid.[2] Secondly, although Saudi Arabia is a developing country, it is nevertheless among the countries providing the greatest scale of aid. The scale of aid is usually calculated by the proportion of the amount of official development financial assistance in the Gross National Income (GNI). With this formula in mind, Saudi Arabia's Gross National Income in 2018 was 790 billion US dollars, and the total financial assistance for the same year was 4.34 billion US dollars, equivalent to 0.55% of its Gross National Income. Compared with developed countries, it ranked eighth after Sweden, Luxembourg, Norway, Denmark, the UK, Germany, and the Netherlands for the scale of aid. In 2018, the average ratio of financial assistance of OECD Development Assistance Committee member states was merely 0.31%.[3] Therefore, despite being a developing country, Saudi Arabia still surpassed most developed countries and the average developed countries in terms of financial assistance.

2 Africa as the Key Recipient of Saudi Aid

Since the 1970s, African countries have been the main recipients of Saudi aid, while African development plans have been the main beneficiaries of that same Saudi aid. The majority of financial assistance made by the multilateral foreign aid agencies in which Saudi Arabia has participated have been provided to Africa.

[1] Saudi Aid Platform, https://data.ksrelief.org, Last Accessed: January 18, 2020.

[2] The website of OECD Development Assistance Committee, http://www.oecd.org/dac/development-assistance-committee, Last Accessed: January 18, 2020.

[3] The website of The OECD Development Assistance Committee: https://www2.compareyourcountry.org, Last Accessed: January 19, 2020.

Africa as Saudi Arabia's Second Largest Aid Recipient

Overall, Africa is the second largest recipient of Saudi aid right after West Asia. According to official Saudi statistics, from 1975 to 2020, Saudi's humanitarian aid and development project assistance to sub-Saharan Africa totaled US$5.37 billion, accounting for 18.4% of Saudi's total aid during the period; there were also 691 aid projects in total, taking up 24.1% of Saudi's foreign aid projects. Including its aid to Egypt, Algeria, Tunisia, and Morocco in North Africa, the total amount of Saudi aid to the whole African continent reached 9.85 billion US dollars, constituting 33.7% of Saudi's total foreign aid during this period. Additionally, there were 822 aid projects to Africa in total, making up 28.7% of Saudi's total aid projects during the period. Africa, next to Western Asian countries, became the second largest recipient of Saudi's foreign aid in terms of the amount of financial assistance or the number of aid projects.[4]

Africa as the Main Beneficiary of Saudi Development Project Assistance

Regarding the number of development projects to which it has provided assistance, Africa is the biggest beneficiary of Saudi aid. The Saudi Fund for Development officially began signing and issuing development project assistance in the form of concessional loans from 1975 on. By 2018, the country had signed 656 aid projects with 83 recipient countries, spending 62 billion Saudi Riyals (16.53 billion US dollars) in total. Among these 83 countries 46 were African countries (including North African countries), accounting for 55.4% of all foreign recipient countries; among all aid projects, 388 of them were in Africa, taking up 59.1% of all foreign aid projects. The amount of loans totaled 35.1 billion Saudi Riyals ($9.36 billion), constituting 56.6% of all loans. Regardless of whether we consider the number of aid projects, aid recipient countries, or the amount of financial assistance, Africa accounted for more than half of the total amount. Saudi's development assistance projects focused primarily on social infrastructure (including water supply and drainage, education, healthcare, housing, and urban development), transportation (including roads, railways, seaports, communications), energy, and agriculture. In addition to the energy sector, Africa has received more than half of the total number of development projects issued by Saudi Arabia in the

[4] Saudi Aid Platform, https://data.ksrelief.org, Last Accessed: January 19, 2020.

other three areas in terms of the number of projects and the amount of aid received. The aid received by the agricultural sector, in particular, exceeded 3/4 of the amount of total aid. To be more precise, Africa received 134 projects (59%) and 10.16 billion Saudi Riyal (54.1%) in transportation and communications, 116 projects (54%) and 11.43 billion Saudi Riyal (60.6%) in social infrastructure, 30 projects (42.9%) and 4.62 billion Saudi Riyal (41.3%) in the energy sector, and 73 projects (77.7%) and 6.71 billion Saudi Riyal (77.5%) in the agricultural sector.[5]

Somalia as the Main Recipient of Humanitarian Aid

Saudi foreign aid mainly focuses on humanitarian and emergency aid. Most Saudi humanitarian aid is channeled to West Asian countries. In comparison, the amount of such aid received by Africa is relatively small, but certain countries receive a greater share. From 1975 to 2020, Saudi Arabia issued a total of US$7.17 billion in humanitarian aid and funded 1,441 humanitarian aid projects. During the same period, African countries received US$241 million in humanitarian aid from Saudi Arabia, making up only 3.4% of Saudi Arabia's total humanitarian aid; 164 humanitarian aid projects were undertaken, representing 11.4% of all Saudi humanitarian aid projects. But Somalia ranked among the top ten countries receiving Saudi humanitarian aid. In total, the country received US$148 million in humanitarian aid from Saudi Arabia, ranking seventh among all recipient countries and constituting 61.4% of the total Saudi humanitarian aid received by Africa. Somalia received 52 humanitarian aid projects from Saudi Arabia, ranking fifth among all recipient countries and representing about one-third of the humanitarian aid projects provided to Africa by Saudi Arabia. Saudi Arabia's humanitarian assistance to Somalia is relatively frequent, and the amount of financial assistance usually ranges from hundreds of thousands of dollars to millions of dollars to even tens of millions of dollars. The largest financial assistance is assistance in kind, valuing US$35.01 million and provided by the King Salman Humanitarian Aid and Relief Center from 2011 to 2012. This was mainly spent on providing food to the famine victims in Somalia during the 2011 food

[5] Saudi Aid Platform, https://data.ksrelief.org, Last Accessed: January 19, 2020.

crisis in the Horn of Africa. In terms of aiding Somalia, the amount of Saudi Arabia's humanitarian aid was second only to Britain and the US, ranking third among the international community.[6]

The Contributions of Saudi Arabia's Multilateral Institutions to Aiding Africa

Saudi aid to Africa is not only bilateral. Since the 1970s, Saudi Arabia has provided donations to multilateral aid agencies and provided aid to Africa through the use of regional and international multilateral agencies with a considerable amount of financial assistance. During 1994–2017, Saudi Arabia provided a large amount of donations to the following multilateral aid agencies: the Arab Fund for Economic and Social Development (US$239 million), the Islamic Development Bank (US$173 million), the Arab Bank for Economic Development in Africa (US$103 million), the International Development Association (US$99 million), the United Nations Development Programme (US$80 million), the OPEC Fund for International Development (US$6,100 trillion), the Arab Monetary Fund (US$55 million), the International Fund for Agricultural Development (US$53 million), the Organization of Islamic Cooperation (US$48 million), the Global Fund to Fight AIDS, Tuberculosis and Malaria (US$40 million), the United Nations Children's Fund (US$40 million), the World Health Organization (US$7.5 million), the United Nations High Commissioner for Refugees (US$6.23 million), the International Committee of the Red Cross (US$6 million), and the African Development Bank (US$2.85 million). Most of the donations were provided to Africa in the form of multilateral aid.

3 Motivations for Saudi Arabia's Aid to Africa

Saudi Arabia takes Africa as its primary aid recipient mainly due to the fact that it has to maintain its foreign relations while also addressing religious and ethnic issues. Saudi Arabia needs the support of African

[6] Saudi Aid Platform, https://data.ksrelief.org, Last Accessed: January 19, 2020.

countries when faced with challenges arising from international relationships and affairs, regional peace and security, and issues related to opening up. Meanwhile, Saudi Arabia's aid to Africa is a one way to put Islamic doctrine into practice.

Foreign Relations Factors

Aid serves as an integral part of Saudi foreign relations as well as an important tool of Saudi diplomacy. This can be observed from the following examples.

Enlisting the Support of African countries in the Arab–Israeli Conflict

As a member of the Arab world, Saudi Arabia has always offered its support to Arab countries in the Arab–Israeli conflict, which has historically been a major agenda of its foreign relations. This position manifests as Africa winning the support of Arab countries through foreign aid. These African countries became tempting targets for Israel after they claimed independence in the 1960s. From 1958 to 1966, Israel provided loans of US$219 million to African countries, dispatched 2485 experts and technicians, signed more than 2000 economic cooperation projects, and trained 5328 African students and military and political officials in Israel.[7] Israel also established diplomatic relations with most sub-Saharan countries. Faced with Israel's aggressive assistance strategy, beginning in the early 1970s, Arab countries also increased their assistance to Africa. Saudi Arabia not only increased bilateral assistance to Africa but in 1973 also jointly established the Arab Bank for Economic Development in Africa with other Arab countries to provide long-term development and technical assistance to Africa. In January 1974, the country also established the Saudi Fund for Development in cooperation with other Arab countries in order to help sub-Saharan African countries cope with the difficulties caused by rising oil prices. These assistance measures greatly contributed to the fact that a large number of African countries broke off diplomatic relations with Israel in the 1970s. By 1976, only three African countries (Lesotho, Swaziland, Malawi) maintained diplomatic relations

[7] Wink: "The Relationship between Black Africa and Arabia after the War", "West Asia and Africa", Issue 4, 1988, p. 33.

with Israel. Nowadays, the relationship between Arab and Israeli is relatively peaceful in comparison with the 1970s. This has caused African countries to adopt a more balanced position between Arabs and Israelis. However, it is acknowledged that as long as African countries keep the same position as Saudi Arabia during voting in the UN, the possibility of their obtaining Saudi assistance will be greater.[8]

Preventing the Risk of Political Unrest During the "Arab Spring"

In 2012, the scale of Saudi aid suddenly increased. It is important to note that before 2012, the scale of Saudi's annual aid never exceeded US$1 billion. However, its financial assistance rocketed to US$23 dollars in 2012. The sudden change is associated with the outbreak of the political turmoil and associated regime changes in many Arab countries since 2010. This turmoil has been referred to as the "Arab Spring" movement by some media. The reason for the outbreak is complicated. It can be understood as an outbreak of problems in Arab countries after years of development. Economic difficulties and people's poor livelihoods also contributed to the unrest. Although no overt inner turmoil broke out in Saudi Arabia, it was rumored that some citizens wrote letters to the king demanding political reforms, which worried the Saudi royal family. To address the turmoil in the Arab world, prevent the further spread of political turmoil, and reduce the risk of political instability that Saudi Arabia potentially faced, Saudi Arabia adopted economic measures to increase assistance to further improve development and people's livelihoods in the countries in which turmoil had broken out or which had suffered economic difficulties. At the same time, the country adopted a "holding together for warmth" policy to prevent any political unrest that may threaten the monarchical countries in the Middle East. In May 2011, the Gulf Cooperation Council, composed of all monarchical countries, welcomed Morocco and Jordan to membership in the council. Saudi Arabia subsequently increased its assistance to these monarchies. What is particularly striking in this case is that, since 2011, multiple large-scale aid projects valuing hundreds of millions of dollars from have passed

[8] ShivitBakrania, Political Economy of Saudi Arabia's Overseas Aid Assistance, Governance and Social Development Resource Centre (GSDRC) website, September 24, 2012. https://gsdrc.org/gsdrc_pub_type/helpdesk/page/27/, Last Accessed: January 22, 2020.

between Saudi Arabia and Africa. Such projects are basically concentrated in the above-mentioned countries. For example, Egypt signed the project on a granary (2012, about US$130 million), a project on irrigation and drainage pumping stations (2012, about US$110 million), a project on drinking water in Nasser City (2012, US$83 million), a project on Tunisian housing (2011, about US$210 million), a project on Sousse power stations (2012, about US$170 million), a network of natural gas pipelines (2012, about US$120 million), Morocco's first high-speed rail project in Africa (2012, about US$2.8 Billion), and more. Oil exports from Sudan were interrupted after it split from South Sudan, leaving the Sudanese stuck in financial difficulties. The country fortunately received large-scale assistance from Saudi Arabia, including the Dam Complex of Upper Atbara Project and a project on the Sister River dam (2011, about US$ 110million; approximately US$140 million was additionally invested in 2012). These large-scale aid projects contributed to the sudden expansion of the scale of Saudi aid.[9]

Addressing the New Challenges of Peripheral Security and Opening Up

There was something unusual about Saudi Arabia's aid after 2014, breaking the concomitant connection between the scale of foreign aid changes and fluctuations in oil prices. Although international oil prices had been sluggish for a long time, the scale of aid broke the record, reaching US$ 4.34 billion in 2018. The total amount of financial assistance from 2014 to the beginning of 2020 reached US$13.38 billion, accounting for 45.8% of total aid of Saudi in the past 45 years. The number of aid projects and recipient countries increased from 139 and 33, respectively, in 2014 to 470 and 110 in 2018.[10] Saudi Arabia made this major change in its foreign aid because it had to address a series of major challenges to its peripheral security and opening up in recent years.

After King Salman came to the throne in 2015, Saudi Arabia was faced with a complicated situation in terms of neighboring political and military relations, especially on major issues such as the Syrian Civil War, the fight against the Islamic State, the Yemen Civil War, the deterioration of relations with Qatar, and challenges from Iran and the Iran-backed Islamic

[9] Saudi Aid Platform, https://data.ksrelief.org, Last Accessed: January 27, 2020.

[10] Saudi Aid Platform, https://data.ksrelief.org, Last Accessed: January 27, 2020.

Shias. In 2015, Saudi Arabia recruited coalition forces and launched a military operation against Iran-backed Houthi forces in Yemen. Economically, the new Saudi King and the new Crown Prince launched the "Saudi Vision 2030" reform and development plan, which included expanding the focus of the country's economic development and economic and trade cooperation to the Red Sea region. Saudi Arabia initiated the establishment of the National Councils along the Red Sea and the Gulf of Aden in January 2020.[11] Therefore, promoting security and cooperation in the Red Sea region became an important strategy for Saudi Arabia. In responding to these major challenges, Saudi Arabia received critical support from some African countries. For example, Egypt, Sudan, and Morocco all participated in the Saudi-led coalition against the Houthi armed forces in Yemen. Sudan also sent a large number of troops to Yemen to participate in ground battles against the Houthi armed forces. Egypt, Sudan, Djibouti, and other countries actively responded to Saudi Arabia's proposal for cooperation in the Red Sea region. Egypt also returned the islands in the Red Sea region to Saudi Arabia.[12] In return for their support, Saudi Arabia greatly increased its assistance to these countries.

From 2015 to 2018, among Saudi Arabia's development project assistance to African countries, large-scale projects valuing hundreds of millions of dollars were mainly signed with the above-mentioned countries, including an expansion of the Western Cairo power plant in Egypt (2016, about US$ 140 million), Castle Aini Hospital (2016, about US$ 170 million), King Salman's Sinai Peninsula development plan (two loans totaling about US$ 200 million in 2016 and 2017), rural rainwater collection and drinking water projects in Sudan (2016, about US$ 170 million), the Monakia power station project (2016, about US$ 180 million), social

[11] Initiated by Saudi Arabia in Riyadh in January 2020, the member states include Saudi Arabia, Egypt, Sudan, Yemen, Jordan, Djibouti, Somalia, and Eritrea.

[12] The Tiran Island and Sanafir Island, located at the entrance of the Gulf of Aqaba in the Red Sea, were included in Egypt in 1905 by the British and the Ottoman Empire, and in 1923 taken over to Saudi Arabia from Britain. In 1950, Saudi Arabia left the two islands under the protection of Egypt. In 2010, Saudi Arabia and Egypt negotiated on the return of the two islands to Saudi Arabia and the maritime delimitation of the two countries. In April 2016, the two countries reached an agreement and decided to return the two islands to Saudi Arabia.

housing and two hospitals in Tunisia (2017, about US$ 170 million), the drinking water supply project in Morocco (2018, about US$ 140 million), a road construction project from the capital of Djibouti to Jalafé (2018, about US$ 170 million), etc.[13]

Religious and Ethnic Factors

Saudi Arabia is an Islamic country with an integrated religion and politics. The royal family and most citizens believe in the conservative Wahhabi sect of Islam, which has exerted a deep influence on internal affairs and diplomacy. Saudi Arabia is one of the Arab countries with a homogenous religious and ethnic identity which, to a large extent, explains why most of its aid to Africa is ultimately aimed at Islamic and Arab countries.

Putting the Islamic Philosophy of Doing Good into Practice

The Islamic classics have many doctrines devoted to the idea that the rich should do good for the poor. It is the belief of Islam that wealth is bestowed by Allāh, so every Muslim must do good and the poor should receive a "fixed share" of the wealth from the rich. The Qur'an, as a Muslim classic, clearly states that it is the right of beggars and the poor to own a part of the property of the rich (51:19). There is "a fixed share" for the beggars and the poor (70:24). As an institutional requirement, Islam also stipulates that Muslims need to pay "Zakah"—in other words, 2.5% of their overall wealth must be spent for good deeds or charity each year. In addition, paying taxes is also one of the "five merits" required by Muslims. According to the Islamic concept of good deeds, Saudi Arabia (a wealthy country with huge oil revenues) should provide aid to Africa (the world's most impoverished region) as a way to fulfill its Islamic obligations. Putting the concept of doing good into practice is crucial for Saudi Arabia to maintain its orthodox image among Islamic countries. On the other hand, as a continent with the most concentrated, least developed,[14]

[13] Calculated based on data provided by the Saudi Aid Platform, https://data.ksrelief.org, Last Accessed: January 27, 2020.

[14] There are 33 African countries in the list of the world's least developed countries: Angola (in the transitional period of graduation review), Benin, Burkina Faso, Burundi, Central Africa, Chad, Comoros, Congo (Kinshasa), Djibouti, Eritrea, Ethiopia, Gambia,

Islamic countries,[15] Africa is stricken with poverty. So the least developed Islamic countries in Africa not only have high expectations for aid from rich countries such as Saudi Arabia, but also confidently put forward their demands for "innate rights" and "reasonable assistance" based on the Islamic philosophy of charity. Such aid obligation arising from the Islamic culture corresponds to aid demands. As a result, most of Saudi Arabia's aid reaches Islamic countries and the least developed countries in Africa. In terms of the development project assistance issued by the Saudi Fund for Development from 1975 to 2018, 27 Islamic countries in Africa (except Libya) were on the list of recipient countries; among the 33 least developed countries in Africa (except Angola, São Tomé and Príncipe, and South Sudan), 30 countries received at least some level of aid from Saudi Arabia.

Supporting the Development of Arab Countries

Although the distribution of Arabs is characterized by wide dispersion across many different countries, their ties are deeply rooted in their history, language, culture, and religion, which forms an important basis for their identity. In the struggle against the rule of the Ottoman Empire and colonialism, the Arabs have formed a "pan-Arab nationalism," which serves as a common ideological foundation for the unity and cooperation of Arab countries. In the early days, pan-Arab nationalism aimed to build a unified Arab country as its final aspiration. After 1945, when the League of Arab States was established, strengthening the close cooperation among the member states became the common purpose of all member states. Saudi Arabia is an important member of the Arab identity group, as well as one of the seven founding members of the Arab League. For Saudi Arabia, aiding other Arab countries is the main way to fulfill its promise of close cooperation with them. Therefore, Arab countries have always been the priority recipients of Saudi Arabia's assistance. The current member states of the Arab League are mostly in

Guinea, Guinea-Bissau, Lesotho, Liberia, Madagascar, Malawi, Mali, Mauritania, Mozambique, Niger, Rwanda, São Tomé and Príncipe, Senegal, Sierra Leone, Somalia, South Sudan, Sudan, Colombia, Tanzania, Uganda, and Zambia.

[15] Of the 57 member states of the Organization of Islamic Cooperation, 27 are African countries: Algeria, Benin, Burkina Faso, Cameroon, Chad, Comoros, Côte d'Ivoire, Djibouti, Egypt, Gabon, Gambia, Guinea, Guinea South Africa, Libya, Mali, Mauritania, Morocco, Mozambique, Niger, Nigeria, Senegal, Sierra Leone, Somalia, Sudan, Togo, Tunisia, and Uganda. In addition, the Central African Republic is an observer country.

the northern and eastern regions of Africa.[16] In addition, about 70% of the Arab population is also in Africa. Therefore, the Arab countries in Africa have become the main beneficiaries of Saudi aid. They were most benefited by development projects that grew out of the Saudi Fund for Development from 1975 to 2018. Among the ten Arab countries in Africa (except Libya, which does not require assistance), nine countries, including Egypt, Sudan, Tunisia, Algeria, Morocco, Mauritania, Djibouti, Somalia, and Comoros, received 23.51 billion Saudi Riyal during this period. The amount constitutes 2/3 of the total aid Saudi Arabia provided to Africa's development projects from 1975 to 2018. In terms of the scale of Saudi Arabia's aid projects, large-scale aid projects of more than US$ 100 million are almost all concentrated in Arab countries. Among non-Arab countries, only Zambia received loan assistance of more than US$ 100 million in 2018 (a construction loan for Abdullah Bin Abdulaziz Specialized Hospital).

4 The Contributions and Prospects of Saudi Arabia's Aid to Africa

The effects of Saudi aid have been positive in promoting the economic development of African countries, particularly in terms of helping countries achieve independent development, obtaining preferential funds, and prompting the international community to assist Africa. However, Saudi Arabia's ability to provide aid is also restricted by the prospects of supply and demand in the international petroleum market.

Saudi Arabia's Active Contribution to the Aid Received by Africa

Saudi Arabia focuses its aid to Africa on connecting with the development strategies of African countries while respecting the sovereignty of recipient countries and has also prompted the international community to aid Africa, which has contributed positively to Africa's development.

[16] Arab League had initially 21 member states. Syria was suspended as a member state in 2011. Qatar was excluded from the organization in 2017. There are 19 member states at present, including 10 African countries: Algeria, Egypt, Djibouti, Libya, Mauritania, Morocco, Sudan, Somalia, Tunisia, and Comoros.

Saudi Arabia's Aid Attaches Great Importance to the Development Strategy of African Countries

Saudi aid places great emphasis on connecting with the development strategy of the recipient country, in contrast with Western countries, which typically promote their own models of economic and political development in the name of foreign aid. The Saudi Fund for Development is aimed at dealing directly with the governments of developing countries so that the funding they provide can be part of the financing of priority development projects.[17] This is the reason that the projects aided by the fund, regardless of how much the amount is at their disposal, are the priority development projects of the recipient countries. In particular, a large amount of funding is expected on projects aimed at transportation, social infrastructure, energy, and agricultural sectors for which the recipient country's own governments and private individuals are incapable of financing independently. These projects thus became key to Saudi assistance. On the other hand, these projects are also significant for breaking the current bottleneck of infrastructure development in Africa and getting rid of poverty, especially in rural populations.

Saudi Arabia Provides More Favorable Funds for Africa

Among the aid offered by Saudi Arabia, its humanitarian aid is basically free of charge. As far as development project assistance is concerned, most of the assistance is in the form of concessional loans. Whether or not such loans are free depends on the level of the grant of the loans—that is, when the loans are due and subsequently repaid, the proportion of the unreimbursed funding needs to be calculated according to the discounted present value. According to the definition of the OECD Development Assistance Committee, only once the unreimbursed funding reaches 25% of all financial assistance can it be viewed as official development assistance. In regards to how much is unreimbursed, this mainly depends on two factors, one of which is the duration of the repayment period (including the grace period) and the other is the interest rate of the

[17] The website of Saudi Fund for Development, http://www.sfd.gov.sa/en/web/guest/home, Last Accessed: January 25, 2020.

loan.[18] In this way, the Saudi Fund for Development provides a favorable loan with a rather large scale for which to measure unreimbursed aid. The current loan repayment period is up to 50 years with an additional 10-year grace period, while the interest rate (or handling fee) is only 1%. Therefore, the level of unreimbursed loans from the Saudi Fund for Development can often reach as high as 60%. It is thus widely known that Saudi Arabia's favorable loans contain "generous repayment terms,"[19] a fact that has earned the country the reputation of being an "extraordinarily generous donor country."[20] In the statistical cooperation between Saudi Arabia and the OECD Development Assistance Committee, all the favorable loans issued by Saudi Arabia besides the Saudi Fund for Development are recognized as official development assistance. In addition, different from the aid from Western countries, Saudi Arabia respects the national sovereignty of the recipient country and generally does not attach conditions that may interfere with their internal affairs, such as political reforms or increased transparency. Saudi Arabia's aid centers around social and economic development projects with no additional requirements for purchasing goods from aid countries in general. Therefore, for African countries, Saudi Arabia often proves to be a more favorable funding source in comparison with Western countries.

Saudi Arabia Prompts the International Community to Aid Africa

To evaluate the scale of Saudi aid to Africa, one needs to assess Saudi Arabia's leading role in prompting the international community to aid Africa alongside its direct aid to Africa. One of the most important measures of aid is the ability to cooperate with regional or international aid agencies to provide joint assistance for various projects in Africa. Up till now, in terms of joint aid projects, the Saudi Fund for Development has cooperated with a large number of foreign and

[18] For the calculation method of the donation level, please refer to Yang Guang: "A Preliminary Study of Development Aid in the Middle East", West Asia and Africa, No. 6, 1990, p. 49.

[19] https://www.devex.com/organizations/saudi-fund-for-development-sfd-44289, Last Accessed: January 25, 2020.

[20] ShivitBakrania, Political Economy of Saudi Arabia's Overseas Aid Assistance, the website of Governance and Social Development Resource Centre (GSDRC), September 24, 2012. https://gsdrc.org/gsdrc_pub_type/helpdesk/page/27/, Last Accessed: February 18, 2020.

international aid agencies, such as The Arab Fund for Economic and Social Development, Abu Dhabi Fund for Development, Kuwait Fund For Arab Economic Development, Japan Bank for International Cooperation, (France) Fonds d'Aide et de Cooperation, (Denmark) The Investment Fund for Developing Countries, African Development Foundation, Economic Community of West African States, the OPEC Fund for International Development, European Investment Fund, International Monetary Fund, International Fund for Agricultural Development, (UK) Overseas Development Institute, Canadian International Development Agency, the Swedish International Development Cooperation Agency, United States Agency for International Development, Export Development Canada, International Development Association, (France) Organization for Economic Co-operation and Development, (Germany) KFW, UnicCredit, the Yugoslav Bank for International Economic Cooperation, African Development Bank, Asian Development Bank, West African Development Bank, the Development Bank of the Central African States, Eastern And Southern African Trade And Development Bank, European Foundation for Management Development, World Bank, the Arab Bank for Economic Development in Africa, Commonwealth Development Corporation Group (CDC Group), Millennium Challenge Corporation, Swedish Development Corporation, the United Nations Development Programme, the World Food Programme, and the Exim Bank of China. In these joint aid projects, Saudi Arabia has often been the main funder. From 1975 to 2018, the Saudi Fund for Development participated in a total of 274 joint-funded projects, investing19.1 billion Saudi Riyal (approximately US$7.07 billion) accounting for 45.5% of all financing in these projects.[21] Many of these projects are jointly financed by two or more of the aid agencies mentioned above. Saudi Arabia's large share of joint aid indicates that it plays a major role as a funder. Without Saudi Arabia as the main party to prompt them, these joint aid projects would not be funded. In this sense, Saudi aid plays a leading role in prompting the international community to aid Africa.

[21] The website of Saudi Fund for Development, http://www.sfd.gov.sa/en/web/guest/home.

Aid Capacity Is Restricted by Oil Prices

For a long time now, the amount of Saudi aid and the number of its aid projects have shown an obvious positive correlation with oil revenues, as well as with the determinants of international oil prices. From the perspective of the relationship between oil prices and Saudi aid, the amount of assistance and the number of aid projects in Saudi Arabia have changed in accordance with the fluctuations in oil prices from the 1970s to 2012. Since 2012, Saudi Arabia's motivation to substantially increase aid despite the low international oil price has been a logical method for maintaining foreign relations. However, such an untenable practice goes against Saudi Arabia's actual aid capacity in terms of the long-term trend of international oil prices.

Despite all odds, the period of low international oil prices that began in 2014 has lasted for six years. Since 2017, joint production restrictions implemented by OPEC and non-OPEC have failed to produce more productive results. Since 2015, major oil exporting countries such as Saudi Arabia, the United Arab Emirates, and Iran have been involved in the Yemeni Civil War; as a result, since November 2018, the US has imposed severe geopolitical incidents such as extreme pressure sanctions on Iran to zero oil exports. However, these incidents neither stimulated oil prices to rebound nor exerted a lasting impact. The key is that the fundamentals of the supply and demand relationship in the international oil market have undergone a radical change. From the perspective of demand, since the 1970s, the progress of global energy low carbonization, the development of alternatives to petroleum, and the improvement of energy efficiency have led to a weakening of the global demand for oil consumption, and particularly for oil in its role as transportation fuel. Some authoritative organizations, including OPEC, predicted that global oil demand may begin to "cease growing any larger" altogether between

2025 and 2040.[22] On the contrary, we contend that there is huge potential for the world's oil supply. US shale oil production has increased while costs fall; Russia's oil production capacity has increased; Africa's oil supply capability has begun to catch our attention. Major oil resource countries such as Iran and Venezuela are temporarily incapable of exporting oil due to geopolitical reasons, but their huge oil production capacities continue to gain momentum. This historical and fundamental structural change in the world's oil market indicates that the international oil market has been locked in a state of long-term oversupply, and the era of high oil prices may be gone forever.

As for Saudi Arabia, which relies solely on oil revenue as a source of aid, this trend in the international oil market surely will pose a serious and long-term negative threat to its aid capacity. Over the last ten years, the impact on the Saudi economy has been obvious. From 2008 to 2017, its per capita GDP growth stagnated at US$20,000.[23] Meanwhile, the government's fiscal revenue was left increasingly short, a situation which made the country supplement its fiscal revenue by reducing subsidies and increasing taxes. Within this milieu, the issue of where to obtain the source of Saudi Arabia's increased aid remains rather questionable (Figs. 1, 2).

5 Conclusion

Since the 1970s, Saudi Arabia, with its huge oil revenues, has rapidly grown to be one of the world's major donor countries and an important source of funding for Africa. African countries are the main recipients of Saudi aid, regardless of the total amount of aid funds or the number

[22] "OPEC: Oil Demand May Peak in 2040", *Economic Information Daily*, November 10, 2016, http://finance.china.com.cn/industry/energy/sytrq/20161110/3978896.shtml, Last interview date: February 17, 2020; David Sheppard, "Peak Oil Demand for 2036", *Financial Times*, July 16, 2018, https://www.ft.com/content/a12af4be-85cf-11e8-96dd-fa565ec55929, Last Accessed: February 17, 2020; Nick Cunningham, "BP Sees Peak Oil Demand in 2030s", Oil Price.com, February 23, 2018, https://oilprice.com/Energy/Crude-Oil/BP-Sees-Peak-Oil-Demand-In-2030s.html, Last Accessed: February 18, 2020; Jiang Xuefeng: "Global Oil and Gas Landscape and China's Oil and Gas Supply and Demand Situation", Natural Gas Industry Website, December 20, 2018, http://www.cngascn.com/homeNews/201812/34331.html, Last Accessed: February 18, 2020.

[23] World Bank database, https://databank.worldbank.org, Last Accessed: February 23, 2020.

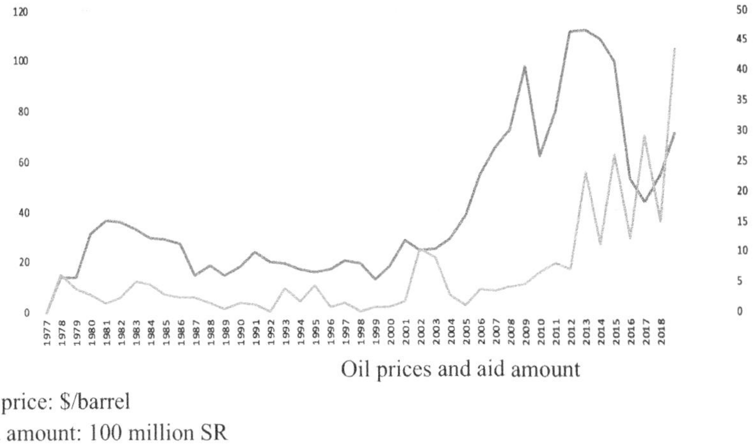

Fig. 1 The relationship between changes in oil prices and Saudi Aid to Africa (*Data source* Saudi Aid Platform, https://data.ksrelief.org, Last Accessed: January 27, 2020)

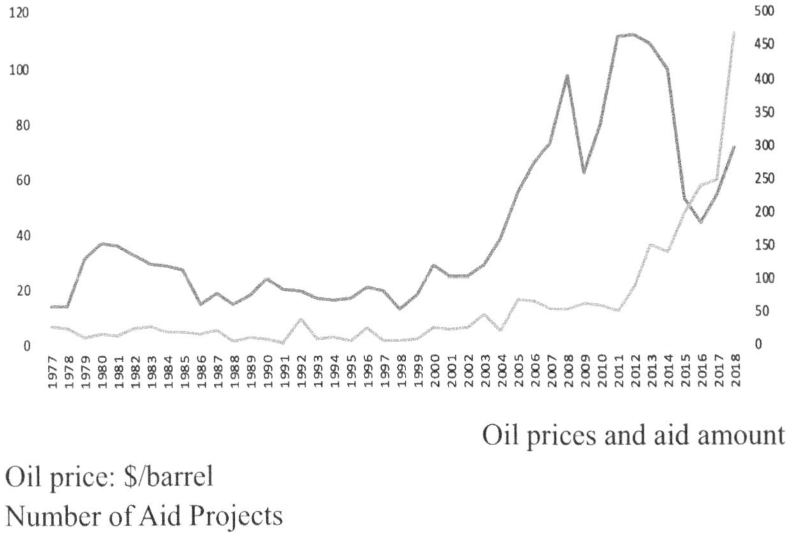

Fig. 2 The relationship between changes in oil prices and Saudi Aid Projects in Africa (*Data source* Saudi Aid Platform, https://data.ksrelief.org, Last Accessed: January 27, 2020)

of development projects. Saudi Arabia's extensive participation in joint assistance programs to and in Africa has also brought more aid funds to Africa. The contribution of Saudi aid to Africa is of significance in terms of encouraging African countries to explore their own independent developmental paths, providing them with relatively more favorable aid, and prompting the international community to aid Africa. Saudi Arabia has hybrid attributes of Islamic countries, Arab countries, and Middle Eastern countries. Its motivation to aid Africa is not only deeply rooted in the combination of Islam's charity and humanitarian spirit, but also out of its own security and development interests. Therefore, no matter what the consideration, Arab countries and Islamic countries are the main beneficiaries of Saudi aid, a fact that is also true of Africa. However, Saudi Arabia is, itself, still a developing country, featuring an obviously homogeneous economy. Its aid capacity depends entirely on oil revenues. A gap between its desire to and capacity for providing aid is emerging, and especially now that major structural changes are taking place in the international oil market, its capacity to substantially increase its aid will be severely constrained.

CHAPTER 13

Brazil's Policy Toward Africa in the Twenty-First Century: Goals, Paths, and Trends

Zhiwei Zhou

1 Introduction

The relationship between Brazil and Africa can be traced back to the slave trade that took place during the period of colonial Brazil. Driven by the slave trade promoted by its suzerain, Portugal, Brazil established racial and cultural ties with sub-Saharan Africa. Statistics suggest that about 12 million African slaves were sold around the world from 1501 to 1866. The number of slaves transported by Portugal was about 5.8 million, among whom about 5.5 million were sold to Brazil. About 3.86 million slaves transported to Brazil were from Central and Western Africa (including today's Angola and the Democratic Republic of the Congo), and about 877,000 were from the Bight of Benin (including today's Ghana, Togo, Benin, and Nigeria). After Brazil's independence and the

Z. Zhou (✉)
Department of International Relations, Institute of Latin American Studies, Chinese Academy of Social Sciences, Beijing, China
e-mail: wxy03@ssap.cn

X. Li and C. An (eds.), *The Changing World and Africa*,
https://doi.org/10.1007/978-981-16-4983-7_13

abolishment of its transatlantic slave trade, Brazil gradually lost contact with Africa.

Brazil's support for continued Portuguese colonialism after the 1950s worsened relations between African countries and Brazil, but since the 1960s, with the implementation of a "pragmatic" form of diplomacy, Brazil's relations with Africa have witnessed a relatively stable improvement. Especially after being the first to acknowledge the government of the People's Movement for the Liberation of Angola before other countries, Brazil has completely reversed its image in Africa's fight against imperialism and colonialism, which is widely regarded as a turning point in the relationship between Brazil and Africa. Due to its debt crisis in the 1980s and the economic downturn in the early 1990s, Brazil shifted its diplomatic focus to improving relations with creditor countries and diversifying trade (including strengthening economic relations with major developing countries). Yet for many years, Africa was still on the margins of its diplomatic strategy. Over the next 20 years, Brazil maintained only a moderate relationship with Africa, and even lessened its diplomatic investment with Africa. For example, the country reduced the number of diplomats in Africa from 34 in the 1980s to 24 in the early 1990s. During Fernando Henrique Cardoso's presidency from 1995 to 2002, Brazil withdrew its diplomatic organs from Africa. By the time President Luiz Inácio Lula da Silva took office in 2003, Brazil had only 18 diplomatic missions in Africa. Brazil's neglect of Africa attracted strong criticism from most African countries.

Since President Lula took the helm in 2003, Brazil has changed its attitude toward Africa. Strengthening relations with Africa has gradually become a more important part of Brazil's diversified diplomacy and South–South cooperation, and the relationship between Brazil and Africa has entered a stage of rapid development. Through a series of policy measures concerning diplomacy, the economy, trade, technical cooperation, and multilateral mechanisms, Brazil has gradually become an important partner with the African continent. That being said, with the dramatic changes to Brazil's political landscape in 2016, the country has made significant adjustments to its foreign policies, noticeably weakening its interaction and cooperation with Africa. This article aims to discuss the goals of Brazil's policy toward Africa, analyze its policy measures as they relate to Africa and its characteristics, evaluate the effects and challenges of its strategies concerning Africa, and explore the current trends in Brazil's relations with Africa.

2 Goals of Brazil's Policy Toward Africa in Its International Strategy

Since the beginning of the twenty-first century, Brazil has continuously adjusted its national identity and altered its diplomatic focus accordingly. After President Lula came into power, the country modified its foreign policy once again. Samuel Pinheiro Guimarães, then-Secretary-General of Foreign Affairs of Brazil and a core member of the country's diplomatic decision-making, clearly defined Brazil as a "periphery country," that is, "an underdeveloped and populous country with vast territories, agreeable climate, economic potential, industrial systems and a huge domestic market." However, he also emphasized that Brazil should not settle for being a periphery country, which often equates to "former colonies" and "neo-colonies," or countries that conform to the current world order and accept existing international political, economic, military, scientific, and technological structures. Although Brazil's economic strength lags far behind that of the United States (US) and Japan, it has advantages in terms of population, territory, resources, and an economic structure unmatched by ordinary peripheral countries. This indicates Brazil's ability to outperform other peripheral countries and change the existing unequal international order. At the same time, Celso Amorim, who served as Minister of Foreign Affairs during President Lula's two terms of office (2003–2006 and 2007–2010), underscored the fact that Brazil is an important international player with advantages in territory, political democracy, and economic strength, in addition to being a developing country faced with serious economic and social vulnerability. Therefore, "Brazil is not a small country, and it cannot and should not only have a foreign policy that suits a small country."

Based on such knowledge, Brazilian diplomatic decision-makers at the beginning of the twenty-first century believed that Brazil, as a developing country with relatively weak level of economic, political, and military strength, must overcome the imbalance of international power and its own vulnerabilities toward the outside world, as countries like Brazil can only benefit from a more balanced and diverse international power landscape. For this reason, it is an inevitable international strategy that Brazil will carry out practical political and economic cooperation with other peripheral countries with similar strategic interests and realize the multipolarization of the international structure. This is why Lula repeated, when he assumed the presidency in 2003, that Brazil should alter its diplomatic strategy, focusing only on the developed countries such as the US and Europe, implement a bolder and wiser foreign policy, and strive to determine a foreign policy that helps maintain a balance between rich and poor countries. Under such a foreign policy, Brazil naturally shifted the focus of diplomatic strategy from the developed countries that had been the focus of its diplomacy in the early 1990s to developing countries. Meanwhile, the strategy of independence proposed by the Lula government indicated that Brazil was determined to place development and the strengthening of its own sovereignty as diplomatic priorities and to resist pressure from developed countries by forming new alliances with foreign and emerging countries. At the same time, strengthening relations with the African continent and consolidating alliances with South American neighbors became the other two important pillars of Brazil's South–South cooperation strategy at the beginning of the twenty-first century.

Brazil's newly adjusted diplomacy has highlighted the status of South–South cooperation in its foreign policy, which was basically in line with the upward trajectory of Brazil–Africa relations in the past decade and reflected Africa's important role in Brazil's South–South cooperation strategy. To a certain extent, Brazil's relations with Africa can better reflect Brazil's above-mentioned diplomatic strategy. Africa's strategic significance to Brazil is mainly reflected in three aspects: historical and cultural connections, market opportunities, and geopolitics.

The slave trade during the colonial period enabled Brazil to establish a direct ethnic relationship with the African continent (especially sub-Saharan Africa) on the other side of the Atlantic Ocean, while Portuguese colonization deepened the historical and cultural bonds between Brazil and Portuguese-speaking African countries. As regards to the current population structure of Brazil, over 50% of the people there share a black lineage (one or more parents). In fact, Brazil is a country with the second largest number of black people in the world, right after Nigeria. In a country with such extensive black ancestry, African culture, religion, language, and customs have played a very important role in the development of Brazilian history and culture. The country's blood ties with Africa have prompted the Brazilian government to prioritize Africa in its international diplomacy, especially after the country shifted the focus of its diplomatic strategy to South–South cooperation. Besides, such a "pro-Africa policy" has gained widespread recognition from the public, thereby strengthening the support of the people (especially the middle- and lower classes) for the Brazilian government. From this perspective,

Brazil's African policy can be seen as an extension of the country's ethnic and social integration policy over the past decade. That is to say, strengthening diplomacy with Africa was actually, to a certain extent, out of considerations for its own domestic politics, although it is difficult to quantify and calculate such a factor.

Economic considerations are also an important factor for Brazil to strengthen its relations with Africa. As the African continent has gradually secured greater political, economic, and social stability, it has begun to release its potential for resource reserves, market demands, and investment opportunities. This was a major reason why developed and emerging nations have increased their exchanges with Africa. However, the economic considerations behind Brazil's policy toward Africa differ from those of other countries in certain aspects. For starters, Africa is a choice made by Brazil to be able to diversify its foreign trade. Since Lula came to power, Brazil has always made reducing its economic dependence on developed countries a focus of its diplomacy, advocated for establishing a new world trade geography, called for the establishment of a new international trade order, and strengthened trade relations with developing countries to diversify its foreign trade. Lula pointed out that in 2003, Brazilian exports to the European Union (EU) and the US accounted for 52% of the country's total exports, adding that the county was in urgent need of expanding the overseas market, which was one way to create jobs and increase foreign currency earnings. Therefore, Brazil must first consider other Latin American countries (or at least start with South America), and then move to Africa, which is relatively close to Brazil. At the same time, the country must also enter the Arab world. It can thus be seen that Africa is an important part of Brazil's overall expansion of its overseas market.

Second, Africa is an important export destination for Brazil's high value-added products. Brazil, one of the countries boasting the world's richest resources, didn't focus on seizing resources from Africa by establishing economic and trade ties, but aimed at exporting high value-added industrial products to Africa, which was another difference between its strategy toward Africa and the policies of developed and emerging countries like China and India. From the perspective of the trade structure between Brazil and various regions, Africa and Latin America are the main export destinations for Brazil's manufactured goods, a fact that could be attributed to the technological advantages Brazil has over the two regions. For example, industrial products (including semi-manufactured products) accounted for 70% (2011) and 84% (2012) of Brazil's exports to the two regions. In 2012, the share of exported industrial products to Africa declined slightly, yet still maintained a relatively high level of about 65%. In 2002, this ratio even exceeded 85%. Therefore, Africa has become the main export market of Brazil's industrial products, which is crucial to Brazil's expansion of its product exportation. Moreover, as Africa has gradually stabilized its economy and society, its consumption potential appeals to Brazil more than ever.

Third, environmental similarities between Africa and Brazil, such as climate and soil, make the continent an important destination for Brazil's technological experimentation and exports. This is also a distinctive feature of Brazil's policy toward Africa, which will be discussed in detail in the next section of this article.

In the past, the African continent was never the focus of Brazil's foreign policy, let alone geopolitical relations. However, in the first decade of the twenty-first century, the strategic role of Africa in Brazil's foreign policy has been greatly enhanced. On the one hand, it already had an inevitable connection with the diplomatic transformation of Brazil toward South–South cooperation. On the other hand, this can be mainly attributed to the power landscape of the African continent. Brazil's "South–South cooperation" strategy not only emphasizes deepening ties and cooperation with developing countries, but also seeks to hold a leadership role among the developing countries. Under such a strategy, the two major priorities for Brazil's foreign policy are to enhance its presence in Africa and to strengthen its participation in African affairs (especially in Africa's developmental issues). Brazil's major goal is to increase its influence on the African continent and win the recognition and support of the various African countries in its pursuit of being the spokesperson for the developing countries. In addition, during Lula's two terms of office, the Brazilian foreign policy formulated by Celso Amorim, Minister of Foreign Affairs, sought the pivot of power changes, new platforms for global participation in related affairs, and a novel space initiated by the developing countries that would enable them to play a greater role. As far as Brazil is concerned, Africa is a region full of uncertainties where multiple forces are engaged in a tug of war. Brazil, a country seeking to rise on the global stage, has many favorable conditions for participating in African affairs. It is also necessary to strengthen Brazil's influence in Africa so as to strengthen its international identity as a spokesperson for the developing countries and to serve its overall international strategic goals. In 1986, Brazil initiated the establishment of the South Atlantic Peace and Cooperation Zone (Portuguese: Zona de Paz e Cooperação do Atlântico Sul, ZPCAS), in a bid to promote regional cooperation in the South Atlantic and maintain peace and security. However, it wasn't until the past ten years that Brazil once again attached any importance to this initiative. Ever since the country discovered a series of deep-sea oil fields rich in reserves, it has significantly strengthened its security awareness of the South Atlantic. Such a transformation is reflected in two important documents published in the past few years: the "National Defense Strategy" in 2008, and the "Defense White Paper" in 2012. As these abundant natural resources may trigger armed conflicts in the Brazilian waters of the South Atlantic, the former document formalized the security of the South Atlantic as a strategic priority, and the latter specified the procedures for ensuring the security of the South Atlantic through Brazil's unilateral actions and international cooperation, not only with South American countries, but also with countries along the west coast of Africa. Therefore, from this perspective, consolidating multi-field cooperation with African countries is of great strategic significance to Brazil's waters and territorial security. It is out of security considerations that military cooperation has become an important part of Brazil's strengthening of its strategic cooperation in defense with Africa. For example, Brazil has helped Namibia with naval construction, developed short-range air-to-air missile A-Darters with South Africa, and assisted several small countries in Africa—including Cape Verde, Guinea-Bissau, São Tomé and Príncipe, and Equatorial Guinea—in training military officers, providing military equipment, and surveying continental shelves.

3 Brazil's Policy Toward Africa and Its Characteristics at the Beginning of the Twenty-First Century

As mentioned earlier, Brazil and Africa have formed many long historical and cultural ties, but these kinds of "blood ties" have not brought long-lasting friendship between the two sides, a fact which mainly stems from the various diplomatic strategies undertaken by Brazil at different times. Since the beginning of the Twenty-first century, Brazil has secured a South–South diplomatic strategy and reversed its long-standing tradition of neglecting Africa. Through diversified policy measures, it has accelerated the ongoing advances in its relations with Africa in a bid to expand its presence and influence on the continent. In terms of its policy measures, Brazil's policy toward Africa is similar to that of other emerging countries, but also has its own distinctive features. Specifically, Brazil's policy toward Africa mainly includes the following aspects:

First, attaching importance to the country's historical and cultural connections with Africa. In the past ten years or so, what Brazilian politicians and scholars have talked the most about when discussing and analyzing Brazil's policy toward Africa was the historical origins between the two entities, which were considered very important factors. In terms of Brazil's policy toward Africa, former President Lula once said, "In a sense, Brazil belongs to Africa" and "Brazil owes debts (to Africa) in history and is now required to pay it back, by establishing a sound relationship with Africa." President Dilma Rousseff emphasized that Brazil regards the African continent as a brother and a close neighbor and that the two parties have similarities and strong attachments to each other. More than half of Brazil's nearly 200 million people consider themselves of African descent. Therefore, African descent is one of the most precious characteristics of the Brazilian nation. When summarizing the policies of the Lula government (2003–2010) toward Africa, Celso Amorim made it clear that in addition to political and economic interests, seeking a closer relationship with Africa is the result of existing bilateral connections in history, population, and culture. Therefore, Brazil is fully prepared in terms of population, race, and culture to play a more important role in the new world of Africa. Emphasizing Brazil's historical and blood ties with Africa will not only bring the two parties closer to each other and help Brazil better fit in with Africa by enhancing its presence on the African continent, but will also give Brazil an edge over other emerging countries in competing for the African market.

Second involves adjusting diplomatic institutions and strengthening diplomacy with Africa. The fact that Brazil has attached importance to Africa is most directly reflected in the institutional adjustments made by the Brazilian Ministry of Foreign Affairs. Since President Lula came to power in 2003, the Brazilian government started to deal with the affairs in Africa and the Middle East separately. It has established a special Department for Africa (DEAF) with three offices and a special coordination office for the Community of Portuguese Language Countries (Portuguese: Comunidade dos Países de Língua Portuguesa, CPLP). At the same time, the number of Brazilian embassies, consulates, and diplomats in Africa has increased significantly. For example, during Lula's eight years in office, Brazil established or restored 17 embassies, bringing the total number of embassies in Africa to 37. Meanwhile, the total number of African embassies in Brazil increased to 34. In addition, the number of Brazilian diplomatic officials in Africa rose from 997 in 2005 to 1397. Professor Cláudio Ribeiro, affiliated with the African Studies Center of the Pontifical Catholic University of São Paulo in Brazil, believes that the restoration and expansion of Brazilian embassies and consulates in Africa and the adjustment of the internal institutions of the Brazilian Ministry of Foreign Affairs reflect on the fact that Brazil has continuously expanded its interests on the African continent and that African countries have played a more important role in Brazil's international strategy. The closer diplomatic relations between Brazil and Africa could be directly seen from the frequent exchanges of visits by Brazilian and African heads of state. During his 2003–2010 presidency, Lula visited Africa twelve times, while African heads of state visited Brazil 48 times during the same period. Since President Rousseff took office, the state visits she paid to African countries decreased in frequency. From taking office in 2011 to being impeached and stepping down at the end of August 2016, Rousseff only visited Africa five times, involving a mere six countries in the region. In contrast, Rousseff preferred to employ "peripheral diplomacy" under the direct or indirect multilateral mechanisms between Brazil and Africa.

Third, introducing incentive policies to encourage trade and investment. Brazil's economic and trade policy toward Africa mainly focuses on expanding exports and increasing investment in Africa. Brazil–Africa trade started from a relatively low level, with only 5 billion USD in 2002, which at the time accounted for less than 5% of Brazil's total foreign trade. In order to accelerate exports to the African market, the Brazilian government formulated special policies to support exports to the African market. For example, in 2008, Brazil's Ministry of Development, Industry and Foreign Trade formulated a program of integration with Africa (Portuguese: Programa Integração com a África) under its productive development policy (Portuguese: Política de Desenvolvimento Produtivo), aiming to expand economic relations between Brazil and Africa and to intensify the presence of Brazilian institutions in Africa. It also set goals for trade, investment, and cooperation with the African continent, such as making sure that Africa accounts for 1.39% or higher of Brazil's total imports and that Brazilian exports to Africa make up 5.1%

or higher of the country's total. This policy also includes establishing a service trade monitoring system and providing funds for the three direct Brazilian investment projects in Africa by 2020. The program aimed to determine specific measures to promote trade and investment relations with Africa through the National Bank for Economic and Social Development (Portuguese: Banco Nacional de Desenvolvimento Econômico e Social, BNDES). So far, the policy mainly targets expanding exports and investing in infrastructure. In 2008, when the program was launched, the supporting funds for exports to Africa stood at 477 million BRL and increased to 649 million BRL in 2009. Regarding investment in Africa, the Brazilian government has largely provided preferential loans for Brazilian companies. For example, Brazil provided a loan of 1.5 billion USD for the ethanol processing plant project in Angola in 2006. Later, in 2010, Brazil established a loan project with a budget of 3.5 billion USD for Brazilian companies in Ghana and Mozambique. In 2012, BNDES provided 682 million USD to Brazilian companies' projects in Africa, an increase of 46% from 2011. In addition, public institutions such as the Federal Savings Bank (Portuguese: Caixa Econômica Federal), Bank of Brazil (Portuguese: Banco do Brasil), and the Brazilian Trade and Investment Promotion Agency (Portuguese: Agência Brasileira de Promoção de Exportações e Investimentos, APEX) have provided important financial and policy support in the process of Brazil's expanding in Africa.

Fourth, greatly promoting development cooperation between Brazil and Africa. Development cooperation is an important part of Brazil's policy toward Africa as well as the core of Brazil's "South–South cooperation" with Africa. Brazil itself used to be a recipient of international development assistance. Since Lula's second term of office (2007–2010), Brazil has gradually strengthened its assistance to other countries while also receiving aid in a bid to serve its international diplomacy, with South America and Africa as the focus of Brazil's development cooperation. Technical cooperation and economic assistance are Brazil's main means of assistance to Africa. According to statistics from the Brazilian Ministry of Foreign Affairs, Brazil's funds for technical cooperation with Africa totaled approximately 22.1 million USD in 2010, three times that of 2009. The funds for Africa accounted for about 57% of the total funds of Brazil for international technical development cooperation that year. In addition, Brazil launched a total of 640 technical cooperation projects with Africa between 2003 and 2014 (including bilateral, trilateral, and multilateral cooperation), of which 77 were initiated during Lula's first term (2003–2006), 406 during Lula's second term (2007–2010), and 157 during Rousseff's first term (2011–2014). Brazil's technical cooperation with Africa involves 40 countries, among which the Portuguese-speaking countries are the primary focus. For example, five Portuguese-speaking African countries accounted for 74% of Brazil's total funds for technical cooperation with Africa in 2010 (Fig. 1).

South America and Africa are, in many ways, like twin continents. Both Brazil and Africa (especially Sub-Saharan Africa) are located in the southern hemisphere and possess very similar ecological and geographical environments, which generates the conditions and space necessary for the two parties to conduct technological cooperation. Brazil's technical cooperation with Africa can be divided into four main fields: tropical agriculture, tropical medicine, vocational and technical education, and social policy promotion. In terms of tropical agriculture technology, the most successful example is the series of technology promotions in Africa by Embrapa, a Brazilian company specializing in agriculture and animal husbandry, and the Brazilian Cooperation Agency (Portuguese: Agência Brasileira de Cooperação, ABC) under the Ministry of Foreign Affairs. At present, Embrapa has set up offices in Ghana, Mozambique, Mali, and

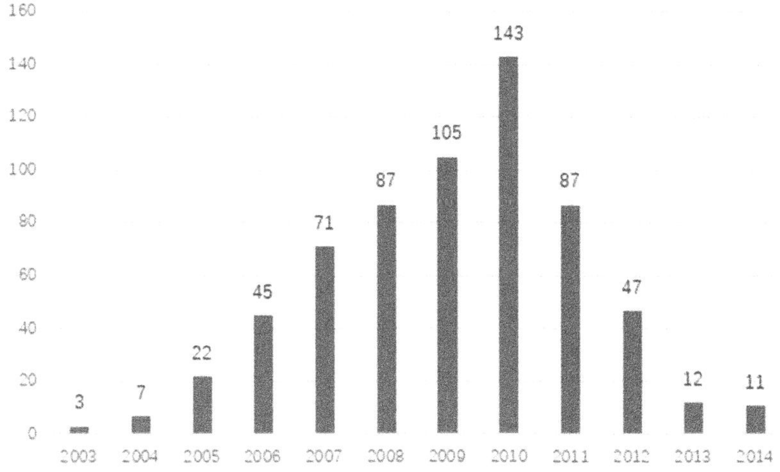

Fig. 1 Number of Brazil's technical cooperation projects with Africa (2003–2014) (*Source* Brazilian Cooperation Agency of Ministry of Foreign Affairs of Brazil)

Senegal, and carried out cooperative projects with 35 African countries. Such projects have been implemented with more than 10 African countries. Major projects include: Brazil's soil improvement project (Projeto ProSavana) with Senegal and Mozambique; Embrapa's cotton project (Projeto Cotton-4) with Benin, Burkina Faso, Mali, Chad, and Togo; Embrapa's land innovation platform project (Plataformas de Inovação Agropecuária), as well as the nutrition and food security program (Programas de Nutrição e Segurança Alimentar) in Mozambique; the rice development project in Senegal, etc. In terms of tropical medicine, Brazil has signed cooperative agreements with nearly 30 African countries in the medical field, mainly focusing on the prevention and treatment of AIDS and other epidemics (including malaria and sickle cell anemia). A representative cooperation project is a biochemical lab established in Mozambique by Brazil's Oswaldo Cruz Foundation (FIOCRUZ) to produce AIDS prevention drugs, which will then be exported to other African countries. In terms of vocational and technical training, Brazil's National Service for Industrial Training (Portuguese: Serviço Nacional de Aprendizagem Industrial, SENAI) has established vocational training centers in African countries like Angola, Cape Verde, Guinea-Bissau, São Tomé and Príncipe, Mozambique, Congo, and South Africa. For example, a center built in 2000 in Luanda, capital of Angola, provides technical training in machinery, civil construction, electricity, clothing, and communications. According to an agreement, the training center was handed over to the Angolan government in 2005. As regards social policy promotion, Brazil's successful experience in poverty eradication (such as the family aid program) has caused many African countries to follow suit. Countries like Angola, Ghana, Kenya, and Senegal have initiated close technical consultations and cooperation with Brazil in terms of poverty reduction policies. A series of policies under the Zero Hunger project (Portuguese: Fome Zero), including the family aid program, has become the Brazilian government's primary policy tools to strengthen ties with African countries and are an important part of Brazil's efforts to build a new image on the African continent (Fig. 2).

Brazil's economic assistance to Africa mainly includes debt relief and concessional loans. In terms of debt relief, both the Lula and Rousseff governments intensified debt relief for some African countries. For example, the Lula government successively canceled the debts of Mozambique (315 million USD), Nigeria (83.1 million USD), Cape Verde (4 million USD), and Gabon (36 million USD). When she attended the

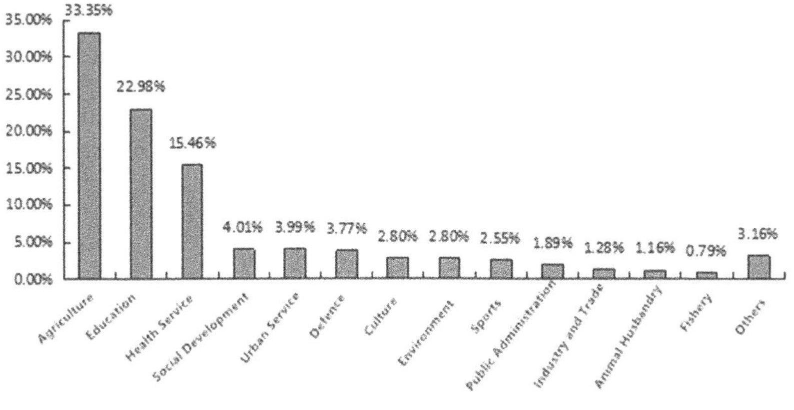

Fig. 2 Distribution of Brazil's technical cooperation with Africa (2000–2014) (*Source* Brazilian Cooperation Agency of Ministry of Foreign Affairs of Brazil)

commemoration of the 50th anniversary of the African Union (AU) in May 2013, Rousseff announced that Brazil would relieve the debts of 12 African countries worth 897.7 million USD. In addition, Brazil has obtained the opportunity to provide loans and financing for African countries through debt relief negotiations and has made it more convenient for Brazilian companies to enter these countries. For example, Brazil provided loans for Cape Verde to support the construction of public facilities. Brazil's Foreign Trade Chamber (Portuguese: Câmara de Comércio Exterior, CAMEX) loaned 95 million USD and 98 million USD to Ghana and Zimbabwe, respectively, to help them import Brazilian agricultural products. BNDES has provided loans to Angola, Ghana, Mozambique, and South Africa for transportation infrastructure, communications, power transmission, water supply facilities, and other projects, among which the loans for Angola alone reached 3.2 billion USD.

Fifth involves advancing relations through multilateral mechanisms. In the past decade or so, strengthening relations with Africa through regional multilateral mechanisms has been an important tool in the Brazilian government's policy toward Africa. The most important multilateral mechanism is the Africa–South America Summit (ASA Summit), established in November 2006. The Brazilian Ministry of Foreign Affairs defines this mechanism as "a special opportunity to strengthen dialogues between the two regions, which share common history, worldview and strategic interests to a great extent." The two regions have extensive consensus on international affairs, especially on issues such as international trade, peace and security, environment, energy, human rights, anti-poverty and anti-hunger, and the reform of multilateral mechanisms. During the first senior officials' meeting, held in Brazil in June 2008, eight working groups were established in line with the cooperation fields of the ASA Summit, with every group chaired by two countries, each country from each of the two regions. As the country with the largest economy and strongest international influence in the two regions, Brazil played a pivotal role in advancing the operation of the ASA Summit. It has not only strengthened its ties with African and South American neighbors at multiple levels, but also enhanced its leading role in the two regions, which is exactly in line with Brazil's diplomatic goal of becoming a leader for the global South. In addition, Brazil and the AU established an institutional dialogue mechanism in 2005 and, in 2007, signed a bilateral technical cooperation framework agreement. In July 2009, then Brazilian President Lula attended the 13th AU Assembly upon invitation. In June 2011, Lula, who had already stepped down, participated in the AU Assembly again on behalf of then-President Rousseff. Later in May 2013, Rousseff attended the 21st AU Assembly upon invitation. By participating in the AU Assembly, the philosophy behind Brazil's policy toward Africa, its appeal for historical and cultural origins with Africa, and the promotion of Brazil's poverty alleviation experience have gained increasing recognition in Africa. Moreover, Brazil has also strengthened relations with subregions and key countries in Africa through multilateral mechanisms. For example, the Southern Common Market (Portuguese: Mercado Comum do Sul, Mercosul) initiated by Brazil and the Southern African Customs Union (SACU), signed a framework agreement and launched negotiations focusing on free trade. In December 2008 and April 2009, respectively, the two sides signed preferential trade agreements. The CPLP serves as a direct bond that helps strengthen ties between Brazil and Portuguese-speaking African countries. Brazil has invested a great deal in its diplomacy with these countries, which is key to the relationship between Brazil and Africa. In addition, mechanisms such as the Summit of South American-Arab Countries, the IBSA Dialogue Forum (India, Brazil, South Africa), and the BRICS are also important in Brazil's policy toward Africa. In short, Brazil–Africa relations have witnessed progress under major multilateral mechanisms involving the two sides (especially those between the two regions), as well as many other mechanisms. Brazil is also good at using these multilateral mechanisms to advance its interests and enhance its image in Africa.

4 Effects and Trends of Brazil's Policy Toward Africa

Through diversified policy measures, the relationship between Brazil and Africa has developed rapidly over the last ten years. First of all, the African countries have generally held a relatively open attitude toward Brazil. Under the support of the Brazilian government, the Brazil–Africa relationship forged through shared cultural inheritance has played a very positive role for their future cooperative endeavors. In addition, Brazil has promoted and exported tropical technologies and provided support and consultation on poverty alleviation social policies in a targeted manner to the African countries, giving Brazil an edge over other countries considering their national image in the African region. Alain Jean Gustave Ilboudo, Burkina Faso's Ambassador to Brazil, once said that "For Africa, Brazil, unlike China, feels closer." Connections due to cultural traditions, as well as Africa's recognition of Brazilian institutions and policies, have brought Brazil closer to African countries, creating a very favorable environment for Brazil to implement its policies toward Africa.

Second, the economic and trade relations between Brazil and Africa have achieved rapid growth. In 2002, the total trade volume between Brazil and Africa registered only 5 billion USD, rising to approximately 28.5 billion USD in 2013. It should be noted, though, that Africa's share of Brazil's total foreign trade didn't increase significantly, only growing from 4.7% in 2002 to 6.3% in 2012. However, the trade between Brazil and Africa has continuously declined since 2013 due to the impact of the global trade environment. In 2019, the scale of bilateral trade shrank to 13.12 billion USD. There has been no specific data regarding the amount of Brazilian investment in Africa. According to statistics drawn from the business of Brazilian multinational companies by the Fundação Dom Cabral (FDC) in 2012, only 27% of Brazilian multinational companies invested in Africa, less than the figures for South America (73%), North America (63.5%), Europe (46%), Asia (44.4%), and Central America (33.3%). However, the growth rate of Brazilian investment in Africa ranked third. Until now, Brazilian companies have only invested in 16 African countries. Among them, Angola and South Africa are the key countries for Brazilian companies' investment: 14% of Brazilian multinational companies have invested in Angola, and the number of Brazilian companies with investment projects in South Africa accounts for 9.5% of Brazil's multinational enterprises. The reasons that Brazil's investment in Africa has increased so rapidly are: first, the past decade has witnessed a rapid expansion of Brazilian companies' internationalization; second, the stability of Brazilian policy toward Africa has greatly improved during this period. In terms of the investing entities, Brazil's companies with existing investments in Africa mainly include Petrobras, Vale, infrastructure manufacturer Odebrecht, Andrade Gutierrez, Marcopolo, WEG, Camargo Corrêa, Queiroz Galvão, etc. As regards the fields of investment, infrastructure, energy, and mining are the major businesses of Brazilian companies in Africa. In specific investment operations, Brazilian companies pay more attention to employing local people, training local employees, fulfilling corporate social responsibilities, and purchasing spare parts and accessories from the local companies. This makes it more convenient for them to integrate into the African market and explore further investment opportunities. In order to make full use of Africa's market potential, Luciano Coutinho, President of BNDES, in 2011 called on Brazilian companies not to be limited to infrastructure, energy, and agriculture during the Brazil–Africa entrepreneurship conference, encouraging them to seize business opportunities in African banks and telecommunications services as soon as possible (Fig. 3).

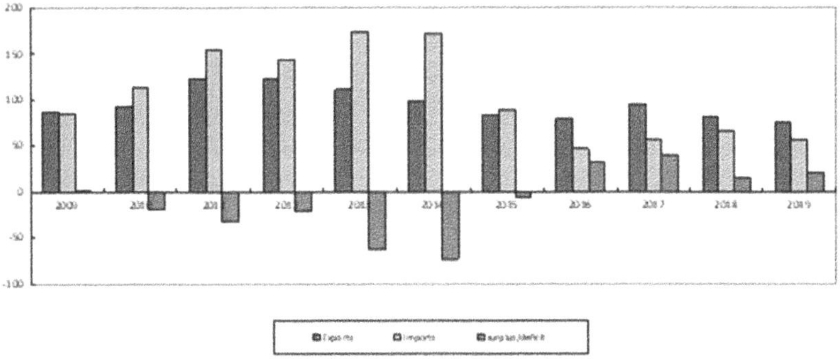

Fig. 3 Trade between Brazil and Africa (unit: 100 million USD) (*Source* Secretariat of the Foreign Trade of Ministry of Economy of Brazil)

The rise of Brazil's economy and international influence, coupled with its reinforced diplomacy in Africa, has rapidly increased Brazil's influence in Africa and gained strong support from the African countries in some major international affairs. For example, on the issue of getting a permanent seat on the United Nations (UN) Security Council, Brazil has gained support from most African countries. In addition, in the 2012 election of the Director-General of the UN Food and Agriculture Organization and the 2013 election of the Director-General of the World Trade Organization, Brazil's two candidates for office, José Graziano da Silva and Roberto Azevedo, respectively, received extensive support from African countries. What's more, for major countries on the African continent, Brazil has gradually become a priority for their diplomatic policies. For example, Brazil and South Africa established a strategic partnership in 2010. Meanwhile, multilateral mechanisms such as the BRICS, the BASIC, and the IBSA Dialogue Forum have brought Brazil–South Africa relations beyond the scope of a bilateral framework and extended such bilateral ties to broader global governance cooperation. In addition, Brazil has established strategic partnerships or strategic dialogue mechanisms with Angola, Mozambique, Nigeria, and other countries. Especially for Portuguese-speaking African countries, Brazil is becoming a priority in their foreign policies. For example, Brazil ranks among the top three in terms of importance and influence in Angola's diplomacy, and enjoys a similar status in Angola's diplomatic policy for the future. Brazil's influence in Mozambique ranks tenth, and has even risen to fifth in terms of top partners for Mozambique in the future.

In October 2018, Jair Bolsonaro, the far-right Social Liberal Party candidate, won the election, which has profoundly changed the political landscape in Brazil and brought great uncertainties to Brazil's diplomacy in the next stage. In terms of foreign policy, President Bolsonaro and Ernesto Araújo, Brazil's current Minister of Foreign Affairs, have repeatedly criticized the country's "ideological" foreign policies during the ruling cycle of the left-wing Workers' Party (2003–2016). However, Brazil's diplomatic preferences are likely to go to the opposite extreme in Bolsonaro's term of office. The new Brazilian government advocates for its foreign trade policy being based on the principle that "partners can bring added value to Brazil." It supports US President Donald Trump's immigration policy, abandons diplomatic relations with autocracies (countries that go against the US, Israel, and Italy), and resolutely refuses to sign trade agreements with such countries. Meanwhile, it also adjusts

Brazil's diplomatic tradition featuring multilateralism and attaches importance to bilateral principles, and so on. Bolsonaro's diplomatic policies are characterized by a strong ideological bias, especially in his criticism of the left-wing ruling countries, socialism, and communism. This is likely to change the priority of South–South cooperation in foreign policies during the ruling period of the Workers' Party. As an important part of South–South cooperation, the relationship between Brazil and Africa is extremely unlikely to continue the development trend in the early twenty-first century in the short to medium term.

First and foremost, this is because the frequency of high-level exchanges between Brazil and Africa has been significantly reduced. In 2019, the first year into Bolsonaro's administration, Brazil's high-level interaction with Africa was reflected only in the visits of Brazilian Foreign Minister Araújo to four African countries (Cape Verde, Senegal, Nigeria, and Angola). During his visit to Angola, Araújo gave a speech titled "Brazil's New Diplomacy" in which he talked about the new Brazilian government's policy toward Africa, including attaching importance to the history and cultural traditions shared with Africa and strengthening economic and trade ties with Africa to participate more widely in Africa's developmental process. Compared with Brazil's aforementioned policy toward Africa at the beginning of the twenty-first century, it can be seen that content such as social development, technical cooperation, and defense cooperation were not included in the Brazilian Foreign Minister's speech, which also reflects how the strategic position of Africa in Brazil's diplomacy has been obviously weakened.

Second, Brazilian diplomatic missions in Africa have been significantly scaled down. In order to strengthen its presence in Africa, Brazil intensified the construction of its missions in Africa at the beginning of the twenty-first century. However, such a policy has shown signs of reversal with the changes to Brazil's political landscape, diplomatic adjustments, and public financial constraints that it has undergone in recent years. In June 2019, the Brazilian Ministry of Foreign Affairs decided to reduce the size of the country's missions abroad, mainly in countries with fewer economic and trade exchanges with Brazil—and fewer Brazilians. Such countries were mainly Caribbean island countries and African countries, including Libya, Liberia, and Sierra Leone. With such diplomatic adjustments, the current Brazilian government has not only weakened its emphasis on South–South cooperation, but will also directly cut its ties with Africa and its influence in Africa.

Third, the potential of Brazil's policy concerning Africa has shrunk. At the beginning of the twenty-first century, Brazil mainly employed economic and trade exchanges and development cooperation as tools of its policy toward Africa. In terms of economic and trade exchanges, bilateral trade has, as mentioned above, been declining since 2013, and the scale of trade in 2019 was even less than half of what it had been in 2013. Considering the economic downturn of Brazil in recent years and the lack of complementarity with African economies, the pulling effect of economic and trade cooperation will gradually be reduced. As many Brazilian multinational companies have been exposed as corrupt, Brazilian overseas investment has gone through a hard time, and many investment projects in Africa have even been directly implicated. In the short term, it is difficult for Brazil to advance its investment business in Africa. In terms of development cooperation, Brazil is also troubled by financial constraints. From 2010 to 2014, Brazil's total development cooperation funds in Africa fell from 19.84 USD million to 5.07 million USD, which also reflected a certain degree of slowed development cooperation between Brazil and Africa.

Finally, the vitality of regional cooperation mechanisms has been affected. At present, the Latin American countries are faced with out-of-beat political cycles. The integration of Latin America is basically at a standstill, with some regional organizations even facing a survival dilemma. For example, the Union of South American Nations now only nominally exists, while the Community of Latin American and Caribbean States is not as vibrant as it used to be over the past several years, which has directly affected the effectiveness of cross-regional multilateral cooperation mechanisms between Brazil and Africa. The ASA Summit has been suspended for many years, and strategic interaction between South America and Africa basically no longer exists.

Generally speaking, since the beginning of the twenty-first century, the relationship between Brazil and Africa has undergone two different stages. In the first stage, that is, the first 15 years of the twenty-first century, Brazil demonstrated its own advantages and characteristics in its policy toward Africa and achieved good results by doing so. Brazil became a new force to be reckoned with on the African continent and began to fully integrate into a more comprehensive global competition in the African market. In the second phase, that is, since the Workers' Party stepped down in 2016, with the corresponding Brazilian adjustments of the country's internal affairs and diplomacy, Brazil has weakened its emphasis on South–South cooperation, and as such the relationship between Brazil and Africa has significantly cooled. Such changes have reflected the limitations of Brazil's own international strategy, the lack of coherence in Brazil's policy toward Africa, and the lack of consensus in Brazilian society as to Africa's importance. Africa is now considered dispensable to Brazil's diplomacy. Although it could become an incremental part of Brazilian diplomacy, the African policy does not have enough weight in the Brazilian diplomatic system. Besides, compared with other major countries (including emerging major countries), Brazil still lacks strong economic means to strengthen its policy toward Africa, which is closely related to the relatively weak economic strength of Brazil itself. In the meantime, the lack of economic complementarity between Brazil and Africa is also an important factor causing the shortage of any internal motivation for pursuing a bilateral relationship.

CHAPTER 14

The Relations Between Cuba and Africa: Characters, Motivations, and Tendencies

Han Han

Cuba is the only socialist country in the Western Hemisphere. Since the success of its revolution in 1959, Cuba has chosen to develop along socialist lines. And with endorsements from an international community that includes Africa, Cuba has firmly and consistently stuck to the socialist system even when encountering many difficulties, such as the five-decade-long trade, economic, and financial blockade instituted by the US, or the collapse of its ally, the former Union of Soviet Socialist Republics (USSR) or the Soviet Union. The United Nations (UN) has put forward an annual resolution draft for the last 27 years—since 1992—calling for the US to remove the economic and financial block and trade embargo on Cuba. Throughout these measures, Africa has been the country's main source of international support.

The relations between Cuba and Africa have passed through several periods and have become a crucial force for diplomatic development in

H. Han (✉)
Chinese Academy of Social Sciences (CASS), Beijing, China
e-mail: wxy03@ssap.cn

X. Li and C. An (eds.), *The Changing World and Africa*,
https://doi.org/10.1007/978-981-16-4983-7_14

terms of South–South cooperation. It is well known that Cuba–US relations have a great impact on Cuba. Africa provides an area of relief for the relations between the two countries. On December 10, 2013, US President Barack Obama and Cuba's leader Raul Castro attended the funeral ceremony for Nelson Mandela at the same time, where they greeted and shook hands together during the official ceremony in Johannesburg, South Africa.

1 Development of the Relations Between Cuba and Africa

Cuba is separated from Africa by the Atlantic but both of them share some similar requirements for their development. Historically, white Europeans dominated the slave trade and originally connected the African continent with Cuba. In the modern era, the common experience of being colonized and oppressed promoted military and revolutionary cooperation between Cuba and the African continent. With the changing international situation and the Cold War, Cuban-African relations, once dominated by ideology and revolutionary exports, also changed. In the new era, Cuba has treated the African continent as an important partner with the third world, sharing common development requirements through its multiple and pragmatic diplomatic channels.

Historical Linkage

The source of Cuba–Africa relations can be traced back to the Spanish colonial period when Cuba pioneered importing coffee plantations in Latin America and the Caribbean region. In addition to coffee, coco, sugarcane, and other agricultural products were also widely cultivated in Cuba, so the plantation economy in Cuba provided the momentum for its development. However, the formation of the plantation economy under colonial rule overworked the local Indian labor population, and together with the diseases brought in by the colonists, the number of local Indians quickly declined on the island. Spanish colonists had to import black slaves in order to solve the problem of labor shortages. This has significantly influenced Cuba's population structure even up through today. The former local residents, the indigenous population, have been diminished in Cuba's population, existing only in the people with mixed ethnicities in Cuba. The immigrants from Africa that arrived in Cuba brought with them African ethnicities, eventually resulting in Afro-indigenous mixed ethnicities. According to the 2017 Cuban population statistics, Cuba's total population was 11.22 million, with 11% of the population made up of black people. In addition, there are many mixed black people in the mixed ethnic groups, making up 22% of the total population. The nation's slave trade history has not only changed the composition of Cuba's population, but also deeply impacted the religion, music, dance, and culture of the island, enriching the existing Cuban culture with that of newly arrived African immigrants. In the more than five centuries since black people began coming into Cuba from Africa, a special black culture with a uniquely Cuban character that differed from the local African style was created, which became the foundation for a common identity between Cuba and African countries. For example, the schools of African religion that are based on the African culture in Cuba have believers from all ethnicities in Cuba. Another example is that in 2019, Cuba held a ceremony celebrating the 55th National Africa Day Festival and invited governmental and national leaders from Ethiopia, Kenya, and Namibia.[1]

Cuba–Africa Relations in the Cold War

The Political Field

Since its successful revolution, Cuba's policy toward African countries has focused on different things in different periods. As a socialist nation, Cuba was naturally against, sometimes even directly confronting, the Western camp in the Cold War. Though Cuba played a key role in the competition between the US and USSR, it was hard to say that Cuba completely followed the USSR's lead. Cuba holds its own independent political views: one stresses the third world to which Cuba belongs; the other is promoting and taking part in the decolonization and national independent movement around the world in the 1960s. So Cuba's diplomatic principles can be defined as the combination of idealism and pragmatism. Its African policy has been going along with the balance between these two principles.[2]

[1] Ministerio de Relaciones Exteriores de Cuba, *Principales resultados de la politica exterior cubana en 2019*, http://www.minrex.gob.cu/es/principales-resultados-de-la-politica-exterior-cubana-en-2019.

[2] Analúcia Danilevicz Peeira, Cuba's foreign policy towards Africa: Idealism or pragmatism? *Brazilian Journal of African Studies*, July/December 2016, p. 108.

In practice, Cuba's African policy has focused more on the efficiency of strengthening the third world, backing anti-colonialist measures, and supporting national independence. After WWII, the West was still trying to extend their sphere of influence even though decolonialization had become a popular trend at that time. And with the super powers' competition in the Cold War, the political situation was becoming complicated in Africa. On one side, the anti-colonialism movement had a new formation of African nations fighting for their basic independence; on the other side, the Organization of African Unity (OAU) was set up in 1963, significantly improving the global influence and power available to the newly independent nations that were emerging as forces in Africa. In the early 1960s, Che Guevara paid a visit to Africa and his trip made Cuba's attitude toward Africa clearer. Since there were many instabilities in Angola, Mozambique, Guinea-Bissau, Congo, Zaire, etc., Cuba predicted the African revolution was coming soon and posited that Cuba should support the de-colonialism process underway in Africa in order to expand the third world forces. This was also the beginning of military cooperation between Cuba and Africa.

Military Field

Apart from providing political support and relational development with Africa, Cuba has committed military forces to aid in the guerrilla wars of some African countries as well, which has gradually become an important part of Cuba–Africa relations. At the initial stage of their military relations, Cuba provided its revolutionary experiences and its well-known familiarity with guerrilla warfare, without particularly stressing Cuba's socialist status. Cuba increased its interest in African affairs rapidly due to third world membership, anti-Western colonialism, dealing with US pressure, etc. In 1959, a republic was established in Cuba and in 1962, Cuba began sending its medical team to support the opposition faction in Algeria, together with guidance for the guerrilla war, beginning Cuba's military involvement in the African area. Monitoring Che Guevara's African trip, the US decided to send their own mercenary army to Africa. Ultimately, Africa became a new staging ground for the US–USSR competition.[3] In 1963, Cuba began sending its guerrilla and medical

[3] Analúcia Danilevicz Pereira, Cuba's foreign policy towards Africa: Idealism or pragmatism? *Brazilian Journal of African Studies*, July/December 2016, p. 109.

teams to Angola to directly join in the Angolan war. This can be understood as an indication that Cuba's military relations with Africa had entered a new phase. Angola used to be a Portuguese colony that became a Portuguese overseas province in 1951. Since the 1950s, three organizations for national liberation were successively established there: the People's Movement for Liberation of Angola (APLM), National liberation Front of Angola (FNLA), and the National Union for the Total Independence of Angola (UNITA). They began their armed struggle for national independence in the 1960s. On January 15, 1975, they reached the "Alvor Agreement" with Portugal on Angola's independence, and on January 31 of the same year, they formed a joint intern government with Portugal. But there was an armed conflict among the three local organizations that lead to an end to the interim government. On November 11, 1975, APLM announced the establishment of the People's Republic of Angola.[4] During the civil war for anti-colonial independence, Cuba kept providing Angola with the military support it needed by sending guerrilla and military forces to take part in various wars. This, then, became the key part of Cuba–Africa relations.

In 1965, more than 400 Cuban soldiers arrived in the Congo, Zaire, and Angola in central Africa. Cuba's Africa strategy at that time could mainly be categorized as one aimed at strengthening the influence of the Non-Aligned Movement, creating more socialist countries, and having more ideological partners in the world.

Since 1975, relying upon its own forces along with material support from the USSR, Cuba transported a large number of military forces from Cuba to Angola to participate in the fight with the military forces from South Africa that were supported by the US, forming a proxy war between the two super powers, the USSR and the US. In 1988, after an agreement was reached among the People's Republic of Angola, the Republic of Cuba, and the Republic of South Africa (the Tripartite Agreement), Cuban and South African troops withdrew from Angola. This was the end of Cuba's military involvement in Africa. In around 20 years of Cuban military involvement in Africa, tens of thousands of Cuban soldiers and medical personnel went to African battlefields both in the

[4] China Ministry of Foreign Affairs website, *Angola Review*, December 2019, https://www.fmprc.gov.cn/web/gjhdq_676201/gj_676203/fz_677316/1206_677390/1206x0_677392/.

front and the rear. In the process of wars and armed threats, Cuban international humanitarian assistance and medical services were also endorsed by Angola and other people in different countries in Africa.

Cuban involvement in the independence movement in Africa has had a significant, long-term effect on South Africa in particular. It not only took military measures against racism, but it also withstood the West's power there. In addition, it provided Namibian people with strong political support for their independence. Its military involvement also dispelled the myth of white invincibility through the Cuban military attack on the apartheid army of South Africa. Therefore, it greatly aided the success of the fight against the apartheid in South Africa and in support of Nelson Mandela's freedom.[5] All in all, Cuba has assisted 17 countries in Africa through multiple means,[6] including the use of soldiers, medical doctors, and social workers in a large scale and number, covering combat, guidance for the war strategies, medical treatment, and other assistance. Cuba has identified with Africa from the point of view of its national development, the Cuba revolution, and national ideology as well.

Medical Field

Medical diplomacy is the most important aspect of Cuban-African relations. Cuba, for all that it is a relatively backward economy and small country, has an outstanding national medical system and conducts world-class medical research, which is well known both in Latin America and the Caribbean as well as in the rest of the world.

"Medical diplomacy" refers to the cooperation between/among nations delivering tangible benefits to the health field. Cuban medical diplomacy is a diplomatic mode with a typical Cuban style, positively benefiting the Cuban international position, goodwill, and prestige as an aspect of Cuba's soft power. Additionally, the success of Cuban medical diplomacy serves to garner a much higher global reputation for Cuba above and beyond its national power, which is the promotion of Cuba's international exchange by providing a constructive health service system abroad.[7]

[5] Nelson Mandela, discurso en Cuba, 26 de julio de 1991, http://www.cubamilitar.org/wiki/Nelson_Mandela,_discurso_en_Cuba,_26_de_julio_de_1991.

[6] Granma: Cuba y Africa unidas por siempre, http://www.granma.cu/mundo/2019-05-25/cuba-y-africa-unidas-por-siempre-25-05-2019-10-05-15.

[7] Julie M. Feinsilver, Cuba's medical diplomacy changing Cuba/changing world, p. 273.

Cuban medical diplomacy formally started in 1959 when the Cuban revolution succeeded. Cuba sent the first medical team abroad when Chile suffered an earthquake in 1960. However, from the early 1960s on, Africa has been the most important recipient of Cuban medical diplomacy. Its duration and scale are internationally impressive. Cuba has also provided free medical training for developing countries, including those in Africa.[8]

Of course, Cuban medical development springs its system, but it is still worthwhile to mention the importance of related ideas and practices in its revolution and construction because these issues have cemented the importance of the country's medical services and healthcare system to its national governance. Therefore, Cuban medical diplomacy has a long history and deep political background. Making medical services and health a priority in the process of the Cuban revolution and the country's subsequent construction can be traced back to the early period of the Cuban revolution. Che Guevara—one of the key early Cuban leaders—used to be a doctor before joining the revolution. He and his colleagues provided medical treatment and service to the farmers and local people in the liberated areas, as well as their wounded colleagues. This greatly benefited the expansion of the Cuban revolution, relying as it did on the rural workforce for overthrowing the Batista[9] regime. So this struggle style was called "the revolutionary medical science." After Che Guevara's death in Bolivia in 1967, many young students from Cuba's new generation followed his model, studying medical science, then starting their medical career in the rural areas to take care of basic medical and health services. When the Cuban revolution was won, drawing from prior Cuban national experience, the new government made it known to the world that "a key task for the revolutionary leadership is to keep people in health. Therefore, the government paid much attention to health and medical service related policies. Few countries in the world could do the same at that

[8] Julie M. Feinsilver, Cuba's medical diplomacy changing Cuba/changing world, p. 274.

[9] Batista came to power again via military coupdétat in 1952 and was turned over by uprising army under Fidel Castro on January 1, 1959. Then revolutionary government was established.

From China Ministry of Foreign Affairs website, *Cuba Review*, January 2020, https://www.fmprc.gov.cn/web/gjhdq_676201/gj_676203/bmz_679954/1206_680302/1206x0_680304/.

time."[10] With the support of the USSR, Cuban leaders set up a national medical network, including everything from the health station or clinic in the community to the general hospitals with a focus on prevention.

Meanwhile, Cuba also provided assistance to some countries against imperialism and colonialism. The main form this took was Cuba sending guerillas together with medical staff or medical teams. Cuban guerillas for the armed struggle abroad were called "medical guerillas" because of the Cuban medical character and tradition. In 1963, when Cuba supported the armed struggle in Algeria, Che Guevara had 58 doctors and nurses with his guerilla forces and brought the seriously wounded Algerian guerilla soldiers back to Cuba for free treatment. The Cuban medical team worked in Algeria for 14 months, setting a record for its medical team operating abroad.[11] Later on, during his guerilla activities in Zaire, Che Guevara also carried out the biggest campaign to date in Africa for epidemic prevention. Afterward, more than ten Cuban guerilla teams arrived in Africa and took part in the armed struggle from Angola to Zimbabwe while the Cuban medical team stationed with Cuba's guerilla forces helped provide for the local health. It is interesting that Cuban medical teams made so many contributions to local health that their influence in Africa was even greater and the social solutions they provided lasted longer even than that of the Cuban guerilla forces. In this way, Cuban-African relations were established. In the 1980s, Cuba began its medical assistance to the African continent by providing related clinical medicine and treatment as a diplomatic trial. At the same time, Cuba also introduced its border management measures for preventing HIV and provided friendly countries with an early medical warning system for identifying infectious diseases. For example, during the annual Non-Aligned Convention in 1986, Cuban leader Fidel Castro notified the Ugandan leader that Cuba found 18 in 60 officials sent from Uganda to Cuba for training had tested positive for HIV, with a warning that this infectious disease could lead to more serious consequences than even a war. This

[10] Andrew Jack, Cuba's medical diplomacy, May 15, 2010, https://www.ft.com/content/debaad0c-5d6e-11df-8373-00144feab49a.

[11] Julie M. Feinsilver, Cuba's medical diplomacy changing Cuba/changing world, p. 274.

directly resulted in the earliest and strictest project for AIDS prevention and control being implemented in Africa.[12]

Cuba had more than 30,000 medical workers in 70 more countries around the world until 2008. And in 2005, Cuba started "Cuba's Comprehensive Health Program," which benefits many countries globally, including 18 African countries: Botswana, Burkina Faso, Burundi, Chad, Eritrea, Gabon, Gambia, Ghana, Guinea-Bissau, the Republic of Guinea, Equatorial Guinea, Mali, Namibia, the Niger, Rwanda, Sierra Leone, Swaziland, and Zimbabwe.[13]

In conducting its medical diplomacy, Cuba mainly has the following considerations in mind: first, Cuba is a socialist country, so Cuba should take responsibility as a member in the socialist camp; second, the Cuban revolutionary successful experience has gradually become synonymous with the Cuban character and its strong point—medical services; third, Cuban values; last but not least, geopolitical reasons. Because of US blockages and hostilities, Cuba believes it needs to have more broad and efficient diplomacy to deal with the imperialist stranglehold. At that time that it was introduced, African countries were in the process of looking for independence, but most interference from outside was in order to invade or prevent Africa from independence. Cuba's "interference" in Africa, on the other hand, was treated as a form of real assistance to African people in the spirit of internationalism and revolution by many African leaders.

Since Cuba's national establishment in 1959, medical diplomacy success has made remarkable contributions to Cuban national interests and offers constructive assistance to developing countries for the public health. At the same time, this also gives Cuban-African relations momentum and cooperative areas in return.

Cuba–Africa Relations After the Cold War

With the end of the Cold War, the international community is increasingly changing from a bipolar structure to a multipolar one. Due to the disappearance of the USSR and the socialist camp, Cuba has gradually run into some difficulties in the new era. So Cuba has to make readjustments for its

[12] Andrew Jack, Cuba's medical diplomacy, May 15, 2010, https://www.ft.com/content/debaad0c-5d6e-11df-8373-00144feab49a.

[13] Julie M. Feinsilver, Cuba's medical diplomacy changing Cuba/changing world, p. 277.

diplomatic policies. The USSR used to be the main supporter of Cuban socialist economic growth. However, due to tremendous changes in the USSR and in Eastern Europe since 1991, Cuba suffered greatly in terms of its trade and economy. Facing such huge changes, Cuba declared that the nation had come into a "special period" and would need to readjust its foreign relations.

In Africa, Cuba gradually withdrew its troops but chose Angola and South Africa to remain as the key links between Cuba and Africa. Cuba–Angola bilateral relations transferred into an equal and common development after the war. From 1960 to 1988, Cuba held ideological dominance in its relations with Angola. Together with the USSR, Cuba sent troops to Angola to join in the civil war there, but Cuba did not fulfill its military goals for the region. When the Cold War ended in 1989, Cuba began to develop an independent and autonomous diplomacy while still adhering to the concept of proletariat internationalism and anti-imperialism.[14] Cuba tried to maintain its pragmatic principles and to improve its exiting domestic economic difficulties through diversified diplomacy, during Cuban adjusted its diplomatic idea. Angola has shared a similar colonial history with Cuba. So Cuba sent troops to Angola and showed to the world that Cuba supported the same ideology and held a common diplomatic orientation. Bilateral relations from this era have steadily transformed into more peaceful and normal state relations over time.

Although Cuba established foreign relations with South Africa later than it did with Angola, the former leaders from both countries shared many common views toward revolution and social movements, which laid the foundation for a good bilateral relation in the new century. Cuba used to be at war with the South African army, which was supported by the US against Angola, but this warfare was in fact a blow to the white political forces prominent in the South African apartheid regime at that time. Thus, Cuba had a relationship with South Africa based on revolutionary friendship even during the Cold War. Cuban participation in the Angolan civil war had an impact upon both the collapse of the white regime and the concurrent apartheid policy in South Africa. In particular, Castro fully supported black people's struggle for their rights when the

[14] Timothy Othieno, Cuba's foreign policy in Angola, Latin American Report, vol. 21, no. 2, January 2005, UNISA Press, pp. 40–55, https://journals.co.za/content/latamrep/21/2/EJC60191.

war between Cuban and South African troops was ongoing. Meanwhile, both in Cuba's own revolution and in Cuba's involvement in the Angola war, Cuban troops and leaders objectively supported South African former President Nelson Mandela and his policy abolishing apartheid from the outside. On May 11, 1994, Cuba established comprehensive diplomatic relations with South Africa. In the same year, former South African President Nelson Mandela noted that South Africa would emphasize and made great efforts to promote trade, culture, and sports relations with Cuba for reciprocal and mutual benefits, based on South African policies toward other countries with normal diplomatic relations. From that point on, top leaders of both countries have frequently visited each other. President Castro paid a state visit to South Africa in September 1998 and attended the Non-Aligned Summit in Durban. Cuba also sent a ministerial-level delegation to President Mbeki's inaugural ceremony in June 1999.[15] In April 2000, President Mbeki went to Havana to attend the South Summit. He paid a state visit to Cuba during March 26–29 that year while in the company of his ministers of foreign affairs, education, sports and recreation, art, culture, sciences and technology, agriculture, and land affairs. During President Mbeki's visit, the two countries signed cooperative agreements on sciences and technology, art and culture, sports and recreation, aviation and merchant marine, and more.

In recent years, medical diplomacy has become a new opportunity to expand Cuban-African cooperation. In 2013, invited by the South African government, Raul Castro attended national leader Nelson Mandela's funeral on behalf of the Cuban people. He also shook hands with the US President at that ceremony, foreshadowing an improvement in bilateral relations. Based on its medical cooperation with Africa, in 2014 Cuba sent several hundred medical workers to West Africa to provide medical assistance when an ebola epidemic broke out. During this epidemic, the amount of Cuban medical assistance provided to West Africa was the largest among the medical teams from all different countries. Cuban leader Raul Castro also expressed the hope that Cuba and the US could at least temporarily shelve the contradictions between the two countries and cooperate in West Africa to address ebola. In June 2018, Cuba also

[15] South Africa Foreign Ministry website, http://www.dirco.gov.za/foreign/bilateral/cuba.html.

had a bilateral medical assistance project with Kenya. Then, Cuba sent a medical team with hundreds of staff and medical services and assistance.[16]

On the basis of supporting the third world against colonialism and racism, Cuba prioritizes South–South diplomatic cooperation in the new era. With its advantages in medical services and training, Cuba stresses public health and cooperation as the basis for its diplomacy in developing countries. Cuba has not only cooperated with the developing countries in Asia, Africa, and Latin America in the areas of medical services and training, but has also increased the medical talent training it provides for developed countries, which has helped garner more support from the Western public for Cuba. Some Spanish and American young people, unable to pay the high educational expenses necessary to pursue medical science in the US, could realize their "doctor dream" through this program. Eventually, they could play a role in improving the West's attitude toward Cuba.

Cooperating on infrastructure and educational projects is another new area of diplomatic momentum for Cuban and African relations. In September 2019, Cuba and South Africa signed an agreement on hydraulic engineering and storage. And in education, Cuba provided scholarship for students from Angola to come to Cuba to study Spanish and the medical sciences.[17]

2 Reasons for Cuba-African Relation Development

The development of Cuban-African relations is a model for South–South cooperation. Apart from the period of military relations in Africa under the influence of the US and USSR's power game, Cuba has cooperated with African countries in both the medium and the long run. Its many achievements in this area are also the model for relations among developing countries. The following reasons are primarily responsible for the development of this relationship.

[16] Cuban Foreign Ministry website, http://www.minrex.gob.cu.

[17] Cuban Foreign Ministry website, http://www.minrex.gob.cu.

Exporting Revolutionary Ideas and Looking for a Common Ideology as a Diplomatic Demand

The foundation of Cuba's diplomatic strategy is sticking to the revolutionary thought from the old generation headed by Castro. Cuban revolutionary leader Fidel Castro chose the socialist road for national development, as well as an open door for development in the world. As a Cuban leader, he realized Cuba's national liberalization and independence and legally endorsed socialism. Under the special geopolitical circumstances in which he held power, Castro also stuck to a diplomatic stance of anti-imperialism, anti-colonialism, and anti-interference, as well as friendly cooperation with other socialist countries. This approach was endorsed by African countries looking for national liberalization and state independence. After Cuba's military involvement in Africa and its partially realized diplomatic target of exporting revolutionary socialist ideology, Cuba, as a developing country, greatly increased its international influence and representation and was identified with African countries at the margin of international power at that time.

Cuba consistently developed its diplomatic relations with Africa over three generations of its national leadership, stressing multiple approaches to pragmatic diplomacy. The Cuban leaders' revolutionary ideas and diplomatic legacy from the old generation are key foundations for Cuba's foreign policy even today. During the time Cuba was under Fidel Castro's leadership, the country focused on developing bilateral relations through leadership and medical diplomacy and laid the basis of further international influence for Cuban diplomatic development. In its multiple diplomatic relations, Cuba has supported the equal international rights of the third world. In his UN speeches, Castro consistently insisted that Cuba was against US trade blockades that placed Cuba in a third world position. This was in line with African countries' own ideas for economic development in addition to strengthening Cuban independent and autonomic diplomacy.

With these international and internal changes and under Raul Castro's leadership, Cuba carried out comprehensive and pragmatic diplomacy even while continuing to stress the socialist road, including the realization of revolutionary ideas and diplomatic principles from the older generation in foreign affairs. This approach strongly matches up with the policy of "updating (actualizar)" Cuba's economic society. After Dias Karner (Díaz Canel) was elected as the new Cuban leader, Cuba's internal affairs and diplomacy underwent an organic adjustment of "changes and the unchanged." Persisting in these principles, Cuba put increased stress on the importance of black culture in its relations with Africa and diluted the negative influence from its previous revolutionary and military link with Cuban-African relations. Cuba is trying to keep its national character in diplomacy and maintain a higher target for Cuban-African relations compared with the link it previously held with Africa and developed countries in peace time.

Domestic Demand for Economic Development

Cuba's revolution developed in a complicated international and internal environment. Its socialism-building needs productive results, as well as consistent social and cultural progress for people to support it. So the relevant political, cultural, and military governmental departments have promoted this development as their foremost working priority. It has also become the core of Cuba's ongoing revolutionary socialist existence.

With Fidel Castro's national principle of autonomic development, Cuba has an interactive relationship between the currently "updating" process and diplomatic development. On one side, Raul Castro, after taking over national power, has initiated and pushed for the Cuban economic model to "update" in order to set up an independent and autonomic image in the international community. For example, the construction of Port Marion has created favorable internal factors and contributed greatly to Cuba's economic diplomacy and bilateral and multilateral cooperation. On the other hand, Cuba's comprehensive and pragmatic diplomacy is in line with the socialist system and is satisfactory in terms of meeting people's requirements. So it gets support and understanding from abroad for furthering the political "update."

Cuba sending troops to Africa left a legacy of military cooperation. More importantly, the African continent has internationally endorsed Cuba's economic model and its developmental path. The domestic "update" process supports Cuba's foreign policy readjustment; meanwhile, it also has new requirements for diplomacy. First, diplomacy must provide domestic development with safeguards. When continuing the consistent public medical service in the process of South–South cooperation, Cuba has also increased its exports and furthered its international cooperation in the medical sciences, medical technology, and education. In its medical diplomacy with African countries, such as Angola,

Equatorial Guinea, etc., Cuba has realized its goals in terms of medical technology exports and cooperation on biological medicine in order to reduce the associated diplomatic costs. And this medical diplomacy can also serve the domestic development strategy as well. Second, the economic model "update" needed the necessary international endorsements through diplomatic efforts for its development of a different process than what had existed previously. Since 2013, Cuba has made a series of new policies and laws indicating that the Cuban economic model would "update" into a more comprehensive form. After the new policies and laws came out, both the scale of foreign capital and the field of investing grew simultaneously. The Mariel special zone building earned support both regionally and internationally, in turn promoting Cuban economic growth. In 2015, Cuba's GDP reached about 4% growth and its per capita increased to US $7,149.5.[18]

The country's medical diplomacy toward Africa has been similarly beneficial to these economic results for development. Some African countries' leaders have commented highly on the Cuban socialist system in both regional and international arenas. The low-cost medical diplomacy has greatly helped develop the international identity Cuba holds in terms of socialist public ownership.

In short, Cuba's diplomatic readjustment as it adapts to the domestic economic reform process firmly guarantees the Cuban "update" of its diplomatic approach. Under the country's socialist principles, Cuban economic reform aims to meet the demand for basic living standards domestically, as well as provide the material basis for upgrading Cuba's international position. That means a foreign strategic adjustment will occur alongside the "update," both with a common guiding purpose. Developing the economy is the priority for Cuba right now.

GeoPolitical Security Demand

Although Africa has ideological differences with Cuba, the sharing of a common black culture is an important link for Cuban foreign relations with Africa. Compared with the European colonization of Africa and then with the US–USSR competition that treated Africa as "a new continent," Cuba considers itself to have a shared black cultural identity and black

[18] Cepal: Porfil of Cuba Nacional economy, http://estadisticas.cepal.org/cepalstat/Perfil_Nacional_Economico.html.

ideology endowed by nature in its relations with Africa. Both sides recognize their equal relations in the international arena. Africa wants to have more attention from the key countries in the world and thus participates in more international affairs. This is the same as Cuba's foreign demand to have more partners and improve its inferior position in its relations with the US.

Cuba's special geographical position and relations with the US are a great challenge for Cuba's economic development; at the same time, they make Cuba a model of anti-neocolonialism for African countries. The Cuban revolution has been affirmed by African countries all along. Cuba only has a national territorial area of 109,884 square kilometers and a population of 11.22 million (2017).[19] From a geopolitical perspective, Cuba is to the south of the US and provides a natural military defense for South America against the US. And in the southern part of the US, there are few ports, but Cuba has several deep and natural harbors. Historically, former US President Thomas Jefferson, through his "Ripe Fruit" policy in 1823, held that Cuba was like an apple on the tree insofar as it has no choice when it becomes ripe. He believed that after Cuba got rid of its colonial status from Spain, it would undoubtedly fall within the sphere of influence. However, history would not vindicate these US political theoretical analyses and predications. Cuba became the only socialist country in the western hemisphere, experiencing several decades of the US's blockade on trade, the economy, and finance while still retaining its national independence and economic progress. Africa, for its part, received the support of medical services and education, but still faced similar interference from the West. Therefore, Africa has always deeply affirmed Cuba's free medical service, education, and culture. They are an important point of assistance for Cuba in the international community.

In conclusion, diversified diplomacy is the new direction for Cuba foreign affairs. Cuba–Africa cooperation is also gradually headed for diversification, but Africa is still the main area for Cuba's own diversified and flexible diplomacy. It is also the main area for Cuban to practice its balanced foreign relations with the unilateral diplomatic relations of other socialist countries. Cuba–Africa relations are not only engaged with a broad range of countries, but also diversify the areas of cooperation. Cuba has upgraded its influence on Africa through multilateral mechanisms,

[19] Oficina Nacional de Estadística e Información de la República de Cuba, http://www.onei.gob.cu.

diplomacy with small inputs and big return, the Non-Aligned Movement, etc. Although Cuba's African policy used to be affected by the USSR and some other countries, Cuba has kept its African policy consistent and relatively independent. Cuba treated its military influence on Africa as a great diplomatic success. There is much diplomatic research on China, India, Brazil, and the Arab countries that studies the big powers' military relations and military influence on Africa. But the research on Cuba remains sparse, even though Cuba is an extremely important outside source of support for Angola's military forces. Today, with both the internal and international law having been more completed, Cuba's political and military involvement in Africa has declined to the lowest level in history. However, among the outside countries with military involvement in the African continent, Cuba is the only less developed country to maintain a significant presence. As a "small" country, Cuba's African policy is more efficient even than that of some of the big powers. Specially, after its alliance with the USSR, Cuba's influence in Africa reached its peak. So Cuba's capacity for involvement across the continents is stronger than what majority of the mid-sized developed countries and most developing countries are capable of.

3 The Future Development of Cuba–Africa Relations

Cuba has African cultural features in the Caribbean region due to its historical resources and ethnic African population. African affairs have come especially significant for Cuba. And both Cuba and the African continent have a common political appeal for getting rid of interventionism. Economically, they also have a common developmental aim for deeper engagement into the international trade and more international investment. Cuba's diplomatic choice of sticking to its own specialties is its main policy for dealing with African relations. Also it is the basement of weakness improvement for small countries. In Africa, Cuba has a long history and consistent diplomatic toolset alongside diversified medical diplomacy. For the developmental level of Cuba, its international influence is far beyond its population and size.

Cuba's revolutionary exports and military personnel abroad during the Cold War not only had a direct impact on the war in Angola, but also earned recognition for Cuba from many African leaders, including former Africa National Council (ANC) leader Nelson Mandela. The

Cuban medical team following along with Cuba's guerilla forces was much appreciated by African society and people in particular. Cuba consistently kept its cooperation in medical services, training, and biological medicine production at a low cost in Africa even during difficult times. Cuba's medical diplomacy is now global. Cuba sends its doctors to places lacking medical services, such as Latin America, China, Africa, etc., while enrolling students from European countries to study medical science in Cuba. Medical diplomacy as the primary Cuban diplomatic character has three policy considerations: (1) realizing bilateral relations between Cuba and other countries; (2) it will help increase the country's revenue from trade in services; (3) improving Cuba's international position by providing medical services and biological media research, education, and exports. Therefore, Cuban medical diplomacy can have more partner countries in line with the diversified diplomatic direction of Cuba. During the period of anti-colonialism and national independence in Africa, its health and living condition were such that disease and poverty became barriers for its further development. So Cuba's political idea has been widely accepted in Africa, and eventually, Cuba's political influence and support have also become accepted. Cuba is not only welcome by African people, but also receives unconditional support from African countries internationally.

In the 1990s, after the USSR collapsed, Cuba lost its main economic assistance and political ally and suffered greatly. During this time, Cuban influence on Africa declined sharply, with Cuba–Africa military cooperation suffering in particular. Until 2000, Cuba–Africa relations were passing through a difficult time, only to later be reinforced and further developed.[20] In this new period of cooperation, Africa is still the key supporting point for Cuba's comprehensive diplomacy. Compared with Latin America and the Caribbean region's shaky stance, Africa is firmly against imperialism and shares a common, consistent, and stable approach to the fields of international cooperation in various regions, supporting unity among third world countries, supporting anti-colonialism and standing against all discrimination in international relations, and opposing the use of threats or force in state–state relations. Since its reform, Cuban

[20] Analúcia Danilevicz Pereira, Cuba's foreign policy towards Africa: idealism or pragmatism? *Brazilian Journal of African Studies*, vol. 1, no. 2, July/December 2016, pp. 106–117.

diplomacy has become more flexible while still sticking to its basic principles. In terms of US relations, Cuba has required an open dialogue that ended with a block on its diplomacy. With US policy readjustments, Cuba has actively joined in the negotiation for more opportunities in national development while keeping Africa as an important platform for its diplomacy. Under the terms of globalization, Cuba has also readjusted its cooperative stance with the African region, which has made both Cuba and those African countries supported by Cuba more focused on their own development and regional cooperation. In fact, this has diluted the Cuba–Africa traditional relationship. In addition, economic relations have been left behind in the Cuba–Africa relationship to the extent that bilateral relations can hardly have extra momentum in the short term under the new situation. Currently, Cuba mainly maintains its relations with Africa via education, training, technological support, and medical assistance.

Since Trump became US President, Cuba–US relations have once again run into difficulties. In 2019, Cuba experienced more serious blockades in trade, the economy, and finance, leading to the erection of further barriers to Cuban economic development. It is difficult to predict how Cuba will even maintain, much less expand, its medical and educational cooperation in Africa with its existing capital shortage.

Under the leadership of Dias Karner, Cuba may still have the continuing support of current African leaders. Karner is the top Cuban leader from the new generation after the revolution. Furthermore, the old military generation of Cubans had outstanding military exploits in Africa, but are becoming irrelevant with the disappearance of the old African political generation, headed by Nelson Mandela. In abolishing South Africa's apartheid policy, Mandela made a concession to the white political forces at the cost of giving up a social revolution. It is not so splendid as it was before for Cuba's work and fight as it was in Angola during 1975–1991. In the new global situation, what is the link between Cuban-African relations if not war and revolutionary friendship?

Though Cuba and Africa have a degree of South–South cooperation characterized by high efficiency and low costs, both of them are still developing areas relying on foreign investment. In the international market, Europe is the main investor for Cuba and Africa, though with different objectives for each. Unfortunately, the European economy has long been at a low level of development, which is weakening its investment capability. With less investment, Cuba and Africa may have some continued demand for foreign capital, as both of them have infrastructure and economic upgrades that demand foreign investment.

CHAPTER 15

Russia's Pivot to Africa: A Tactical Move or Strategic Calculation?

Xiaoyun Qiang

In October 2019, Russia hosted the first-ever Russia–Africa Summit since the dissolution of the Soviet Union, held at its Black Sea resort in Sochi, where President Vladimir Putin welcomed leaders and representatives from all 54 African countries, including Egyptian President Abdel Fattah el-Sisi (co-chairman of the summit) and South African President Cyril Ramaphosa. Attracting worldwide attention, this historic event was perceived to be a symbol of Russia's grandiose return to Africa based upon the decades-long diplomatic heritage the Soviet Union left on the continent. Over the course of the two-day summit, a wide range of issues—including bilateral economic cooperation, regional stability, and joint responses to international challenges and threats—were discussed. Meanwhile, US$12.5 billion worth of deals were struck, and Russia announced that it would set up a US$5 billion online platform for tapping the country's trade potential with Africa.

X. Qiang (✉)
Shanghai Institutes for International Studies (SIIS), Shanghai, China
e-mail: wxy03@ssap.cn

X. Li and C. An (eds.), *The Changing World and Africa*,
https://doi.org/10.1007/978-981-16-4983-7_15

Notably, in his opening remarks, President Putin emphasized the "strategic" and "long-standing" nature of Russia-African cooperation and vowed to intensify this cooperation across various fields. According to the 2016 Concept of the Foreign Policy of the Russian Federation (CFPRF), Russia would "enhance its multifaceted interaction with African states on a bilateral and multilateral basis with a focus on improving political dialogue and promoting mutually beneficial trade and economic cooperation and contribute to settling and preventing regional conflicts and crises in Africa." This policy paper also recognized that developing partnerships with the African Union (AU) and other regional organizations are important elements of the Russia–Africa relationship.[1] However as the CFPRF and the National Security Concept of the Russian Federation (NSCRF) also reveal, Russia's long-standing relationship with Africa seems to be less strategic than it could be, for Africa has remained at the bottom of Russia's diplomatic priorities since 2000. As a matter of fact, Russia did not substantially expand its footprint in Africa until 2014, when the Ukraine crisis pushed Russo-Western relations near the freezing point and the Russian economy was hit hard by international sanctions.

What's the driving force behind Russia's return to Africa? Was it a strategic calculation—as President Putin described at the Russia–Africa Summit—or is it just a tactical move to counterbalance Western sanctions? This paper will try to explore answers to these questions in three steps. The first part elaborates on the achievements and new developments that have taken place in Russia–Africa relations since 2014. The second part presents a critical assessment of Russian-African cooperation under the weight of years-long Western sanctions. The third part examines challenges and opportunities for Russia to build up strategic relations with Africa in the long run. The paper concludes by discussing the potentiality of advancing Russia–China cooperation in Africa.

1 Russian-African Cooperation Since 2014

For a long time now, Russia's presence in Africa has been inconspicuous when compared with other parts of the world. However, the eruption

[1] Концепция внешней политики Российской Федерации (утверждена Президентом Российской Федерации В.В.Путиным 30 ноября 2016 г.). http://www.mid.ru/foreign_policy/official_documents/-/asset_publisher/CptICkB6BZ29/content/id/2542248.

of the Ukraine crisis in 2014 prompted Russia to scale up its diplomatic efforts on the continent, with more bilateral visits and meetings attended by top leaders and foreign ministers, more dialogues and new mechanisms between Russia and regional organizations in Africa, more African students studying in Russia, and the initiation of enhanced cooperation in some traditional fields such as peacekeeping, military technology, and development assistance.

First and foremost, Russia has strengthened its political coordination with African countries through bilateral visits by heads of state and senior officials. 2014 witnessed the most frequent interactions between Russian and African political elites, although similar exchanges had existed for a long time. In that year, heads of state and senior officials from 16 African countries, including South Africa, Algeria, and Uganda paid a visit to Russia for bilateral or multilateral consultation, improving their mutual understanding of hot topics at the global and regional levels. Similarly, President Putin made two state visits to Egypt in 2015 and 2017, respectively, and met with his African counterparts on various multilateral occasions such as the leaders' summits of the Group of Twenty (G20) and BRICS countries. In 2018, a new record was set for over 20 bilateral visits in one year.

It was easy to discern the key targets of Russian diplomacy in Africa from the visits by the Russian president and senior officials. In terms of individual countries, Egypt, Ethiopia, and South Africa were been picked as the three anchors, thanks to their prominent geographical locations and regional influence. It's worth mentioning that President Putin has been holding annual meetings with his Egyptian and South African counterparts since 2014 and has also maintained good communication with Ethiopian leadership.

In terms of international organizations, Russia has established sound relations with the AU and the Southern African Development Community (SADC). It was given AU observer status as early as 2006 and signed the Memorandum of Understanding (MoU) on the mechanism of political consultations with the AU in 2014, which injected new impetus into bilateral ties. Actually, since 2014, Russia's Federal Council has signed a number of cooperation agreements with its counterparts in some African countries. In October 2017, the 137th General Assembly of the Inter-Parliamentary Union (IPU) was held in St. Petersburg in Russia and was attended by speakers from eight African countries. One year later, Russia signed the MoU with the SADC on the basic principles of relations and cooperation. Additionally, Russia-African cooperation at the global level functions well within the frameworks of the United Nations, the G20, and the BRICS.

Second, Russia has further institutionalized its relations with Africa, launching an all-dimensional, multi-layered diplomacy on the continent. On October 31, 2014, a new position called the "Special Envoy to the Middle East and African Countries" was created in accordance with the Russian presidential decree No. 701, and Russia's Deputy Foreign Minister Mikhail Bogdanov was appointed to this position. About two months later, the Russian Ministry of Foreign Affairs nominated Valery Utkin, then head of the Department of African Affairs, as Special Envoy to African Regional Organizations. Besides these Russia–Africa joint intergovernmental committees, a periodically held Russia-Egyptian two-plus-two dialogue mechanism, which involves foreign and defense ministers of the two countries, was also initiated. Moreover, at the end of 2013, Russia and South Africa agreed to set up a Russian business office in Johannesburg. In October 2018, the Russia–Africa Social Forum was successfully held in Moscow and has become a new platform of communication for civil organizations from both countries.

In addition, Russia has largely expanded its engagement with international organizations in Africa, making its voice heard in the meetings of not only the AU and the SADC, but also the Intergovernmental Authority on Development (IGAD) in Eastern Africa, the Economic Community of West African States (ECOWAS), the Common Market for Eastern and Southern Africa (COMESA), and the Eastern African Community (EAS), in order to mediate conflicts in Libya and South Sudan, jointly fight piracy in Somalia, and combat the threats of terrorism and drug trafficking in Africa more broadly. Hopefully, Russia's cooperation with these organizations will be extended to the economic field following the 2019 Russia–Africa Summit.

Third, Russia has been seeking more robust trade ties with African countries, though the overall trade volume remains relatively low. On the one hand, African countries are given trade preferences and tariff and tax relief to facilitate the export of their products to Russia; on the other hand, Russian exports to Africa still remain one of the few areas where Russia enjoys established advantages such as geologic exploration, exploiting and processing oil, gas, and other raw materials, equipment and technological service for hydroelectric and nuclear power plants, and weapons and military technologies. Specifically, big Russian companies (instead of small and medium-sized ones), play a major role in the country's export trade to Africa. For instance, Russia's State Atomic Energy Corporation (ROSATOM) has written a success story in Africa by building centers for nuclear science and technology in Zambia and Nigeria, and is currently exploring opportunities for cooperation with other African countries including Tanzania, Ethiopia, and Ghana on the peaceful uses of nuclear energy.

In the meantime, Russia has also attempted to export some cutting-edge technologies in its military industry, digital economy, and information sciences that can be applied in African markets. As Anton Kobyakov, Advisor to the Russian President and Executive Secretary of the Organizing Committee of the Russia–Africa Summit put it, Russia would be happy to share its products and experience of innovation with African partners.[2]

What's more, the establishment of Russian industrial zones in Africa has also been put on the agenda of bilateral cooperation. By the end of 2020, the first Russian Industrial Zone (RIZ) will come into operation in the Suez Canal Economic Zone in Egypt. And so far, eight Russian enterprises have signed up for the projects in the RIZ and more companies manufacturing agricultural machinery are expected to join in the coming months.

Over the decade from 2005 to 2015, Russian investment in Africa has increased 185% and reached the amount of US$ 17 billion. In 2018, the trade turnover between Russia and Africa totaled more than US$ 20 billion. In the same year, Russian export to the African continent saw 18% year-on-year growth totaling US$17.5 billion. Particularly among Russia's exports to Africa, non-energy products reached the value of US$14.3 billion. In this light, experts reckoned that the Russia–Africa trade volume will double in the next three years.[3]

Fourth, foreign assistance has long been used by Russia as a diplomatic tool in Africa. As a permanent member of the United Nations

[2] "Подведены итоги Ежегодного собрания Афрэксимбанка – 2019". Российская газета, 25 June 2019. https://rg.ru/2019/06/25/podvedeny-itogi-ezhegodnogo-sobraniia-afreksimbanka-2019.html.

[3] И.О. Абрамова. "Ежегодное собрание Афрэксимбанка в Москве: новые горизонты российско-африканского сотудничества", *Азия и Африка сегодня,* 2019, №9, с.7–10.

Security Council (UNSC), Russia has actively been taking part in the peacekeeping actions in Africa and delivering various kinds of assistance to African countries. For example, Russia has been committed to forgiving debt for African nations under the "Heavily Indebted Poor Country" (HIPC) Initiative jointly launched by the World Bank and the International Monetary Fund (IMF) in 1996, writing off African debt worth over US$20 billion. Meanwhile, Russia is also carrying out its intergovernmental agreements with Madagascar, Mozambique, and Tanzania under the Debt-for-Development Initiative. This innovative method has unlocked a commitment of US$40 million, which will be used by the United Nations World Food Programme (WFP) to support the government of Mozambique to provide school meals for children with nutritional deficiencies.

By funding a multitude of international organizations including the World Health Organization (WHO), the WFP, the International Civil Defense Organization (ICDO), and the United Nations Industrial Development Organization (UNIDO), Russia has delivered a large amount of humanitarian aid to African countries.

Concretely speaking, in coordination with the WFP, Russia sent US$21.5 million in food assistance to African countries from 2014 to 2017, among which Kenya got US$5 million, Somalia acquired US$4 million, Guinea, Zimbabwe, Sierra Leone, Ethiopia, and Namibia each received US$1.5 million, and Liberia, Madagascar, Sudan, and the Democratic Republic of the Congo (DRC) each obtained US$1 million. During the program year of 2018–2019, Russia gave African nations another US$9 million as food assistance, among which the Republic of the Congo and Somalia each got US$2 million, Guinea and Uganda each obtained US$1.5 million, and Burkina Faso and the Republic of Chad each received US$1 million.[4] Since 2011, Russia has begun to supply the WFP offices in Africa (Ghana and Uganda) and the Middle East with Kamaz trucks and accessories. By April 2019, these WFP regional offices had already received a total of 355 Kamaz trucks from Russia.[5]

In 2017, Russia donated US$2.26 million to the UNIDO to fund the organization's project called "technical and institutional capacity building

[4] "Роль России в решении проблем Африки к югу от Сахары", МИД РФ, 2019, http://www.mid.ru/ru/rol-rossii-v-resenii-problem-afriki-k-ugu-ot-sahary.

[5] Ibid.

for the increase in production and development of the aquaculture and fisheries value chains in Ethiopia," in order to develop the fisheries and aquaculture in Ethiopia under the Programme for Country Partnership (PCP).

In terms of medical aid, the WHO's two leading programs in Africa, namely, "Facilitating Individual Countries to Formulate Domestic Policies on Combating Noncommunicable Diseases" and "Launching Multi-agency Response in the Countries Most Affected by Tuberculosis," have both benefited greatly from Russian funding. In November 2018, through its funded WHO projects, Russia handed over to Zimbabwe multiple batches of medicines to fight a cholera outbreak in that country. Also, Russia was among those countries to be earliest invested in research aimed at defeating the Ebola virus, pumping over US$600 million into this cause. In Guinea, the United Company RUSAL funded the building of a clinical and diagnostic research center for epidemiology and microbiology that has since been put into operation. In 2017, this center was further developed into the Russian-Guinean research center of epidemiology and prevention of infectious diseases.

In August 2017, Russia decided to allocate about US$1 million from the Russia-UNDP Climate Window Trust Fund Program for the implementation of the "Support for Zimbabwe" project that would run from 2018 to 2020. Furthermore, Russia's humanitarian aid was usually available when African countries were ravaged by natural disasters. In April 2019, for example, Russia delivered humanitarian relief to Zimbabwe, Malawi, and Mozambique after tropical hurricanes devastated these countries.

Fifth, financial cooperation remains a key component of Russia–Africa relations, and Russia's involvement in the African Export–Import Bank (Afreximbank) as a shareholder is a case in point. Notably, the Afreximbank held its 2019 annual meetings, including the 26th meeting of shareholders, in Moscow, one of only two exceptions to the meetings being held on the African continent (the other was its 2012 annual gatherings in Beijing). In particular, then Russian Prime Minister Dmitry Medvedev was present at the official opening of the Afreximbank shareholders' meeting. In the frameworks of the shareholders' meeting, as well as the Russia–Africa Economic Forum, more than 20 side events were held to discuss a variety of topics such as how to facilitate the collaboration of small and medium-sized enterprises (SMEs) between the two sides, Africa's integration into the world economic system, how to empower economic and trade activities in Africa under protectionism and sanctions, and how to develop the digital economy as well as enhance cyber, energy, and food security in Africa.

Needless to say, Russia's being a shareholder of the Afreximbank is conducive to boosting Russia–Africa trade cooperation. Up through the end of 2017, the Russian Export Center had been projected to be the third largest non-African shareholder of the Afreximbank, which would promote the export of Russian commodities to the African market, reduce financial risk for Russian exporters in the region, and simplify the trade settlement system between Russia and Africa. As George Elombi, Executive Vice President of the Afreximbank commented, the bank's annual gatherings in Moscow would be a great opportunity for Russia to restore its influence among African countries—a situation that also called for diverse sources of foreign investment.[6]

Sixth, Russia has deepened its cooperation with Africa in education and training on the foundation of close people-to-people exchanges that was previously established during the Soviet era. For one thing, Africans have been seeing a steady increase of scholarship offers from the Russian government in recent years. The number of African students studying in Russia has jumped from mere thousands in 2013 to over 15,000 in 2018, and the number of African students with Russian government scholarships has risen from 968 in 2013 to over 4000 in 2018. Currently, around 17,000 African students are studying in Russian universities, over 4000 of whom have won a Russian government scholarship to be there. For another thing, more and more police and peacekeeping officers in African countries have been invited to visit Russia in order to attend short and long-term training programs. In 2017, for the first time, Russia hosted 28 female police officers from African countries for training. In the same year, 144 police officers from 18 African countries participated in training courses in the universities affiliated with the Russian Interior Ministry, and 135 police officers from 23 African countries were studying in different undergraduate programs in Russian universities.[7]

[6] "Afreximbank: Ties with Russia Open Way to European, Central Asia Markets," *Sputnik News*, October 1, 2018, https://sputniknews.com/africa/201810011068475731-afreximbank-russia-africa-europe-asia/.

[7] "Внешнеполитическая и дипломатическая деятельность РФ в 2017г.: обзор МИД РФ", *МИД РФ*, 2017г.

Since 2014, Russian-African cooperation has entered into a new stage, with salient progress being seen in almost all sectors. Politically, Russia has attached great importance to the establishment of multilateral institutions. Economically, Russia broadened cooperation on innovative technologies in addition to the traditional business of natural resources and military equipment. When it comes to foreign assistance, Russia has gone beyond the bilateral approach in order to properly coordinate with relevant international organizations to deliver aid to Africa. More importantly, Russia has doubled its efforts on training African students and law enforcement staff. That is, Russian influence has continued to grow in Africa over the past few years and has virtually extended to every corner of the continent, even though Russian-African cooperation looks much weaker in terms of economic figures alone.

2 Africa: Russia's Friend in Need?

Doubtlessly, Russia's return to Africa was quite remarkable, though this did not surprise many observers, for Russia has long been famous for its employment of the element of surprise when faced with rivals. After all, the Russian military intervention in the Syrian Civil War amid the Ukraine Crisis showcased the flexible use of tactics, which not only consolidated Russian interests in the Middle East but also, to some degree, forced the US to retreat to the negotiation table.[8] Therefore, Russia's diplomatic attempts to court Africa seemed to be a tactical move to counterbalance Western sanctions following the Ukraine Crisis, and Africa was regarded by Russia as a friend in need when the country found itself isolated in a worsening international environment.

To begin with, Russia has long been an advocate for a multipolar world system and diversified models of development, in order to better safeguard international security and strategic stabilities. Given Africa's political weight in global affairs—especially its voting power at the United Nations—gaining political support from African countries was timely in terms of Russia being able to manage the fallout of the Ukraine Crisis.

Next, Russo-Western relations have been deteriorating over recent years, which has compelled Russia to seek new diplomatic directions.

[8] Feng Yujun, "New Order in Eurasia (volume 2) The Transformation of Russia: Foreign Policy and China-Russian Relations," China Social Sciences Press, December, 2018, p. 48.

Africa proved to be just such a significant breakthrough in Russia's diplomacy toward non-Western countries. From Russia's perspective, its security has been under grave threat due to continuous Western expansion. In particular, the eastward expansion of the North Atlantic Treaty Organization (NATO) with the US working behind the scenes, was listed as Russia's No. 1 security threat in the 2015 NSCRF.[9] Similarly, the 2016 CFPRF also pointed out that the expansion of NATO, as well as the European Union (EU), has generated severe crises for Russo-Western relations. Specifically, "the containment policy adopted by the United States and its allies against Russia, as well as political, economic, information, and other pressure Russia is facing from them, undermine regional and global stability, are detrimental to the long-term interests of all sides, and run counter to the growing need for cooperation and addressing transnational challenges and threats in today's world."[10]

Furthermore, developing closer ties with Africa could, to a greater or lesser degree, help Russia reduce the negative impact of Western economic sanctions. Until the end of 2015, the US and the EU had imposed 18 rounds of sanctions against Russia, from blocking individuals and entities through travel bans and asset freezes and suspending Russia's membership in the Group of Eight (G8), to targeting Russia's energy, defense, and financial services sectors and cutting off military cooperation with Russia. As a result, in 2015 the Russian economy slipped into recession. According to economic data released by the Russian Federal State Statistics Service on January 25th, Russia saw a 3.7% contraction of its GDP in 2015 and the country's fiscal deficit widened to US$25 billion. Meanwhile, domestic demand for consumption and investment in Russia declined sharply with drops in quite a few indicators including retail sales (down 10%), capital investment (down 8.4%), industrial output (down 3.4%), and household income (down 4%). Around 4.26 million Russians unfortunately lost their jobs, adding 7.4% to the unemployment

[9] "Стратегия национальной безопасности Российской Федерации", Президент РФ, 31 декабря, 2015г., http://www.kremlin.ru/acts/bank/40391.

[10] Концепция внешней политики Российской Федерации (утверждена Президентом Российской Федерации В.В.Путиным 30 ноября 2016 г.). МИД РФ, 30 ноября 2016 г., http://www.mid.ru/foreign_policy/official_documents/-/asset_publisher/CptICkB6BZ29/content/id/2542248.

level of 2014.[11] Another report published in 2019 also disclosed that Russia's economic loss in terms of currency amounted to over US$ 11 billion, if calculated with the price standard of 2011. Western sanctions have restricted the opportunities available to get foreign loans for Russian companies, whose corporate debt shrank by 25% on average during 2014–2015.[12] All in all, since 2014, Russia has never seen its GDP growth reach the level of 6% owing to the Western sanctions as well as structural changes and the slowdown of the global economy.

Finally, Russia's return to Africa was accelerated by US diplomacy toward the continent. It's notable that top diplomats of both the US and Russia visited Africa at almost exactly the same time in 2018. From March 6–13, then US Secretary of State Rex Tillerson made his one-week official trip to five African countries including the Republic of Chad, the Republic of Djibouti, Ethiopia, Kenya, and Nigeria. Apparently, these countries are of great significance to American interests in Africa. Specifically, as the largest African economy by GDP, Nigeria has been a major importer of US oil and military equipment. For the US, the Republic of Chad and Kenya are at the forefront of the battle against international terrorism, while the Republic of Djibouti and Ethiopia are both influential states in East Africa, with Djibouti even hosting the only US permanent military base in Africa (Camp Lemonnier). Coincidently, Russian Foreign Minister Sergey Lavrov also paid a short visit to five African countries from March 5–9, including Angola, Namibia, Mozambique, Zimbabwe, and Ethiopia, all of which have received political and economic support from the Soviet Union in the twentieth century in their national independence and liberation movements and have since maintained traditional friendships with Russia. To a certain extent, Tillerson's journey to Africa seemed to be a wake-up call for Russian political elites to pay more attention to Russian-African cooperation.

In short, Russia's pivot to Africa since the outbreak of the Ukraine Crisis can be considered a reflexive reaction to the ensuing Western economic sanctions, which were intended to undermine its economic development and squeeze Russia's international presence. Keeping in

[11] "Western Sanctions Drugged Russian Economy into Recession," *Sina*, January 26, 2016, http://mil.news.sina.com.cn/dgby/2016-01-26/doc-ifxnuwfc9560356.shtml.

[12] А.Гальчева, "Экономисты оценили потери России от санкций в 800 млрд. Руб.", RBC, 14 августа, 2019г., https://www.rbc.ru/economics/14/08/2019/5d5178 0c9a7947cd5eb6eb56.

mind the unfavorable international environment, Russia strengthened its engagement with non-Western countries and reprioritized its diplomatic focuses. Countries in the African, Middle Eastern (particularly Syria), and Asia-Pacific regions (particularly China) have gained prominence in Russia's overall diplomatic focus. Consequently, Russia's push for stronger ties with Africa was widely interpreted as both a tactic to hedge against containment from the West, and also a supplement to the country's Asian diplomacy.

3 Challenges and Opportunities for Strategic Russian-African Relations

In view of the tactical nature of Russia's return to Africa, it still remains to be seen how far Russian-African cooperation could go in the coming years. Obviously, challenges lie ahead for Russia to further deepen its ties with Africa. Today's Russia is no match for the former Soviet Union in terms of economic power. In the African market, Russian products, technologies, and financial resources do not enjoy absolute advantages over those of other competitors like the US, the EU, China, and India. Also, Russia has failed to successfully unleash its trade potential with Africa by making full use of their traditional friendship. To date, Russia–Africa trade has been sluggish despite their close economic partnership. In 2017, the trade volume between Russia and sub-Saharan Africa only reached the amount of US$3.6 billion, and figures from the Russian trade agency showed that the country achieved a total volume of US$17 billion from its trade with all 54 African countries. In contrast, the same year witnessed trade turnovers of US$170 billion between China and Africa, US$70 billion between India and Africa, US$53 billion between the US and Africa, and US$20 billion with Africa by Brazil and Turkey, respectively.[13] It is thus evident that Russia lags far behind the West and even some emerging powers in regard to trade with Africa.

In the face of these challenges, Russia can find some opportunities to make its relations with Africa more strategic in the long term. On the one hand, Africa as a continent of hope, promise, and potential has become increasingly attractive to Russia. Since the beginning of the twenty-first

[13] И.О. Абрамова., "Африканское турне С.В. Лаврова: новые вызовы и перспективы российско-африканского сотрудничества", *Азия и Африка сегодня*. 2018, №5, с.2–10.

century, African countries have undergone fundamental transformations in all aspects. Peace and security, especially, were partially realized on the continent, and civil wars and regional conflicts were terminated in most African countries except for a few including Libya, South Sudan, the Central Africa Republic, and the DRC. Also, Africa is ready for the economic take-off that has seen its annual growth grow by 3–5% from the period of 1995 to 2017, as a World Bank study disclosed.[14]

Thanks to these political and economic achievements, Africa has now turned into a popular partner for both the West and for emerging economies. In the eyes of traditional colonial powers in Europe, Africa has long been vital to their geopolitical interests and socio-economic stability, while for theUS, Russia, China, and India, Africa presents an enormous number of developmental opportunities that all these economies would like to seize. Russia is no exception. Today, no single power in the world can completely dominate the African continent, which is good news for Russia as it attempts to renew its policy to make inroads into Africa. In order to balance their diplomatic ties with other big powers like theUS, the EU, and China, African countries are more than happy to see a growing Russian presence, and they neither take sides with any particular power or power bloc, nor extend their full support to one single power.[15] Undoubtedly, this neutral attitude has boosted Russia's confidence in pursuing cooperation with Africa.

More significantly, Africa owns a huge market that Russia badly needs to diversify its export structure and nurture its non-energy sectors. Similar to Russia, Africa has abundant supplies of natural resources (30% of the world's total) including platinum (83% of the world's output), cobalt and manganese (55–60% of the world's output), diamond (46% of the world's output), palladium (43% of the world's output), chromium (42% of the world's output), gold (40% of the world's output), and bauxite (30% of the world's output).[16] Therefore, Russian minerals can hardly

[14] "Sub-Saharan Africa Trade Statistics: Exports, Imports, Products, Tariffs, GDP and Related Development Indicator," World Intergrated Trade Solution, World Bank Group, December, 2018, https://wits.worldbank.org/CountryProfile/en/SSF.

[15] "Лукоморья больше нет. Цели и задачи России в Африке", Валдай, 19 декабря, 2018, http://ru.valdaiclub.com/events/posts/articles/lukomorya-bolshe-net/.

[16] И.О. Абрамова, Л.Л. Фитуни, "Потенциал Африканского континента в стратегии развития российской экономики", *Мировая экономика и международные отношения*. 2017, №9, с.24–35.

find any market space in African countries. The collapse of international oil price has also taught Russia an important lesson about no longer being able to rely too much on the export of oil, gas, and other raw materials. Instead, non-energy sectors should be transformed into sources of export for high value-added products. The new "May decree," signed by President Putin in 2018, ambitiously set the goal for Russia to jump into the list of the top five global economies by the end of 2024. More concretely, the decree required the development of much stronger manufacturing, agricultural, and service sectors with global competitiveness in order to make Russia get rid of its heavy dependence on oil and gas exports. By 2024, Russia's annual export of non-energy products should reach the amount of US$250 billion (US$50 billion from manufactured goods, US$45 billion from agro-industrial complex, and US$100 billion from services).[17]

On the other hand, Russia can still offer some technologies to the continent, as well as the financial resources necessary for the development of Africa. Over the last couple of years, regional economic integration in Africa has been gaining momentum since the 2063 Agenda laid down the strategic framework for transforming Africa into a global powerhouse in the next 50 years. In July 2019, the African Continental Free Trade Area (AfCFTA) was launched to further advance regional economic integration. Meanwhile, many African countries have also set their own national priorities for development such as strengthening the digital economy, encouraging industrial innovation, upgrading infrastructures, and safeguarding energy and food security, as well as protecting the environment. With this background, Russia's advantage in the above areas can fit in well with the development priorities of African countries. Besides this complementarity, the diplomatic heritage left by the Soviet Union can help catalyze more opportunities for Russia-African cooperation. For lots of Africans who experienced national independence and economic rehabilitation, Moscow's helping hand is still vivid in their minds. In other words, nice memories about the Soviet Union in history have increased Africans' positive emotions toward today's Russia. In practical matters, Russia and African countries can also find a common language when

[17] "Указ Президента Российской Федерации от 07 05 2018г. № 204: О национальных целях и стратегических задачах развития Российской Федерации на период до 2024 года", Президент РФ, 7 мая, 2018, http://www.kremlin.ru/acts/bank/43027.

fighting US hegemony at the global and regional levels. Notably, more and more countries in Africa consider Russia an "alternative"[18] to China and theUS, and thus welcome technological and financial assistance from Russia.

4 Conclusion

Given its geopolitical significance, rich resources, and growing economic weight, Africa has been projected to be the focus of powers around the world in the coming years. Since 2014, Russia has doubled down on its relations with Africa as a tactical move to counterbalance political containment and economic sanctions from the West that resulted from the Ukraine Crisis. The first-ever Russia–Africa Summit held in Sochi in 2019 was a historic breakthrough for bilateral relations and marked Russia's great pivot to Africa. In the long run, however, Russia's return to Africa may go beyond being a tactical move and finally lead to a strategic partnership. To achieve this end, Russia needs to collaborate well with African countries in order to best implement what has been agreed upon at the Sochi summit, align its African policies with Agenda 2063, and formulate visionary plans in multiple areas of cooperation. Only by doing so will Russia be able to render its African ties more sustainable and better serve the country's diplomatic interests.

As a major stakeholder in the African continent, China has been Africa's largest trading partner for ten consecutive years. On the basis of their robust economic connections, China-African cooperation has achieved many tangible results in many fields including politics, security, and people-to-people exchanges, which also makes it possible for Russia and China to explore cooperative opportunities in Africa. Russia can learn from the experience and lessons provided by China's peacekeeping actions, as well as its involvement in regional infrastructure construction. As members of the UNSC and the BRICS, both Russia and China shoulder the responsibility of promoting peace, security, and prosperity in Africa. Despite their differences in economic aggregate and investment philosophy, Russia and China can still try to be in tune with each other

[18] И.О. Абрамова, "Африканское турне С.В. Лаврова: новые вызовы и перспективы российско-африканского сотрудничества", *Азия и Африка сегодня*. 2018, №5, с.2–10.

in order to make their cooperation in Africa another highlight of the coordinated China–Russia strategic partnership for a new era.

PART III

China and Africa Relations

CHAPTER 16

China–Africa Relations and the Belt and Road Initiative

Xinfeng Li

China and Africa are both the cradles of human civilization. China and Africa have a long history of cultural exchange, which can be traced back to ancient times. Supported by a deep-rooted friendship, China–Africa cooperation is growing and showing great potential. In modern times, both China and Africa have greatly suffered from colonialism—China was turned into a semi-colonized state and Africa, a colonized continent by imperial powers. The common experience of suffering has bound Chinese and African people together. By supporting each other in the struggle against imperialism and colonial powers, China and Africa have established a deep and genuine relationship.

The original version of this chapter was revised: Affiliation of the chapter author has been updated. The correction to this chapter is available at
https://doi.org/10.1007/978-981-16-4983-7_23

X. Li (✉)
Institute of West-Asian and African Studies, Chinese Academy of Social Sciences (CASS), Beijing, China
e-mail: wxy03@ssap.cn

X. Li and C. An (eds.), *The Changing World and Africa*,
https://doi.org/10.1007/978-981-16-4983-7_16

After the founding of the People's Republic of China (PRC) in 1949, modern-day China has been continually and highly emphasizing increased cooperation with African countries, showing deep understanding of and strong support for African countries and African people as they fight for and safeguard their national independence. Because both sides have fought side by side with each other, the relationship between China and Africa has made great achievements, and China–Africa friendly cooperations and their traditional friendship continue to deepen.

At present, under President Xi Jinping's proposal of the Belt and Road Initiative (BRI), China and Africa are expected to speed up the development of an unprecedented relationship in the new era. The success of the Beijing Summit of the Forum on China–Africa Cooperation (FOCAC) has highlighted the importance of China–Africa relationships as the basis of the BRI, which will guarantee a higher, deeper, and wider China–Africa cooperation.

1 The Long-standing China–Africa Relationship Has Close Ties with the BRI

Africa is among the major focuses of the BRI blueprint. In ancient times, Chinese envoys, headed by His Excellency Zheng He, visited Africa four times during his seven ocean voyages. It is a marvelous chapter in the story of the ancient maritime Silk Road and is considered a high peak of ancient China–Africa relations.

Zheng He's voyages, whose westernmost destinations were coastal regions in East Africa, can be divided into two phases. In the first phase, the farthest place that he visited was the Guli Kingdom, or Calicut (an ancient kingdom in Southwestern India), during which he reached Southeast Asia and South Asia. In the second phase, he crossed the Indian Ocean and reached the countries along the East African coast. This is undoubtedly an unprecedented achievement of world geography at that time.

In 1431, Zheng He and his fellow explorers erected a stone tablet to commemorate his journey to Africa. The tablet was named "The Record of the Divine Manifestation of the Goddess of Sea." The opening remark on the tablet made the purpose of his voyages to the other sphere of the globe very clear. It reads,

> Our Ming Dynasty (1368-1644) has unified the whole world. Our emperors have made more and grater achievements than those of the Xia (2070 BC-1600 BC), Shang (1600 BC-1046 BC), or Zhou (1046 BC-256 BC) dynasties and those of the prosperous Han (202 BC-220) and Tang (618-907) dynasties. In areas at even the ends of the earth, people have pledged allegiance to His Majesty. There are regions to the west of China's westernmost lands and to the north of China's northernmost lands. They are remote from China, but the distance of the journey can still be measured. As to the countries on the other side of the seas, they are really faraway places. Despite of the long distance, their envoys, accompanied by interpreters that they found by all kinds of means, came to meet His Majesty, bringing all kinds of treasures and gifts. His Majesty deeply commended them for their loyalty and sincerity. I, together with several other officials, was designated by His Majesty to visit them by leading a fleet consisting of hundreds of ships and tens of thousands of officials, officers, and soldiers. We have brought a large number of treasures and gifts in order to reward them. The purpose of our journey is to demonstrate our good virtues and cultures, and attract those people in remote areas.

This document illustrates that the purpose of Zheng's maritime voyages was to spread the country's reputation to the places wherever a ship may reach. This was an ambitious activity surpassing those of all the former dynasties in Chinese history, even the Tang Dynasty (618–907) at its peak.

During Zheng He's voyage to Africa, he brought massive amounts of gold, silver, silk, fine porcelain, tea, and other goods in exchange for local specialties such as ambergris and ivory. This fair and complimentary trade accelerated the development of China–Africa marine trade.

Through a trade network consisting of maritime transportation centers and strategic trade bases and headquarters, Zheng He's voyage integrated the East Africa region to the Maritime Silk Road trading system and strengthened the geopolitical importance of the countries along the East African coast. This action improved the development of transportation and trade between Asia and Africa, cementing the important role of the maritime Silk Road in world history.

History has shown that direct trade between China and Africa started very early. Today's China–Africa trade and cooperation, as well as the new wave of Africa "looking east," is an inherited and coherent reaction to the long-standing historical connection between China and Africa.

Kenya's President Uhuru Kenyatta remarked that, "President Xi's Belt and Road Initiative reinvigorated the long-standing ties between China and Africa, connected China, Southeast Asia, and the South Asian subcontinent with the Mediterranean Sea and East African coast and linked the Western Pacific Ocean with the Indian Ocean. Zheng He's several visits to the ancient Malindi Kingdom in Kenya six hundred years ago are particularly meaningful."

With the purpose of setting up a stable foundation for the development of the BRI today, we dig deep into the historical significance of the Zheng He's sea voyage to Africa, to help a global audience understand China's peaceful foreign policy with a clear lens.

Zheng He visited Africa four times. According to the aforementioned stone tablet, the ships of the fleets "with sails that can blot out the sky like clouds, sailed day and night as fast as shooting stars. They rode on the crest of the raging waves as steadily as walking on an avenue." What a magnificent image! There is no doubt that the Ming Dynasty had a powerful military force. Nevertheless, China had not occupied the land of any African country, nor taken a single penny out of African people's pocket, nor trafficked African people for slavery, nor threatened any African country. It is not because China was incapable of doing so, but because China chose not to, since its relations with Africa complied with traditional Chinese virtuous moral principles. The underlying reason for this is that the Chinese are a people who love peace, honor harmony, choose to be kind, and to live in peace with other people. Therefore, the purpose of Zheng's voyages was not to capture lands, grab property, enslave people, or intimidate other countries, but to explore opportunities for exchange, friendship, and trade.

This reminds us of President Xi Jinping's speech at the symposium held on the 5th anniversary of the BRI: "Building a community of a shared future for humanity through the BRI platform is based on China's reform and opening-up and its long-term development goal. It also echoes China's philosophy of living with great harmony with countries that are far away from us. China's actions stand in high moral ground in the international community."

Zheng He's four voyages to Africa in ancient times and the joint effort between China and Africa to build the BRI today share a similar philosophy. They demonstrate the continuity and development of the peaceful concept of Chinese culture. Former South African President Thabo Mbeki said with deep affection: "History tells us that hundreds of years ago, no matter Africans or Asians, they did not regard each other like barbarians. Although far from the ocean, both sides believed that our very own well-being depends on the happiness of each other." The humanity of all humankind shines in the basic truth revealed in this belief. It is based on this belief that the Chinese ships visited the African port during the fifteenth to sixteenth centuries. This belief brings beneficial exchanges, not the destruction and despair brought by the slave trade and colonialism that came with Arab and European conquests.

Zheng He's voyages to Africa have greatly promoted China's friendly exchanges with countries and regions along the way. Ancient African countries sent representatives and envoys to visit China with Zheng He's fleet several times. At the same time, China was hosting two "Grand Gatherings." The first was on the Dragon Boat Festival in 1413. Ming Dynasty Emperor Zhu Di hosted a brilliant opening ceremony and invited Chinese government officials and foreign national representatives to attend. The imperial court welcomed the visiting kings, tribal chiefs, and envoys with equal respect and received them in a respectful, high-standard protocol. It was so successful that the emperor named it the "Grand Gathering of Good Faith from around the World."

The second "Grand Gathering" took place in 1423. This time, the envoy Zheng He invited more than a thousand Asian and African representatives to visit China via ship so that they could experience the beauty of the Chinese culture and the presence of the Chinese military power at that time.

The two "Grand Gatherings from around the World" deeply impressed the emperor's grandson, the Xuanzong Emperor Zhu Zhanji. After he ascended to the throne, he sent the envoy Zheng He to Africa (1430) for the fourth time.

It should be pointed out that the "Grand Gathering" in the Chinese feudal era was constrained to a relatively small scope and was not sustainable. This is fundamentally different from the situation in today's China, the PRC. Nowadays, China–Africa cooperation is more inclusive and practical, with the people as the major participants, bringing concrete benefits to the people of both sides.

When considering the long-standing China-African relationship, from Zheng He's visits to Africa to the subsequent friendly China–Africa exchanges, we can see that a predominant theme running through the whole process is the peace-loving gene of the Chinese nation. President Xi Jinping pointed out that the fundamental reason for the long-standing friendship between China and Africa is that the two sides always adhere to the virtues of equality, sincerity, cooperation, mutual benefit, and mutual development. This profoundly reveals the core spirit of China–Africa cooperation.

2 China–Africa Cooperation Is Enhanced and the BRI makes Great Achievements

After Zheng He's voyages to Africa, Western colonists dominated the world's oceans, and further direct communication between China and Africa was blocked. In 1956, the PRC first established diplomatic relations with the Arab Republic of Egypt, which symbolized the beginning of a new era for the China–Africa relationship. Ever since then, China has been selflessly providing assistance to Africa, from aiding the construction, to dispatching medical teams, to sending peacekeeping officers, to training all kinds of talents, all reflecting Chinese people's friendship with the African people. Just as Mr. Liao Chengzhi, Chairman of the Chinese Asian-African Solidarity Committee said, "Chinese people were, are, and will always be the most reliable friend to the African people." Likewise, Chinese people will never forget who brought the PRC back to the seat in the United Nations.

If Zheng He's visit to Africa is the peak of ancient China-African relations, then China's aid to construct the Tanzania-Zambia Railway (TAZARA), the construction of the Mombasa–Nairobi Standard Gauge Railway (SGR), and the establishment of the FOCAC represent the new peak of the contemporary China–Africa relationship.

The TAZARA is an artery linking East Africa and Central-South Africa. It originates from the port city of Dares Salaam of Tanzania in the east and stretches to Kapiri Mposhi in Zambia in the west, with a total length of 1860.5 kilometers. This world-famous railway is financially funded and technologically assisted by China and co-constructed by China, Tanzania, and Zambia.

The TAZARA project began in October 1970 and was put into testing in June 1975, then officially handed over to the Tanzanian government in July 1976. During this period, workers from China, Tanzania, and Zambia worked side by side to accomplish a project that Westerners considered impossible. The success of the TAZARA has proven a strong support to the liberation movement of the people in southern Africa and accelerated the economic development of the newly independent African countries. It has been praised as "the road of friendship," "the road to freedom," "the road of the spirit," and "the road to development." The impact and significance it represents have long exceeded the railway itself.

Since China's reform and opening-up, the relations between China and Africa have maintained a healthy development, which can be illustrated by the Mombasa–Nairobi SGR.

As early as the late nineteenth century, the British government built the narrow-gauge (1.0 m) Uganda Railway to the Great Lakes region from Mombasa via Nairobi. The railway was built to strengthen colonial domination over East Africa and grab more resources and wealth. More than 2,400 local workers lost their lives during the construction process, which killed an average of one person per kilometer; this is now called a bloody gauge in the truest sense of the term.

After Kenya's independence, due to the technical barriers it faced and its limited financial resources, the narrow-gauge railway was deteriorating day by day. In recent years, it has not even been able to compete with the parallel highway in terms of transportation capability. It could not sustain the development of Kenya's social-economic development, to say nothing of being able to support the further development of East African integration. Considering this, when formulating its "Vision 2030," Kenyan leaders decided to build a modern SGR (1.435 m) as a starting point to tackle the bottleneck of transportation issues.

With the full support of China on finance, design, construction, equipment, and models of operation and management, the new Mombasa–Nairobi Railway was constructed. The Mombasa–Nairobi Railway is a modern SGR that runs 480 kilometers, with passenger trains traveling at a top speed of 120 km/h, while freight trains run at a maximum speed of 80 km/h.

The construction of the SGR started in December 2014 and was officially in operation by May 2017. Since the opening of the operation, passenger and cargo transportation remain at high operational levels, and the number of freight trains has been continuously growing. For example, at the Nairobi Inland Container Depot (ICD), the terminus of the SGR, cargo trains were busy with unloading containers transported from Mombasa Port.

During Kenyan President Uhuru Kenyatta's railway inspection trip, he stated that "the SGR will contribute 1.5% to Kenya's economic growth, help Kenya to achieve economic transformation." The SGR directly created over 46,000 local jobs, trained more than 45,000 local employees, and sent a large number of Kenyan technicians to China to study railway operations. In the long term, the SGR will, as a major route, connect Uganda, Rwanda, Burundi, South Sudan, and other Eastern African states. As an extension of the SGR, the first phase of the Nairobi–Malaba Railway is now under construction.

With the SGR as a starting point, a modern railway network bringing greater connections among the countries of East Africa is emerging. The integration of East Africa on the railway track has injected a new vitality into the economic and social development of the African continent.

The SGR is the flagship example of the comprehensive integration of China's BRI with Kenya's Vision 2030 and the African Union's Agenda 2063.

The unification of Kenya and China through the BRI will surely improve the liberalization and facilitation of trade and commerce in Eastern Africa, boost Eastern Africa's social-economic development from the coast to the inner land areas, and from the Great Lakes region to surrounding areas. It will form a new development pattern of linkages between the land and the sea and between the east and the west.

3 China—Africa Ties Promote the Concrete Implementation of the BRI and Safeguard Peoples' Well-Being

China is the largest of the developing countries and Africa has the largest number of developing countries. China–Africa cooperation has the right reasons, a solid foundation, and promising prospects. At present, the world is facing changes unseen for a century. Global governance and international order are undergoing accelerated reforms. The fate of the people of all countries in the world has never been as closely related as it is today. At the same time, hegemony and power politics still exist, and wars, terrorism, famines, and epidemics are still occurring despite this greater global connectivity. Under such new circumstances, Chinese-African common interests, mutual benefits, and needs are expanding, which adds even more importance to the development of China–Africa relations.

The FOCAC is an important platform for developing China–Africa relations. The 1st FOCAC ministerial-level meeting in 2000 in Beijing set China–Africa relations on the right track and established a direction for China–Africa to develop a new long-term, stable, equal, and mutually beneficial partnership.

This is the first platform for institutionalized dialogue that China has established with developing countries, and is also the first cooperation mechanism that China established with an entire continent. The FOCAC has held three summits and seven ministerial-level meetings. The China–Africa cooperation mechanism is gradually improving, significantly elevating the general level of China–Africa ties. China-African trade value has grown from USD 10.6 billion in 2000 to USD 204.19 billion in 2018. China has been Africa's largest trade partner for many consecutive years now.

In 2015, when delivering a speech at the Johannesburg Summit of the FOCAC, President Xi Jinping emphasized that by adhering to the principles of sincerity, real results, affinity, and good faith in its African policies and maintaining an approach of upholding justice and pursuing shared interests, China will work with its African friends to achieve a new era of win–win cooperation and common development.

Therefore, President Xi Jinping upgraded the China–Africa New Strategic Partnership into a Full Comprehensive Strategic Partnership, and called for efforts in the fields of the "Five Pillars:" the principle of political equality and mutual trust; the win–win principle for economic cooperation; actions taken toward civilization exchange; the security partnership; and coordination and cooperation in international affairs.

President Xi further expressed that, in order to promote the Full Comprehensive Strategic Partnership, China is willing to make great efforts with the African side to further enhance the "Ten Major Plans:" industrialization, agricultural modernization, infrastructure, finance, green development, trade and investment facilitation, poverty alleviation, public health, humanities and cultural exchanges, and peace and security issues. Since the Johannesburg Summit, China and Africa have fully implemented the "Ten Major Plans." China has supported Africa's efforts to tackle the infrastructure bottleneck, its shortage of talents, and the shortage of financial funds to accelerate the process of industrialization and agricultural modernization in Africa, eventually aiming at reaching its own sustainable development with an independent spirit.

In September 2018, the Beijing Summit of the FOCAC was held successfully. The leaders of China and Africa focused on the year's theme, "*Work Together for Common Development and a Shared Future*," to strengthen their partnership, discuss further cooperation, and their shared future. This is a great reunion for China and Africa and another peak since the 2006 Beijing Summit and the 2015 Johannesburg Summit. The 2018 Summit was marked by three characteristics regarding its scale, level, and achievements. The number of attendees of African leadership and delegations reached a record high. The UN Secretary-General Antonio Guterres and delegates from 26 African and international organizations presented, with a total of over 3,200 people. Chinese President Xi Jinping chaired and delivered a keynote speech in the opening ceremony, and attended over 70 bilateral and multilateral meetings—a record high for Chinese leaders' intensive diplomatic meetings at home with foreign leaders.

Four fruitful achievements were reached as a result of the summit: (1) the keynote speech delivered by President Xi Jinping set up a clear goal, and pointed out the direction for China–Africa Fully Comprehensive Strategic Partnerships; (2) the summit passed two fruitful papers: the *Beijing Declaration—Toward an Even Stronger China–Africa Community with a Shared Future* and the *Forum on China–Africa Cooperation Beijing Action Plan (2019–2021)*, declaring a critical consensus had been reached by all parties and the concrete action plans for the upcoming three years of China–Africa cooperation; (3) China signed BRI cooperation documents with 28 countries and the African Union Commission, which set off another wave of support for participation in BRI construction and expanded the zone of BRI friends; (4) it injected positive energy and a new sense of vitality into, and created a favorable international public opinion of, the environment for the development of China–Africa relations.

China–Africa cooperation not only needs to continue to develop in breadth but also needs to develop in depth, actively dealing with new challenges and solving new problems. Specifically, China–Africa economic and trade cooperation must achieve transformation and upgrading, expand and strengthen its cooperation, and continuously increase the quality of that cooperation. In other areas, China and Africa should play up their comparative advantages, release their potentials, overcome the shortcomings, and achieve the fully coordinated, better, and faster development of China–Africa relations.

Based on a comprehensive summary of the successful experiences of China–Africa's "Ten Cooperation Plans," the 2018 Beijing Summit focused on people-oriented cooperation, paying attention to the development needs of African countries, and valuing the aspiration of African people. For the next three years and in the next period of time, China proposed the joint implementation of the "Eight Major Initiatives," including industrial promotion, facility connectivity, trade facilitation, green development, capacity building, health and hygiene, people-to-people exchanges, and peace and security. This will bolster China–Africa cooperation to an even higher level, and help achieve the fully coordinated, sustainable development of China–Africa cooperation.

The "Eight Major Initiatives" not only covers the traditional advantageous fields of China–Africa cooperation but also explores new development possibilities for China–Africa cooperation, establishing a complete cooperation system. To promote the smooth implementation of the "Eight Major Initiatives," China has provided USD 60 billion to Africa through government aid, investments, and financing from financial institutions, enterprises, etc.

In his congratulatory letter to the establishment of the China–Africa Institute, President Xi Jinping pointed out that, under these new global dynamics, deepening China–Africa traditional friendships, enhancing communication and cooperation, and promoting mutual learning among different civilizations, will not only benefit Chinese and African people, but will also make greater contributions to world peace and development overall.

China–Africa will keep working on the good soil that has been fertilized by their historical friendship along the ancient Silk Road in order to open up the right path of cooperation, mutual benefit, and common development. There is a saying in Africa: "A single hand does not lift a calabash to the head." Only two hands working together can lift heavy objects. The same philosophy could be applied to China–Africa relations, which call for a win–win success through cooperation between the two entities. Upon the comprehensive integration between the BRI and the AU's Agenda 2063, China–Africa friendly cooperation will surely usher in even more vigorous development in the new era.

4 Unique Characteristics of the Africa–China Relationship Establishes the Foundation for the Sustainable Development of BRI Cooperation

Recalling the long history of the development of China–Africa relations, the work the two sides have been cooperating on together over the past years has been done on the basis of similar experiences and missions, united in the Chinese dream of the great rejuvenation of the Chinese nation and the African dream of uniting and rejuvenating the African people. China and Africa, by advocating for each other and helping each other, have explored a form of win–win cooperation.

China–Africa relations have witnessed a trend of incremental growth with both ups and downs. In their long history, this growth has experienced several peaks, particularly in the early Ming Dynasty and after the founding of the PRC. This process is influenced by both internal and external factors. The internal factor is the respective internal situation of China and Africa and how this internal situation changes. The external factor is the evolution of global dynamics. Despite the setbacks these global dynamics occasionally pose to China–Africa relations, the development of China–Africa relations is in the fundamental interests of the people. The general trend of a deepening friendship between the Chinese and African people has not changed.

The Chinese people and the African people have a natural social rapport with each other. Since ancient times, both China and Africa have shared an innate understanding and commonality of thought. This kind of rapport, understanding, and commonality has laid the emotional foundation for mutual respect and equal treatment between China and Africa. This is reflected in the peaks of China–Africa exchanges. In the Tang Dynasty, Du Huan went to Africa as a war captive lost in the Battle of Talas and left the earliest accounts of a Chinese experience in Africa. Africans came to China at the same time, and the black pottery discovered in the tombs in Xi'an that can be traced back to the Tang Dynasty is a solid witness to this fact. In the Yuan Dynasty (1271–1368), Chinese traveler Wang Dayuan visited various places in Africa and wrote "A Brief Account of Foreign Islands." Meantime, Ibn Battuta, the African scholar, came to China, leaving the book "The Travels of Ibn Battuta" to the world as a history of exchange and communication between nations.

In the modern era of national independence movements against imperialism and colonialism, African and Chinese people have supported each other in fighting for the independence of their respective nations. That precious spirit and the comradely experience of resistance both strongly demonstrate that "China and Africa are separated by oceans, but our hearts are connected." Once the right time comes, this spirit and friendship will blossom and spur the further rapid development of China–Africa relations. Since the founding of the PRC, the rapid, comprehensive, and healthy development of China–Africa relations has explained that "China–Africa relations not only embody a profound traditional friendship, a close bonding of interest, but also a close link of dreams."

Advancing with the times is among the major characteristics supporting the development of China–Africa relations. For more than half a century, China has continuously adjusted its policies toward Africa in the hopes that in every critical period of China–Africa relations, both sides can find the best policies to promote the continuous development of China–Africa relations. President Xi Jinping's keynote speech at the opening ceremony of the 2018 Beijing Summit of the FOCAC was magnificent and insightful on this point. He put forward the slogan that states the two entities must "Work Together for Common Development and a Shared Future," stressing that China and Africa should work together to build a China–Africa community with a shared future that assumes joint responsibility, pursues win–win cooperation, delivers happiness for all, enjoys cultural prosperity, enjoys common security, and promotes harmony between humans and nature. This points out the direction of the development of China–Africa relations, clarifies the undertone of the new era and the path of progress for building a shared future, and shows a brighter future for China–Africa cooperation.

Currently, the development of China–Africa relations is often disrupted and sabotaged by Western imperial powers. For example, China–Africa exchanges have made significant contributions to world history, but these contributions have been concealed, ignored, and often forgotten. The direct interaction between China and Africa in history has been interrupted by Western colonists who broke into the Indian Ocean, bringing with them hundreds of years of unabated, devastating history. For this reason, there is still a great need to constantly strengthen the mutual understanding between China and Africa. This is particularly important for contemporary China–Africa relations, which are constantly being twisted by forces in the West with all kinds of rumors, groundless criticism, and even fake information, which aims to have a destructive influence on China–Africa relations.

The history of China–Africa relations reflects the developmental trend of world history. Zheng He's four-time visit to Africa spread China's peaceful foreign policy out of China, establishing a traditional friendship between China and Africa. The independence and national liberation of China and Africa marked the end of the Western colonial era. Since then, a vast number of developing countries have stepped onto the world stage and opened a new chapter of world history.

Current China–Africa relations are full of vitality and flourishing bilateral flows, and have in fact become models for solidarity and cooperation among developing countries. Great Poet Han Yu in the Tang Dynasty said that "it is the solid, firmly ground roots that make the fruit grow and thrive; it is the fat and nutrition inside the body that make the outer skin glow."

The long-standing China–Africa relations that are so deeply rooted in mutual respect and shared interest regarding the cooperation on the BRI and the building of China–Africa community with a shared future are all valuable experiences for humanity. These are valuable experiences from which we can draw wisdom and set an example for our aspiration to build a global community with a shared future for humanity. May our dream come true!

CHAPTER 17

The Shared Future of a China–Africa Community: Ideological Implications and Implementation

Wenping He

A great change that has not been seen this century is the major strategic judgment necessary to observe how the world and China have changed over the last hundred years. On June 23, 2018, at the Central Committee's Foreign Affairs Work Conference, General Secretary Xi Jinping pointed out: "At present, China is in the best period of development since modern times, the world is in a hundred years of unprecedented changes, the two are intertwined and stirring each other."[1]

[1] "Xi Jinping: Efforts to Create a New Situation of Major Powers Diplomacy with Chinese Characteristics," Xinhua News Agency Beijing, June 23, 2018; also see Xinhua net website on June 24, 2018, http://pic.cnr.cn/pic/nativepic/20180624/t20180624_524280223.shtml.

W. He (✉)
Institute of West-Asian and African Studies, Chinese Academy of Social Science, Beijing, China
e-mail: wxy03@ssap.cn

X. Li and C. An (eds.), *The Changing World and Africa*,
https://doi.org/10.1007/978-981-16-4983-7_17

Over the course of the twentieth century, humankind has experienced two world wars. A hundred years ago, after the first world war, the so-called "nations of all nations" that took place at the 1919 Paris Conference was actually only attended by 27 countries. Today, in the twenty-first century, the membership of the United Nations has reached 193 countries. As time passes, not only have the number of sovereign countries increased, but many former oppressed and colonized developing countries have also begun down the path of development and have formed a powerful bloc of emerging countries, such that world affairs are no longer dominated by the "seven powers."

The same great changes that have also taken place in China over the last hundred years are the same ones that have driven the world. From a poverty-stricken country to the world's second largest economy, China's magnificent turn and its entrance to the center of the world stage have transformed China while also driving changes in global governance. Yet a hundred years of unprecedented changes have also been accompanied by a hundred years of uncertainty. The changes in the pattern of new, old, and major powers will inevitably bring about great shocks and adjustments. The rise of power politics in international politics and diplomacy, the prevalence of unilateralism, the rise of Cold War thinking, the increasing uncertainty of the international political economy, and so on are the reflection of this period of great shock and adjustment of anti-globalization. Therefore, as the strongest forces of solidarity among the "New South" developing countries, it is more urgent than ever for China and Africa to strengthen their communication and coordination, show the strength of unity, speak with a common voice, and grasp the strength of the developing countries as a whole in a changing world by building a closer China–Africa community with a shared future.

1 Analysis of a Hundred Years of Unprecedented Change from Three Dimensions

Although we can elaborate upon the last hundred years of unprecedented changes from many perspectives—including the newest round of large-scale scientific and technological revolutions, the impact of globalization on the world's political, economic, social, cultural, ecological systems, etc.—but govern the length and the subject of this article, the author here mainly observes the last 100 years of unprecedented change from three dimensions based on changes in the international structure.

Changes in the International Balance of Power: The Loss of Western Dominance and the Overall Rise of Emerging Powers

The most significant changes in the balance of international power over the course of the last hundred years are the "East up and the West down" and "new rise and old fall" of the international system and the world's balance of power. In terms of the number of sovereign states, a hundred years ago, at the end of the First World War, the so-called "universal" representatives of the post-war 1919 Paris Conference actually came from only 27 countries, and only five powers—Britain, France, theUS, Japan, and Italy—actually formulated the rules of procedure for the Paris Conference behind the scenes.

After the end of World War II in 1945, the Bretton Woods system, led by theUS, began to govern the global economy. By the 1970s, after the first oil crisis hit Western economies, developed Western countries formed the Group of Seven (*G7*) to dominate global economic governance and then went on to dominate global security and political affairs for more than 30 years. With the growing power and role of developing countries, the Group of 20 (G20) was established in 1999, which included China and other developing countries. But for a longer period of time, the G20 seemed to only play a role as a bridge between the West and the rest, functioning as just a ministerial mechanism between finance ministers and central bank governors of developed and developing countries.

The most important turning point of the increasingly significant international status of emerging and developing countries and the "upgrade" of the G20 role was the 2008 world financial crisis. The year following the crisis, when the leaders of the world's 19 major economies and the European Union gathered for the G20 summit in Pittsburgh, US, the G20 announced that the G20 would replace the G8 (which had previously invited Russia to join the Group and become a member G8), as the premier global forum for international economic cooperation and coordination. Going forward, the G8 will take a back seat in international economic affairs and focus on international security, diplomacy, and other issues. Western media has commented that the G20 "upgrade" reflects the current state of the world and highlights the importance of emerging economies such as China, India, and Brazil.[2]

Indeed, since Jim O'Neill, chairman of Goldman Sachs Asset Management, first introduced the concept of the BRIC in 2001, consisting of the initials of Brazil, Russia, India, and China, this acronym was originally just a business concept that identified these four countries as the most growth-oriented and attractive emerging markets for TNCs and was based solely on economic indicators such as population, resources, market potential, and level of economic growth. But with the increasing overall developmental strength of developing countries and the strengthening trend of South–South cooperation, the BRIC appellation has gradually developed into a multilateral consultation and cooperation mechanism among leading emerging market countries, and recently it has even ascended to the highest level of the annual leaders' summit.

The accession of South Africa, Africa's largest economy, in 2010 not only expanded the BRIC countries into BRICS, making it a South–South cooperation mechanism covering Asia, Europe, the Americas, and Africa, and further strengthening this important channel for cooperation with Africa as a whole, but also further extended the very concept of the BRICs beyond their initial economic scope to encompass the broader political, security, and social spheres. Judging by the weight of the world's political and economic landscape, the concept of the "BRICS" countries has sufficient weight. At the political level, it is not only Russia and China that are permanent members of the UN Security Council, but Brazil, India, and South Africa are also the major powers of their continents and regions; at the economic level, the BRICS countries account for 26.46% of the world's total territory and 42.06% of the world's population. It is estimated that in 2018, these five countries accounted for about 23.53% of the world's economy and 16.30% of the world's total trade. The five countries have 13.39 percent of the World Bank's voting rights and 14.84 percent of the total share of the International Monetary Fund.[3] In addition, the five countries' foreign exchange reserves accounted for 75% of

[2] "The G20 Replaces the G8 as the World's Premier Economic Forum and the G20 'Upgrades'," *Southern Metropolis Daily*, September 26, 2009, http://news.sina.com.cn/w/2009-09-26/053516360826s.shtml.

[3] The website of the Ministry of Foreign Affairs of China, "BRICS," https://www.fmprc.gov.cn/web/gjhdq_676201/gjhdqzz_681964/jzgj_682158/jbqk_682160/.

the world's total as early as the end of 2011 and contributed more than 50% to world economic growth.[4]

From the perspective of the global scope and development trends, the gap between traditional developed countries, emerging economies, and the vast number of developing countries is shrinking, while emerging economies and other developing countries' "late comer advantage" and "catch-up advantage" continue to develop. Emerging markets and developing countries accounted for about 40% of the world economy in 2018 and contributed about 80% to world economic growth. If the current pace of development is maintained, the total economies of emerging economies and developing countries will be close to half of the world total in ten years, which will make the map of global development more comprehensive and balanced. Additional United Nations data predict that by 2050, the global population will reach about 10 billion, of which 8.5 billion will belong to current emerging markets and developing countries.[5] The gradual loss of the dominance of the Western powers and the overall rise of the emerging powers can be said to be the most revolutionary, historic, and even irreversible change to the international power balance in modern times. This change is what President Xi Jinping described as the three "unprecedenteds," namely "the rise of emerging market countries and developing countries is unprecedented, the new round of scientific and technological revolution and industrial change brought about by the metabolism and fierce competition unprecedented, the global governance system and the international situation changes are not adapted, asymmetric unprecedented."[6]

Changes in China's Power and Status: China's National Power Grows as It Continues to Step into the Center of the World Stage

What is driving the great changes in China is also what has driven the world over the last hundred years. From a poverty-stricken country to the world's second largest economy, China's magnificent turn and its

[4] He Wenping, "The BRICS Countries Have Crowed History," *Global Times*, March 31, 2012. https://world.huanqiu.com/article/9CaKrnJuNwr.

[5] "Why Now Is the Great Changes in a Hundred Years?," The Website of the CPC Party Members, August 29, 2019, http://www.12371.cn/2019/08/29/ARTI1567071473915983.shtml.

[6] Ibid.

entry into the center of the world stage have transformed China while also driving changes in global governance. If the clock were turned back for 40 years, the average gross domestic product of China before its period of reform and opening-up would be even lower than that of the small African states of Malawi and Burkina Faso today.[7] But now, 40 years later, China has not only succeeded in lifting more than 700 million poor people out of poverty,[8] but it has also grown to become the world's second largest economy after theUS. According to *The 2019 National Economic and Social Development Statistics Bulletin* issued by China National Bureau of Statistics on February 28, 2020, China's gross domestic product (GDP) in 2019 is close to RMB 1000 billion, ranking second in the world; gross domestic product per capita exceeded the $10,000 mark for the first time. China's GDP is expected to account for more than 16% of the world's GDP in 2019, and China's economic growth contributes about 30% of the world's economic growth, continuing to be the main driving force for world economic growth overall.[9]

Since the beginning of the twenty-first century, China's development has been gradual process of stepping toward the center of the world stage. In 2000, China's gross domestic product exceeded trillions of dollars, marking the country's ascendance to the ranks of the big powers. In 2001, China joined the World Trade Organization, accelerated the pace of accession of the international economic system and the division of international production, and vigorously promoted economic development through its participation in globalization. In 2008, China successfully hosted the Beijing Olympic Games, highlighting China's peaceful rise to the world stage. In 2013, China launched the "Belt and Road" cooperation initiative, a cooperative concept characterized by "co-consultation,"

[7] According to the World Bank data, China's per capita GDP was only $156.396 in 1978, Burkina Faso per capita GDP was $225.942, and Malawi is $163.416. http://databank.shihang.org/data/reports.aspx?source=2&series=NY.GDP.PCAP.CD&country=CHN, http://databank.shihang.org/data/reports.aspx?source=2&series=NY.GDP.PCAP.CD&country=BFA, http://databank.shihang.org/data/reports.aspx?source=2&series=NY.GDP.PCAP.CD&country=MWI

[8] See the White Paper, "China's Poverty Reduction Action and Human Rights Progress," published by the Information Office of the State Council on October 17, 2016, http://news.xinhuanet.com/politics/2016-10/17/c_1119730504.htm, May 19, 2017.

[9] Wang Enbo, "China's Contribution to World Economic Growth in 2019 Is About 30%," China News Network, http://www.chinanews.com/cj/2020/02-28/9107603.shtml.

"co-construction," "co-sharing," and being a "win–win situation" for all participants. Since the BRI was launched more than six years ago, it has increasingly become a bright business card for China's peaceful development diplomacy and a global public product that is popular with the international community. Given its openness and inclusiveness, being able to participate is no longer confined to countries along the physical Belt and Road for which it was originally designed. By the end of January 2020, in fact, China had signed 200 co-construction "Belt and Road" cooperation documents with 138 countries and 30 international organizations.[10]

After 40 years of hard work since the country's reform and opening-up, the Chinese people not only lead the world in new economic fields such as high-speed railways, new energy, and the network economy, but China's diplomacy has also increasingly moved closer to the center of the world stage and entered a "new era of making continuous greater contribution to [hu]mankind." Unlike in the past, when the major powers held the center of the world stage by means of forced conquest, China, which instead advocates "peace and harmony" and having a "shared culture," has now entered the world stage with its Belt and Road Initiative, shared development, and win–win cooperation, and the thus effected the concept and vision of a human community with a shared future.

To this end, we put forward a new concept of international cooperation marked by "co-consultation," "co-construction," "co-sharing," and a "win–win situation." China's Belt and Road Initiative aims to further strengthen multilateralism, promote international development cooperation and inclusive globalization, and fully embody the spirit of equal participation and joint construction. China hopes that by bringing together all countries to work together, all countries involved in the Belt and Road construction will benefit from the process of participation and share developmental results.

We also put forward the idea of a "community for humanity with a shared future" as an international global development goal. The reference to "community for humanity with a shared future" excludes neither the maintenance nor pursuit of national interests, nor does it advocate for a so-called "unconditional internationalism." It is aimed at pursuing

[10] See "Belt and Road Network": "A List of Countries That Have Signed a Document on Co-building the Belt and Road Cooperation with China," https://www.yidaiyilu.gov.cn/info/iList.jsp?tm_id=126&cat_id=10122&info_id=77298.

national interests with a balance between the legitimate concerns of other countries and promoting the common development of all countries in the pursuit of their own development. Because humankind has only one earth, with all nations living together in a single world, the development and security of any country cannot be based on the sole development and exclusive security of a single country. The development of globalization has objectively and naturally made the world we live in an interdependent community characterized by a shared human destiny. From the perspective of international relations and international development cooperation, the global value of a "human community with a shared future" actually contains the advanced concepts of global development, such as the interdependent view of international power, the concept of common interests, the concept of sustainable development, and the concept of global governance.

It should be said that the concept of a "human community with a shared future" has a greater historical understanding, structure, and worldview than the "America First" approach proposed by US President Trump, and is therefore more easily recognized and accepted by the world at large. In 2017, President Xi Jinping delivered keynote speeches entitled, "To Promote Global Development with Shared Responsibility" and "Building a Human Community with a Shared Future" at the World Economic Forum in Davos and the United Nations headquarters in Geneva that attracted widespread attention and resonance from the world. A year later, in early 2018, the World Economic Forum in Davos identified the forum's theme as "building a common destiny in a divided world," a recognition and continuation of the theme of President Xi Jinping's 2017 speech. Worldview determines methodology. The "America First" worldview determines that the Trump administration has taken a "retreat diplomacy" and evasion of global responsibility over the past few years, including at the Paris Agreement on climate change, its withdrawal from UNESCO, etc. So the worldview of building a "human community with a shared future" is accompanied by the implementation of the Belt and Road Initiative and China's continuous strengthening of its contribution to international peace, security, and development.

In short, the contemporary change in China's power and status and the great rejuvenation of the Chinese nation is undoubtedly one of the most unique and astounding chapters of this once-in-a-century period of great change. As Larry Summers, former President of Harvard University, put it: "The focus of world history in the first half of the twenty-first century will be on China."[11]

A Century of Uncertainty: Great Shocks and Adjustments to the International System

Any major change of international power in history will inevitably bring about major shocks and adjustments to the international system. After the end of the Cold War, the US took advantage of what it saw as its victory to not only vigorously promote the neoliberal economic paradigm, multi-party democratic competition, and free political development model in the world, but also demonstrated its superior military power by waging wars in Afghanistan and Iraq. The dominant international structure moved away from the "bipolar pattern" seen during the Cold War period and toward one characterized by "one superpower and many strong powers." Moreover, the power represented by the only "Superpower," theUS, is not on an equal scale with other powers such as Europe and Russia, etc. That is, the "one superpower" takes the leading position, and the "other strong powers" are listed in parallel.

However, with the outbreak of the Wall Street financial crisis in 2008 and the strategic and resource constraints caused by the United States' long-standing global strategy, the will, determination, and ability of the US to control the regional and international situations independently have declined significantly. The trend of equilibrium of international status changes between "many strong powers" is becoming increasingly prominent. Especially after Mr. Trump took office, the US wanted to shift the cost of its security expenditures to allies and other emerging market countries and to remove the elements and constraints that were detrimental to theUS. To this end it engaged in unilateralism, authoritarianism, and bullying under the banner of "America First," politicizing and weaponizing tariffs. This has led to chaos and disorder in the international order, and the process of globalization has also suffered countercurrents. At the same time, the threat of terrorism has not been eliminated, regional

[11] See Xinhua News Agency, Beijing, August 28, 2019, "Special Report: A New Era for the Great Change in the World," http://www.xinhuanet.com/world/2019-08/28/c_1124931634.htm.

conflicts have become driven by war, and the competition and games between the great powers have become increasingly opaque.

On one side is theUS, busy with "retreat" (unilateralism) and "building walls" (protectionism), moving further and further away from the path of globalization; on the other side is China's Belt and Road Initiative, which holds high the banner of free trade, advocates inclusive globalization, and promotes multilateralism and regional cooperation. China and theUS: one is the world's largest developing country in a rising state of power, the other is the world's largest developed country in a declining state of power. The "China moves on—the US retreats" situation as it relates to global governance will also be reflected in the strategic competition between China and the US in Africa. In recent years, senior US government officials, including US Secretary of State Mike Pompeo, have been clamoring for African countries to move back from the "China model" to the "American model" during their visits to Africa. Therefore, in order to seize this once-in-a-century opportunity for development, we must continue to develop and consolidate the friendly cooperation between China and Africa and build a China–Africa community with a shared future.

2 China and Africa: Building a Stronger Community with a Shared Future

At the Beijing Summit of the Forum on China–Africa Cooperation held in early September 2018, Chinese and African leaders shared friendship and cooperation on the theme of "win–win cooperation and working together to build a closer China–Africa community with a shared future." Building a closer "China–Africa community with a shared future" is not only conducive to the unity and cooperation between China and Africa, but can also play a leading role in the construction and benchmarking of a wider scope of human destiny.

In the course of building a "China–Africa community with a shared future," China and Africa draw not only on their solid historical foundation, moral foundation, and common values, but in the new era, because the BRI has empowered new development, the idea of building a "China–Africa community with a shared future" has a more solid and broad condition for its rise and development than ever before.

The Historical Foundation and Common Values Supporting the "China–Africa Community of a Shared Future" Through Tough Times

The concept of the "China–Africa community of a shared future" was proposed by President Xi Jinping during his visit to three African countries, namely, South Africa, Tanzania, and the Republic of the Congo in March 2013. During his visit to Africa, President Xi pointed out that China and Africa share a common historical experience, a common developmental task, and a common strategic interest, and that they are in the same boat: a "community of destiny." Indeed, both China and African countries are developing countries, both belong to the third world, have had similar historical experiences and common trajectories, and have experienced colonialism and aggression by foreign powers. Therefore, the mutual solidarity and level of support between China and Africa has stood the test of the international climate and has remained strong on the basis of common political and development aspirations, which constitute the most solid foundation for building the China–Africa community and its shared future.

The historical basis of China–Africa friendship is reflected in mutual political trust and the long-standing mutual "support" and mutual "concerns." This kind of political and diplomatic mutual support can be traced back to the mutual support of Chinese and African people in the struggle against imperialism and colonialism during the struggle for national liberation and national independence that has been a feature of the areas in modern times, as well as the firm support African countries have given to China's core interests in recent years (for example, in relation to the South China Sea dispute and the "one China" principle). Similarly, China is constantly drumming up support for and calling for Africa's development in the G20 and on the international stage. And every New Year's visit by China's Foreign Minister since 1990 has been to Africa, a diplomatic tradition that has lasted nearly 30 years and is fully reflected with the keywords of mutual "support" and mutual "concern."

The "all-weather" China–Africa friendship also highlights the shared weal and woe of difficult times. Like with a friend, lover, or couple, it is easy to enjoy a happy moment together, but to be able to share difficulties can test true love and true feelings. Only remaining hand-in-hand and never giving up, even when times are difficult, can best reflect the depth of friendship. Chinese people will not forget that when they suffered from earthquake disasters in Wenchuan, Yushu, and other places in China in May 2008 and April 2010, African countries, even with their relatively lagging economic development and modest national strength, were eager to support China. For example, Equatorial Guinea, a small central-western African country with long-term poverty and a population of just 1.01 million, donated 1 million euros to China. Although the total is small, it is equivalent to 1 euro per national.

Likewise, African people will not forget that when African countries were hit by famine and disease outbreaks, China was quick to lend a hand and work hand-in-hand with its African siblings. In the summer of 2011, when countries and regions such as Somalia and the Horn of Africa suffered severe drought, the Chinese government announced twice in a short period of time that it would provide emergency food and cash assistance to the affected countries totaling RMB 443.2 million. The speed and the amount of assistance were unprecedented in the history of China's emergency humanitarian assistance to Africa. In March 2014, when Ebola broke out in West African countries such as Guinea, Liberia, and Sierra Leone, China provided aid and funds totaling RMB 750 million in four batches. Instead of fleeing, China's medical teams in the affected countries choose to stick close in order to share in the suffering of their African friends and fight the outbreak together. At the height of the epidemic, as many as 700 Chinese experts and healthcare workers worked in the affected areas, surpassing Cuba as the world's largest contributor of aid. Regarding this, the Minister for Foreign Affairs of Sierra Leone praised China's help for the "genuine friendship" it had shown during the disaster. He said, "China's generous assistance shows that China is a true strategic partner for Sierra Leone and African countries" and that "China's image has become taller and more upright."[12]

Around the 2020 Lunar New Year in China, when places such as Wuhan, China, faced serious challenges from the outbreak of the new coronavirus epidemic, African countries were the first to send support and encouragement. On February 4, 2020, President Obiang of Equatorial Guinea chaired the Council meeting of Ministers and decided to contribute $2 million to the Chinese government to support China in its fight against the new outbreak of coronavirus. On February 7, 2020, the

[12] See "Foreign Media: China's Aid to Africa Anti-Ebola Harvest Diplomatic Dividend, Being Praised Image High," Reference News Network, October 2014, http://china.cankaoxiaoxi.com/2014/1022/536350.shtml.

36th AU Ministerial Meeting issued a special communique announcing that it stood with the Chinese government and people to support China's efforts to combat the outbreak, expressed confidence in China's ability to deal with it, and called on all countries in the world to work together with China to strengthen cooperation and jointly deal with the impact of the outbreak. In addition, the leaders of Ethiopia, Tanzania, Sierra Leone, Kenya, Zimbabwe, Angola, and many other African countries sent letters of *sympathy* and support to President Xi Jinping, saying that while they recognized that the outbreak is a challenge for China and the world, they are sure that Chinese people will surely win the fight against it.

In addition, the "China–Africa community of a shared future" is based on the common concept of being "true, honest, close, sincere" and having the "correct view of righteousness and benefit." During his visit to Africa in March 2013, President Xi Jinping not only put forward the concept of building a "China–Africa Community with a shared future," but also commented on the basic concepts and principles that needed to be adhered to for the future development of China–Africa relations—that is, being "true, honest, close, and sincere." He pointed out that we speak a "true" word to our African friends, speak the word "honest" when we take cooperation with Africa, speak the word "close" when we strengthen our friendship with Africa, and speak a "sincere" word when we solve the problems together in cooperation.[13] In addition to the concise "true, honest, close, sincere" four-word maxim, when they visited Africa over the last seven years, President Xi Jinping, Premier Li Keqiang, and Foreign Minister Wang Yi have repeatedly pointed out that China needs to establish a correct "righteous view" in its policy toward Africa, especially in China–Africa economic and trade cooperation, and that the relationship between "benefit" and "righteousness" should be correctly positioned.

For example, in January 2014, Chinese Foreign Minister Wang Yi stressed during a visit to Africa that "the right view of righteousness is a banner of Chinese diplomacy in the new era." In a joint meeting with reporters after talks with Senegalese Foreign Minister Ndiaye, he said that "righteousness" means "morality." There is an ancient Chinese saying that claims, "gentlemen love wealth, but take it in a moral way."

[13] International online report (Reporter: Ding Ning, Shi Ran), "Xi Jinping: China and Africa Have Always Been a Community of Destiny, and China Will Step up Cooperation with Africa," March 25, 2013. http://news.cri.cn/gb/27824/2013/03/25/6651s4064306.htm.

China should take "morality" as the number one priority in its dealings with African countries, insist on treating its African siblings equally, be sincere and friendly, be faithful, and speak up for the defense of Africa's legitimate rights and legitimate demands. "Profit" here means "mutual benefit." China will never follow the old path of colonial plundering in its dealings with African countries, will never follow the capitalists' profit-making approach, nor will it be like some countries that are only trying to achieve their own self-interests; it is instead willing to develop and prosper together with its African siblings. In this process, China will take greater account of the legitimate needs of African countries and strive to make Africa gain and benefit through increased cooperation. If needed, we have to re-take the benefits, and even the benefits. The Tazara Railway project in 1960–1970s was a model of what we mean when we take the correct righteousness view, and the AU Conference Center in Addis Ababa today is another example.[14]

In addition, it should be noted that, in order to enhance Africa's voice in international affairs and the attention paid to African affairs on the international stage, China, as the Chair of the G20 held in Hangzhou in 2016, gave full play to the platform setting and to making the meeting agenda being about host country diplomacy. Not only did it set "support for the industrialization of African countries and least developed countries" as an important topic for the summit, for the first time it included support for the industrialization of African countries in the G20 communique. It also provided a platform for communication and exchange between African countries and G20 countries. In addition to South Africa, already a member of the G20, China also invited Chad, the African Union's rotating chair country at that time, Senegal, the Chair of the New Partnership for Africa's Development, and Egypt, an important North African country, to join the G20 in Hangzhou. This not only reflects the fact that China has attached great importance to Africa's development, but also provides African countries with urgent and necessary opportunities for global governance reform, contributing to Africa's rising international status.

[14] "Wang Yi: The Right View of Righteousness Is a Banner of Chinese Diplomacy," Xinhua Network, http://world.people.com.cn/n/2014/0111/c157278-24087373.html.

Realistic Conditions and Empowerment: Building a China–Africa Community with a Shared Future for Development, Cooperation, and Mutual Wins

China and Africa are both developing countries, and after completing the great cause of national independence and national liberation in the twentieth century, in the twenty-first century both are now faced with the demands of economic development and improving people's well-being and realizing their "dreams of rebirth." Judging from the elements of their respective economic developmental paths, China and Africa have a strong basis for complementary advantages, mutual benefits, and win–win cooperation and resource endowment. The African continent is rich in natural and human resources, with 54 countries and vast markets for more than 1 billion people. However, as a result of its prolonged colonial plunder and the turbulence caused by local conflicts, the economy is still relatively backward and lacks the funds, technology, and experience necessary for development. China has accumulated rich development experience, capital, and applicable technology after more than 40 years of reform and opening-up. Therefore, on the one hand, African countries hope to learn from China's development experience to find the key to their own development. On the other hand, China has the capacity and strong will to help Africa's development, and is able and willing to work with African countries to forge forward with a partnership based on South–South cooperation that is mutually respectful and cooperative to bring about a win–win situation.

It is said that "one flower alone is not spring, a hundred flowers in full spring garden." The rapid development of emerging economies in a few developing countries, such as China and India, does not mean that the entire global South has achieved development; only through the guidance of emerging economies and the deep integration of the economies of the developing countries as a whole can the substantive development and further promotion of South–South cooperation be achieved. Regarding this issue, Chinese Premier Li Keqiang said during his visit to Africa in May 2014 that China and Africa together have a population of more than 2.3 billion people, that China has helped Africa develop, and that through sincere, open, mutually beneficial cooperation, nearly one-third of the world's population will truly benefit from economic growth and improve their lives. This will greatly contribute to the balanced development of the world economy and the overall progress of human society.[15]

Despite the fact that China's own economic growth rate is no longer rapidly increasing and has in fact stabilized at a "new normal" pace of development, China has still invested huge amounts of resources in its pursuit of developing China–Africa cooperation. By promoting Africa's development through strengthened South–South cooperation even as international cooperation more broadly declines, China and Africa are mutually committed to facing the challenges of structural adjustments and adapting their production capacity. For example, at the China–Africa Cooperation Summit in Johannesburg, South Africa, in December 2015 and the Forum on China–Africa Cooperation Summit in Beijing in September 2018, respectively, China has twice proposed the "Ten Plans for China–Africa Cooperation" and the "Eight Actions," together totaling US$60 billion. The "Ten Plans for China–Africa Cooperation" refers to promoting all-round cooperation in ten areas, including industrialization, agricultural modernization, infrastructure, financial cooperation, green development, trade and investment facilitation, poverty reduction and people-friendly cooperation, public health cooperation, people-to-people cooperation, and peace and security cooperation. The "Eight Actions" refers to the eight major areas of industrial promotion, facilities connectivity, trade facilitation, green development, capacity building, health, people-to-people exchanges, and peace and security.

Whether considering the "Ten Plans" or the "Eight Actions," its connotations are similar to the "Five-Connectivities" construction (i.e., policy communication, facility connectivity, smooth trade, financial communication, and people-to-people connectivity) advocated by the Belt and Road Initiative launched in 2013. In fact, the Belt and Road Initiative has become a huge engine for building the China–Africa community with a shared future, which is being implemented through the "Ten Plans" and "Eight Actions" between China and Africa.

Since the Chinese government launched the Belt and Road Initiative in 2013, the African continent, which has abundant resources, huge market potential, and a strong demand for infrastructure construction,

[15] See He Wenping, "China-Africa Cooperation Has Been Comprehensively Upgraded," *China Today (Chinese Edition)*, April 4, 2014, http://www.chinatoday.com.cn/ctchinese/chinaworld/article/2014-06/04/content_622416.htm.

has actively participated in the construction of the Belt and Road Initiative. Africa has seized important historical development opportunities in the process of participating in the BRI, with a view toward realizing the leapfrogged development of its economy from a pre-industrial society to full industrialization. At present, 44 of the 54 African countries on the continent have signed an intergovernmental memorandum of understanding (MoU) with China to develop the BRI. The Belt and Road Development Drive has two important areas: infrastructure construction and industrialization, which have become the two wings of Africa's economic growth.

In terms of infrastructure construction, China–Africa infrastructure cooperation has accelerated significantly since the Belt and Road Initiative was launched in 2013. During his visit to Africa in 2014, Chinese Premier Li Keqiang proposed helping Africa build the "three major networks"; that is, the railway network, the road network, and the African regional aviation network. From 2011 to 2016, China invested an average of $12 billion a year in infrastructure in Africa, making it the largest source of investment in infrastructure in Africa, according to a 2017 report by the African Infrastructure Consortium (ICA).[16] In the past five years, China's investment in the infrastructure of railways, roads, bridges, and ports in Africa has blossomed, changing the continent's landscape and effectively improving the investment environment in African countries while improving the living standards and transportation facilities of local people. For example, the Kenya Standard Gauge Railway, which connects the Kenyan port city of Mombasa and its capital city, Nairobi, has had 2.7 million passengers since it opened in May 2017, with an average occupancy rate of more than 90 percent and a punctuality rate of 99.9 percent. It also delivered 4.21 million tons of cargo. The railway created more than 50,000 jobs during construction and operation, boosting Kenya's GDP by 1.5%.[17] Even then British Foreign Secretary Jeremy Hunt told a

[16] "China Is the Largest Investor in Infrastructure Development in Africa," AFP reported on October 19, 2017. Quoted from the website of the Ministry of Commerce of China, October 20, 2017, http://www.mofcom.gov.cn/article/i/jyjl/k/201710/20171002661137.shtml.

[17] Hou Jieru and Zhang Mei, "The Kenya Standard Gauge Railway All-Round Helps Kenya Revitalization—Interview with Mr. Dai Yunjie, the Vice-President of the Railway Operations Company," *China Investment, Africa Edition*, April Issue, 2019, pp. 62–63.

press conference that, when he visited Kenya, Britain very much appreciated the great contribution made by Chinese constructed major projects such as the Kenya Standard Gauge Railway to the economic and social development of Kenya and Africa as a whole.[18]

With regard to cooperation in the fields of productivity capacity and industrial promotion, by the end of May 2018, China had established about 25 economic and trade cooperation zones and hundreds of industrial parks in Africa, making positive contributions to Africa's industrialization, job creation, and export growth. According to initial statistics, a mere 25 economic and trade cooperation zones have created about 40,000 jobs, with nearly $1.1 billion in taxes being paid to the host country.[19] In June 2019, the first China-Africa Economic and Trade Expo was successfully held in Changsha, the capital city of Hunan province. During the Expo, 15 China–Africa cooperation project agreements were formally signed, involving the construction of agricultural cooperation zones, investment in factories in cooperation zones, automobiles, medical facilities, textiles, engineering machinery, mines, cross-border e-commerce, and more. All these projects, which amount to nearly $300 million, will promote Chinese enterprises to invest in production capacity cooperation in Africa and to upgrade the China–Africa economic and trade cooperation zone in Africa. The "two-wheel drive" model of "infrastructure construction and industrial park development," which is being promoted by China, will boost Africa's economic development and transform Africa's resources and energy endowments into a more sustainable development capacity while benefiting African people's livelihoods.

[18] "British Foreign Secretary: Appreciated the Great Contribution Made by Chinese Constructed Major Projects Such as the Kenya Standard Gauge Railway," *China Economic Network*, May 5, 2019. https://baijiahao.baidu.com/s?id=1632644627904673826&wfr=spider&for=pc.

[19] "The Ministry of Commerce of China (Mofcom): 25 Economic and Trade Cooperation Zones Have Been Established in Africa, Creating About 40,000 Jobs," *China Net Finance*, June, 4, 2018, http://finance.china.com.cn/news/20190604/4997221.shtml.

3 Looking to the Future and the Direction of Development: Six Dimensions to Build a Closer China–Africa Community with a Shared Future

In his opening address to the 2018 Beijing Summit of the Forum on China–Africa Cooperation, Chinese President Xi Jinping elaborated upon the profound connotation of building the six dimensions of China–Africa community with a shared future, which points to the direction for the future construction of a China–Africa "community of destiny."

The first is to work together to build a stronger China–Africa community of a shared future that assumes joint responsibilities. China and Africa should not only strengthen the political dialogue and policy communication between each other, but also strengthen cooperation and coordination in regional and international affairs in order to safeguard the common interests of China and Africa and the developing countries in general. From the view of building a new type of international relations, China–Africa solidarity and cooperation also reflects the firm determination of one-third of the world's population and one-fourth of the world's countries to follow the path of peaceful development, while also making a great contribution to the cause of world peace and development. China–Africa cooperation and win–win development not only leads South–South cooperation overall, greatly enhancing the overall strength and influence of developing countries, but also promotes North–South dialogue, promotes the reformation of the international pattern and governance system toward a more balanced and rational direction, and makes greater contributions to the construction of a new type of international relations with win–win cooperation as its core.

Historically, due to the weakness and poverty of African countries, the continent has lacked a voice in the international community and has long been marginalized globally. China's Forum on China–Africa Cooperation sets high standards to receive African leaders visiting China, such that not only has the importance of African countries on the international stage has been unprecedentedly improved, but African countries have also received the attention that has not historically been paid to North–South relations. The international community is now paying increasing attention to Africa. Today, both traditional powers and emerging economies are also increasingly paying attention to Africa, with a wide variety of African summits forthcoming as a result. Following the first China–Africa summit in Beijing in 2006, the major countries all held their own respective summits on Africa. In August 2014, for example, the first US–Africa Summit was held in Washington. In October 2015, the third India–Africa Summit also attracted the attention of the world on a massive scale. There were also the 2019 Russia–Africa Summit, the Japan–Africa Summit, and so on.

The second is to work together to build a stronger China–Africa community with a shared future that pursues win–win cooperation and a mutual development strategy. The link between the Belt and Road Initiative and Africa also requires a three-dimensional connection process to make the strategy connect from "general" to "detailed." The three-dimensional connection process is: the connection between the BRI and the UN 2030 Agenda for Sustainable Development, the connection with the African Union's 2063 Agenda, and the connection with the development strategies from different African countries. In particular, the connection with each African country's development strategy should be precise, so that our development advantages and each African country's areas of priority development can be matched, so as to achieve win–win cooperation. Specifically speaking, precise connections between countries requires us to do a solid job of studying the regional and national circumstances of each African subregion and each country, not only to understand the political and social stability of the region and countries themselves, but also to fully understand the medium- and long-term national development strategies and priority development areas there. At the same time, China also needs provinces and cities to do their own homework, to find out which enterprises in the provinces and cities are suitable for China–Africa ventures, and which extra production capacity is suitable to transfer and invest in which African countries. That is to say, we should also know ourselves in a detailed and precise way.

In addition, as Africa's economic development is still relatively underdeveloped in terms of capital, technology, and talent, the continent not only need a "blood transfusion" of technology, but also need to improve its "blood making" capacity. Special attention needs to be paid to technology transfer, talent development, and job creation in Africa through the process of China–Africa industrialization cooperation. This should be aimed at strengthening the existing financing support for African SMEs as a part of the China–Africa financial cooperation process, to address financing and debt problems with innovative financial instruments such as developmental finance, and to focus on sustainable project development in the process of strengthening interconnected infrastructure construction. The China–Africa community of a shared future can be indestructible only if it helps Africa to realize its own "African dream" of poverty reduction and development.

The third is to build a stronger China–Africa community with a shared future that delivers happiness for all of us. The word "happiness" is the key here. China–Africa cooperation projects should focus more on improving people's living standards, rather than just on some "face-saving projects," and to promote people's well-being as the starting point and primary purpose for the development of China–Africa relations. Only by bringing tangible results and benefits to the people of China and Africa in the first place can China–Africa cooperation ensure the further sustainable development and strengthening of the most powerful basis of people-to-people relations for the China–Africa community with a shared future. As an old Chinese saying goes, "the good relation of the country lies in the friendship between the people" and, further, "mutual affinity between the people lies in the mutual understanding of the heart." Only by making the Chinese and African people have a sense of happiness in the construction of the Belt and Road Initiative will a "China–Africa community with a shared future" become more than a slogan and empty words.

In fact, the Forum on China–Africa Cooperation, which meets every three years, lists a number of specific projects for the benefit of the people in the action plan that guides the work of the next three years. In October and November 2019, China's "thousand villages connectivity" project was completed in Nigeria and Cameroon. The completion of the project has made it important for villagers in Nigeria, Africa's most populous country, to receive clear, informative satellite television programs and to provide children with better educational resources through satellite television education programs. The "thousand villages connectivity" project was implemented mainly by the Chinese company Sida Times Group and is only one of the China–Africa cooperative initiatives aimed at strengthening people-to-people connectivity. First proposed at the Johannesburg Summit of the Forum on China–Africa Cooperation in 2015, it was aimed at bringing satellite television to 10,000 villages in African countries. Furthermore, in order to increase vocational and technical training in Africa, the 2018 China–Africa Beijing Summit decided to set up ten "Luban workshops" in Africa over the next three years to provide vocational skills training and a quota of 50,000 Africans to study in China. The investment in human resources training in Africa is a real solution to the lack of vocational training and employment for Africans, especially young people. In short, it is only through the improvement of the people's livelihoods and the improvement of their living standards that more people can benefit from the fruits of China–Africa cooperation, and the more solid the basis of public opinion for China–Africa common development can be.

The fourth is to work together to build a stronger China–Africa community with a shared future that promotes cultural prosperity. Both China and Africa are the birthplaces of human civilization and have their own splendid civilizations and cultural heritage, which makes the cultural exchange between China and Africa have its own rich heritage and intercultural connotations. To promote the exchange, mutual learning, and integration of Chinese and African civilizations will not only provide lasting assistance to the revival of each other's civilizations, cultural progress, and literary and artistic prosperity, but also make greater contributions to the diversification of world civilization as a whole. In recent years, there have been many new ideas and new phenomenon for cultural and people-to-people exchanges between China and Africa. People-to-people exchanges are not limited to cultural group visits, tourism cooperation, or museum exhibitions, but are aimed at expanding and deepening the exchange of ideas and common research.

China has decided to set up the China Africa Research Institute to build an enhanced version of the existing China–Africa Joint Research Exchange Program. The China Africa Research Institute was officially established on April 9, 2019, and President Xi sent a special congratulatory letter at that time expressing the hope that the China Africa Research Institute would pool the resources of the China–Africa academic think tank, enhance mutual understanding and friendship between the people of China and Africa, pool ideas and suggestions for cooperation between China and Africa and other parties, and contribute to the development of China–Africa relations and the building of a China–Africa community of a shared future.[20] At present, the China Africa Research Institute focuses on 18 research topics about China–Africa cooperation; these 18 projects are already underway, with another 32 cooperation topics to be launched. A

[20] Xinhua News Agency, Beijing, April 9, 2019, "Xi Jinping sent a congratulatory letter to the establishment of the China Africa Research Institute", http://www.xinhuanet.com//mrdx/2019-04/10/c_137964344.htm.

series of other China–Africa academic exchanges and cooperation projects will also be launched at the same time.[21]

The fifth is to work together to build a stronger China–Africa community with a shared future that enjoys common security. With China–Africa economic and trade cooperation and people-to-people exchanges becoming increasingly frequent, peace and security in Africa are inextricably linked to China's investment in Africa and the security of Chinese in Africa. Supporting Africa's peace and security construction and building an "Africa to silence the sound of guns" will not only create a more peaceful environment for China–Africa economic and trade cooperation, but will also be an important part of maintaining world peace. China–Africa security cooperation has also expanded in recent years in several areas and become more involved, from participating in United Nations peacekeeping missions to Africa, to supporting African counterterrorism and piracy, to providing military assistance to the African Union, to training African peacekeepers, and to strengthening security cooperation seminars and exchanges.

From June 26 to July 10, 2018, China's Ministry of Defense held the first China–Africa Defense and Security Forum in China. Representatives of defense and military forces from 50 African countries and the African Union participated in the forum, including twelve chiefs of staff and deputy chiefs of staff from Sierra Leone, South Sudan, and other countries. On the theme of "Working Together and Helping Each Other," the participants deeply discussed the Belt and Road Initiative and the China–Africa Community with a Shared Future, the security situation in Africa, the capacity building of African independent security, and the military relations between China and Africa in the new era. They also visited China's land, sea, and air forces. Over the next three years, China will also set up the China–Africa Peace and Security Cooperation Fund and the China–Africa Forum for Peace and Security, as well as promote the implementation of 50 security assistance projects in the areas of the Belt and Road projects construction, social security, United Nations peacekeeping, anti-piracy, and counterterrorism.

[21] Wu Chuanhua Li Xinfeng, "Strengthen Academic Research and Deepen mutual learning of China-Africa civilization - written on the occasion of the establishment of the China Africa Research Institute", China Africa Research Institute website, http://cai.cssn.cn/xjdt/201905/t20190505_4874179.shtml.

The sixth is to work together to build a stronger China–Africa community with a shared future that promotes harmony between humans and nature. President Xi often said that only clear waters and lush mountains can be labeled "golden mountains and silver rivers." Not only should China's domestic development pay attention to the concept of green, environmental protection, and sustainable development, but it should also uphold the idea of environmental protection, wildlife protection, and "the unity of heaven and humans" in China–Africa cooperation, so as to realize the organic combination of economic development and environmental protection. As the last frontier of human development, the African continent enjoys the advantages of later-comer development, which can completely avoid the path of pollution—development—governance that some former industrialized countries have taken. It can also advance its industrialization process from a relatively advanced starting point and incorporate new ideas about green, low-carbon, circular, sustainable development.

China has many lessons and experiences to share with Africa as it pushes Africa forward on the path of green industrialization. Over the next three years, China will implement 50 green development and ecological and environmental assistance projects in Africa, steadily promote the construction of the China–Africa Environment Cooperation Center, explore projects such as waste power generation and the "toilet revolution" with African countries, and strengthen exchanges and cooperation with Africa in the fields of combating climate change, applying clean energy, preventing desertification and soil erosion, and protecting wildlife and other ecological and environmental protections. In addition, the initiative is also in accordance with the characteristics of Africa's prolific bamboo; in particular, China has indicated it would like to help Africa develop its bamboo and rattan industry, to build a bamboo center in Ethiopia, and become involved in the in-depth development of bamboo for food, furniture, construction, textiles, paper, medicine, and other aspects of comprehensive value. This focus on bamboo will help create more African jobs and promote economic diversification in Africa.

In the face of unprecedented global changes, developmental opportunities, and challenges, China must mobilize its unique national characteristics to continue rising in power and to build a "human community with a shared future." The diplomacy among the great powers with Chinese characteristics is not only reflected in how we deal with our relationships with developed countries such as the US and the associated response to internationally hot issues, but is also shown in our attitude and commitment to Africa in the face of international cooperation and development. The African continent, with the most concentrated and development-oriented developing countries, is the stage that best embodies this global responsibility for building a human community with a shared future.

CHAPTER 18

Inception, Development, and Achievements of the Forum on China–Africa Cooperation

Xiaolei Shen

Between 3 and 4 September 2018, the Forum on China–Africa Cooperation (FOCAC) was held in Beijing. Working together under the theme "China and Africa: Toward an Even Stronger Community with a Shared Future through Win–Win Cooperation," China and Africa unanimously agreed to build a "China–Africa Community with a Shared Future" and advance the Belt and Road Initiative. Thus China–Africa relations opened up a new chapter for future cooperation. This new chapter, inaugurated on the occasion of the 20th anniversary of the founding of FOCAC, sought to review its inception and analyze its developmental history. This not only helped sum up its historical lessons and promote the ongoing stability and development of China–Africa relations, but also helped to

The article was originally published in *Pacific Journal*, Vol. 28, No. 3, 2020, pp. 25–37

X. Shen (✉)
Institute of West-Asian and African Studies, Chinese Academy of Social Science, Beijing, China
e-mail: wxy03@ssap.cn

X. Li and C. An (eds.), *The Changing World and Africa*,
https://doi.org/10.1007/978-981-16-4983-7_18

refute the arguments that FOCAC is exclusively dominated by China and that China's cooperation with Africa is promoting "neocolonialism" by plundering Africa's resources and increasing Africa's debt burden.

Since its inception, the FOCAC has highlighted the importance of promoting Africa's own development and the development of China–Africa relations. Scholars began to systematically study it only after the Beijing Summit in 2006 or, more precisely, after 2008. In that year, Zhang Zhongxiang wrote an article on the positive role played by the FOCAC in Africa's development. South African scholars Garth Shelton and Farhana Paruk published their book, *The Forum on China–Africa Cooperation: A Strategic Opportunity*, which provides a detailed overview of the first three forums and argued that the FOCAC provides a valuable and strategic opportunity for China–Africa cooperation.[1] Thereafter, British scholar Ian Taylor published his book, *The Forum on China–Africa Cooperation*, in 2010, which argues that the FOCAC is driven and dominated by China. While this book has attracted global attention, its argument is biased[2] and has since been refuted by Chinese scholars. In 2012, Li Anshan wrote an article analyzing the origins of the FOCAC, pointing out that it was Africa itself that promoted the establishment of the forum. This article proved valuable in academic circles both in China and abroad.[3] After 2012, research on FOCAC has increased and has emerged as an important theme of China–Africa relations. For example, *The China Monitor*, sponsored by the Center for Chinese Research at Stellenbosch University in South Africa, even dedicated a special edition specifically on the FOCAC in 2015.[4]

Chinese scholars also continue to study the FOCAC from various perspectives. Such scholarship includes Zhou Yuyuan's analysis of the

[1] See Zhang Zhongxiang, "The Role of the Forum on China-Africa Cooperation in Africa's Development," *Exploration and Free Views*, No. 12, 2008, pp. 83–85; Garth Shelton and Farhana Paruk, *The Forum on China-Africa Cooperation: A Strategic Opportunity*, Institute for Security Studies, 2008.

[2] See Ian Taylor, *The Forum on China-Africa Cooperation (FOCAC)*, Routledge, 2011.

[3] Li Anshan, "Origin of the FOCAC: Reflections on China's Africa Strategy," *Foreign Affairs Review*, No. 3, 2012, pp. 15–32.

[4] See Young-Chan Kim, ed.,*China and Africa: A New Paradigm of Global Business*, Palgrave Macmillan, 2017; Arkebe Oqubay and Justin Yifu Lin, eds., *China-Africa and an Economic Transformation*, Oxford University Press, 2019; Centre for Chinese Studies, Stellenbosch, *The China Monitor* (Special Edition: FOCAC VI: African Initiatives towarda Sustainable Chinese Expressionship), 2015.

achievements and challenges of FOCAC and Zhao Chenguang and Zhang Chun's research examining the forum as an example of collective dialogue diplomacy and a supplier of global public goods.[5]

On the basis of these studies, this article relies on data collected through interviews with some of the main officials of the Chinese Ministry of Foreign Affairs (MFA) who have participated in the FOCAC. This is further substantiated by field research undertaken in South Africa, Zimbabwe, Zambia, and Lesotho. It also incorporates the new changes and developments that have emerged since the 2018 Beijing Summit. Taken together, this research aims to offer a more comprehensive analysis of the inception, development stages, characteristics, and contribution of the Forum, as well as briefly discuss its challenges and related policy recommendations.

1 Origins of the FOCAC

The establishment of the FOCAC is the product of historical inevitability. On the one hand, it was the result of the long-term friendship and cooperation between China and Africa; on the other hand, it stemmed from the common needs of both China and Africa. As Tang Jiaxuan, China's former Foreign Minister, said: "In the new situation, China needs more cooperation and support from Africa in order to achieve common development; Africa needs more aid and support from China in order to keep political stability and promote national development."[6] The FOCAC, which was first held at the turn of the century, is inextricably linked to the advancement of African countries insofar as Africans exercised their collective agency and ownership in the process of its inception and institutionalization.

The first proposal to establish a mechanism for multilateral cooperation between China and Africa came from African officials. During a visit to China in September 1997, Albert Tévoédjrè, the Minister

[5] Zhou Yuyuan, "Fifteen Years of Forum on China-Africa Cooperation: Achievements, Challenges and Prospects," *West Asia and Africa*, No. 1, 2016, pp. 4–21; Zhang Chun, "Global Public Goods with Chinese Characteristics: A Case Study of FOCAC," *Foreign Affairs Review*, Vol. 36, No. 3, 2019, pp. 1–28; Zhao Chenguang, "The Chinese Collective Dialogue Diplomacy: The Case of the Forum of China-Africa Cooperation," *International Forum*, No. 5, 2018, pp. 23–28.

[6] Tang Jiaxuan, *Heavy Storm and Gentle Breeze*, World Affairs Press, 2009, p. 433.

of Planning, Economic Rehabilitation and Employment Promotion of Benin, proposed to Chinese Deputy Prime Minister Zhu Rongji that the two sides should establish an "Asian-African Development Forum" composed of scholars, entrepreneurs, and businessmen to conduct dialogue and seminars on cooperation projects aimed at promoting cooperation between China and Africa specifically and between Asia and Africa more broadly. Zhu instructed the MFA and the Ministry of Foreign Trade and Economic Cooperation (MOFTEC) to study the feasibility of such a project. After some discussion, MOFTEC concluded that the forum would involve too many countries and would likely encounter difficulties; it could only have set up an irregular "China-Africa Economic and Trade Council" to exchange information and strengthen cooperation.[7]

A similar proposal was put forward in 1998 by a number of African diplomats attending the "Workshop for African Diplomats on Modern China" sponsored by China Foreign Affairs University, in the hope of establishing a "one-to-multi partnership." During his visit to China in January 1999, Ahmed Haggag, then Assistant Secretary-General of the Organization of African Unity (OAU), met with Vice Premier Li Lanqing and Assistant Foreign Minister Wu Donghe and suggested that China and Africa should establish a multilateral cooperation mechanism. This proposal was not accepted, however, because China believed that multilateral cooperation would not be an easy task.[8] From April 14–28 of the same year, the "Fourth Delegation of Senior Diplomats to China," comprising division levels from the OAU, Mauritius, Egypt, Eritrea, Nigeria, Zambia, and Zimbabwe was held. In discussions with their Chinese counterparts, African officials suggested that China and Africa should establish bilateral (OAU and China) or multilateral (African countries and China) mechanisms for consultation and dialogue. They said that most countries in Africa already had long-term friendships with China and hoped to establish a more regular forum with China to exchange experiences. In garnering Chinese support for a forum, it was explained that even though China-African friendship was much closer than relations

[7] The author's interview with former China's Ambassador to Eritrea and Rwanda, Beijing, April 8, 2019.

[8] Li Anshan, "Origin of the FOCAC: Reflections on China's Africa Strategy," *Foreign Affairs Review*, No. 3, 2012, pp. 23–24.

between Africa and the West, most major Western countries had already established dialogue mechanisms with Africa.[9]

Although ultimately none of the proposals by African officials were adopted by China, they undoubtedly played a subtle role in shaping China's multilateral diplomacy with Africa, and eventually prompted the Chinese Government to give practical consideration to the proposal by Madagascar's Foreign Minister, Lila Ratsifandrihamanana. In May 1999, when she visited China and talked with China's Foreign Minister Tang Jiaxuan, Ratsifandrihamanana pointed out that African countries had already established institutional cooperation platforms with Western countries, such as the France–Africa Summit, the Commonwealth Heads of Government Meeting, and the Tokyo International Conference on African Development. In view of the friendship between China and Africa, African countries were also eager to establish a partnership with China based on common areas of concern related to peace and development. She made it clear that the two sides should establish the China–Africa Forum. Mr. Tang attached great importance to this proposal and asked Ji Peiding, Vice-Minister in charge of African Affairs, and Liu Guijin, Director-General of the Department of African Affairs, to study it immediately. Liu Guijin convened a conference among the officials of the Department of African Affairs to discuss it, but they were divided on this question. Those in agreement thought that the rapid development of China–Africa relations required a multilateral mechanism that could be used to coordinate domestic work on Africa in order to solve bilateral problems such as debt relief for the continent. Those who disagreed were mainly concerned that the forum would become a mere formality without practical results; however, the conclusion was that they should be innovative. This culminated in the 2000 Ministerial Conference of the Forum on China–Africa Cooperation.[10]

In August 1999, MFA and MOFTEC jointly sent a report to the Central Committee of the Communist Party of China that was approved. In the report, they suggested holding the Ministerial Meeting of the Forum on China–Africa Cooperation in Beijing in October 2000. Participants included the ministers for Foreign Affairs and ministers of Trade,

[9] Shu Zhan and Zeng Aiping's interview with Ambassador Liu Guijin, Shanghai, March 30, 2018.

[10] Tang Jiaxuan, *Heavy Storm and Gentle Breeze*, World Affairs Press, 2009, p. 433; interview with Liu Guijin on March 30, 2018.

Industry, and Economic Affairs that were responsible for international cooperation with those African countries with diplomatic relations with China. In October of the same year, Chinese President Jiang Zemin addressed letters to the heads of African countries that had diplomatic relations with China, as well as the Secretary-General of the OAU. In this communication, he extended an invitation to the meeting and also introduced its background, purpose, and themes. In February 2000, Foreign Minister Tang Jiaxuan and Minister of MFTEC Shi Guangsheng also sent a joint letter to their African counterparts, inviting them to attend the conference.

In January 2000, China set up a preparatory committee for the conference, led by MFA and MOFTEC and composed of 20 ministries, including the State Planning Commission, the Ministry of Finance, the International Department of the Central Committee of the CPC, and the Ministry of Education. In the preparatory process, China committed to extensive consultation with and joint contribution from African countries. In this way, African countries were able to demonstrate their agency in the process.

The spirit of extensive consultation and joint contributions with China is mainly reflected in two ways: first, officials of MFA and MOFTEC personally visited African countries to hear their opinions. In January 2000, Tang visited the leaders and foreign ministers of Nigeria, Namibia, Zimbabwe, Mozambique, and Seychelles to explain China's proposed FOCAC, and to invite their contributions as to how to hold the meeting.[11] In June and August, Vice Minister for Foreign Affairs, Ji Peiding, and Vice Minister for FOCAC, Sun Guangxiang, also led delegations to Morocco, Algeria, Egypt, Ethiopia, South Africa, Nigeria, and Côte d'Ivoire to get feedback.

Second, in drafting the documents of the Conference, China advocated for and followed the principles of "democracy and openness, equality and consensus, and highlighting the focus," while also extensively seeking and fully respecting the suggestions of African countries. From April to September 2000, the department of African Affairs of the MFA held at least eight rounds of consultations with African ambassadors to China. The contact group was comprised of eleven African envoys organized in consultation of the outcome documents of the conference, which were

[11] Ma Wuye, "The Fate of Thousands of Miles often Meet——Foreign Minister Tang Jiaxuan Talk about the Trip to Africa," *World Affairs*, No. 4, 2000, pp. 11–13.

then revised three times. Chinese embassies in African countries also generally sought advice from African countries. Lu Shaye, then Director-General of the Department of African Affairs, who participated in the drafting of the documents, said that "We have paid special attention to equal consultation with the African side, so long as the problems were raised by African friends, we would try our best to resolve them."[12] As a result, the final document was nearly three times as long as the original, and many of the concerns and pieces of advice from African countries were included.

The agency and ownership of African countries are mainly reflected in the presentation of advice and the promotion of institutionalization represented in the FOCAC. During the preparations for the conference, hundreds of written comments and verbal suggestions were made by African states and especially South Africa, which prepared two written texts for the outcome documents.[13] The advice given by African countries can be categorized into four aspects: first, that FOCAC should be different from the France–Africa Summit and the Tokyo International Conference on African Development. It should be a meeting of developing countries and should create new paths for global South–South cooperation and produce specific results. Second, the FOCAC should have tangible follow-up actions, and the two sides should establish a mechanism for implementing conference documents and related initiatives. Third, China–Africa cooperation should primarily focus on development, and the FOCAC should formulate certain concrete measures to promote economic and trade cooperation between China and Africa, such as setting up a special fund for Africa's development, offering assistance toward infrastructure construction, and tackling the issue of trade imbalance between China and Africa. Fourth, Africa will attach great importance to China's unique role and contributions to the development of Africa's human resources.[14]

[12] Shu Zhan's interview with Lu Shano, Chinese Ambassador to France and former Director-General of The Department of African Affairs of MFA, Beijing, October 22, 2018.

[13] Interview with Liu Guijin on March 30, 2018; Interview with Shu Zhan on April 8, 2019.

[14] Zeng Aiping and Shu Zhan, "Achievements, and Prospects of the Forum on China-Africa Cooperation," *China International Studies*, September/October 2018, pp. 91–92.

With regard to the institutionalization of FOCAC, China was initially cautious, considering it inappropriate to institutionalize it immediately. China's suggestion was that they should hold a senior officials meeting two years after the first Ministerial Conference to determine whether or not to institutionalize it, depending on the implementation of the outcome of the meetings. However, from the very beginning and during initial preparations for the conference, senior officials from Africa unanimously insisted on institutionalization. Ambassador Liu Guijin later said in an interview: "Their enthusiasm is very high. They think only one meeting is really meaningless. They think since China-Africa relations are very good, why not institutionalized (sic) it from the beginning? We talked about it for several days, including, of course, other issues, but we cannot reach an agreement on institutionalization. Finally, we report to the superior leaders, from vice minister to minister, then to deputy Prime Minister Qian Qichen. Qian made a final decision and decided to institutionalize it immediately: the Ministerial Conference will be held every three years in China and African countries alternately."[15]

Following the joint efforts of China and Africa, the first Ministerial Conference of FOCAC was held in Beijing in October 2000. The leaders of China and presidents of Algeria, Togo, Zambia, and Tanzania, 79 ministers from 44 African countries, representatives of 17 regional and organizations—including OAU, the United Nations Development Programme, and the Economic Commission for Africa—as well as representatives of African business communities all attended the conference. Malawi and Liberia, which had not yet established diplomatic relations with China at that time, also sent observers.[16] The conference adopted the Beijing Declaration of the Forum on China–Africa Cooperation and the Program for China–Africa Cooperation in Economic and Social Development, leading China–Africa relations into a new stage of historical development.

[15] "Speech by Ambassador Liu Guijin at the 8th China-Africa Think Tanks Forum," Fangfei Yuan, Diaoyutai State Guesthouse, Beijing, August 26, 2019.

[16] Tang Jiaxuan, *Heavy Storm and Gentle Breeze*, World Affairs Press, 2009, pp. 434–435.

2 Development of FOCAC

Since 2000, seven Ministerial Conferences have been successfully held, of which the third, sixth, and seventh have been upgraded to summits. After more than 20 years of development, the FOCAC has grown into the most important platform for China–Africa cooperation in politics, the economy, culture, and security. Taking China's financial support to Africa as an example, the 2006 Beijing Summit provided US$3 billion of preferential loans and US$2 billion preferential export buyer's credit to African countries. The 4th and the 5th Ministerial Conference increased these preferential loans to $10 billion and $20 billion, respectively. The Johannesburg Summit in 2015 continued to significantly increase the amount of financial support being offered, up to US$60 billion. This amount was retained at the Beijing Summit in 2018.

While the growing scale of economic cooperation is being deployed in response to the needs of both sides, the areas of cooperation between China and Africa under the framework of FOCAC have also been expanding. As a result, new content has been highlighted or added at different periods. In terms of development, the FOCAC can be divided into three phases, with each phase focused on a different aspect of cooperation. The first phase runs from the 1st Ministerial Conference to the 4th Ministerial Conference in 2009; the focus of cooperation in this phase was on economic areas, which indicated a shift from solely focusing on trade to including investment. Promoting economic and trade cooperation between China and Africa was one of the most important reasons for the establishment of the FOCAC in the first place. MOFTEC and MFA jointly submitted the report for holding the conference and took the lead in setting up a preparatory committee; the ministerial conference itself was attended by ministers for foreign affairs and those responsible for international economic cooperation. This demonstrated the importance of economic cooperation in the framework of the FOCAC. In 1999, the trade volume between China and Africa was only $6.49 billion, of which China exported $4.15 billion to Africa and imported $2.375 billion from Africa.[17] For this reason, increasing the volume of China–Africa trade, and especially enhancing the export capacity of African countries to China, became an important task in the first phase of FOCAC.

[17] China National Bureau of Statistics data, http://data.stats.gov.cn/easyquery.htm?cn=C01.

The Program for China–Africa Cooperation in Economic and Social Development that was adopted in 2000 stated that China would help Africa to diversify the composition of its exports and overcome Africa's export dependence on primary commodities, single products, and raw materials. In order to achieve this goal, China would encourage its enterprises to give preference to the importation of African products in light of market demands and conditions.[18] From the 2nd Ministerial Meeting in 2003 on, China began to grant zero-tariff treatment for some commodities from Africa's least developed countries (LDCs) in order to access the Chinese market. By the eve of the 2006 Beijing Summit, the number of export items to China eligible for zero-tariff treatment from African LDCs reached 190. At the 2006 Beijing summit, China pledged to further open its market to Africa and increased the number of export items to China that were eligible for zero-tariff treatment from 190 to more than 440. It also sought to properly address trade disputes and friction and to gradually improve the mechanisms regulating the China–Africa Joint Chamber of Commerce and Industry.[19] Thanks to these measures, China became Africa's largest trading partner in 2009, with a bilateral trade volume reaching $91.65 billion, of which Africa exported $43.331 billion to China, 18 times as much as in 1999.[20] In the 1st Ministerial Conference, China expressed its willingness to increase investment in Africa, and during the 2nd Ministerial Conference, China held the China–Africa Business Conference to boost investment and cooperation between the two sides. By the time of the 2006 Beijing Summit, China had taken on a series of important initiatives, including the establishment of the China–Africa Joint Chamber of Commerce and Industry and the establishment of the China–Africa Development Fund, which was officially launched in June 2007 and encouraged well-established Chinese companies to set up overseas economic and trade cooperation zones in Africa. This kicked off the establishment of overseas economic and trade cooperation zones and industries zones by Chinese companies in Africa. After the 2006

[18] The Program for China-Africa Cooperation in Economic and Social Development, the website of FOCAC, https://www.focac.org/eng/zywx_1/zywj/t606797.htm.

[19] Forum on China-Africa Cooperation Beijing Action Plan (2007–2009), the website of FOCAC, November 5, 2006, https://www.focac.org/eng/zywx_1/zywj/t280369.htm.

[20] China National Bureau of Statistics data, http://data.stats.gov.cn/easyquery.htm?cn=C01.

Beijing summit, China's net foreign direct investment increased dramatically, from $1.574 billion in 2007 to $5.49 billion in 2008. Although it fell sharply in 2009, it has steadily increased since then, and China's stock of foreign direct investment has begun to increase rapidly.[21] It is from this phase that China–Africa economic cooperation has developed from trade-oriented to focusing on both trade and investment, and this trend has been further consolidated since then.

The second phase is from the 4th Ministerial Conference in 2009 to the Johannesburg Summit in 2015. In this phase, the economic and trade cooperation between China and Africa has been further deepened, with people-to-people exchanges now receiving special attention.

China–Africa economic and trade cooperation increased greatly during this phase. In terms of trade, China further opened its markets to Africa and gradually granted the zero-tariff treatment to 97% of all tariff items from LDCs in Africa that had diplomatic relations with China. At the same time, China implemented the "Special Plan on Trade with Africa" and actively extended aid-for-trade to African countries. As for investment, China set up a special loan of US$1 billion for African SME development, the China–Africa Development Fund was scaled up to $5 billion, and by the end of 2015, China's stock of foreign direct investment had reached $34.69 billion. In the area of agriculture, from 2009 to 2012, China sent 50 agricultural technology teams to Africa, trained 2000 agricultural technicians for African countries, and increased the total number of agricultural technology demonstration centers from ten to twenty. In the area of infrastructure, a joint working group on Africa's transnational and cross-regional infrastructure development was established by China and the African Union (AU). During his visit to Africa in May 2014, Prime Minister Li Keqiang raised the idea of "three major transport networks" for cooperation with African countries through the construction of railways, roads, and regional aviation networks. This included the construction of the Mombasa–Nairobi Railway and the Addis Ababa–Djibouti Railway.[22]

[21] China National Bureau of Statistics data, http://data.stats.gov.cn/easyquery.htm?cn=C01.

[22] Zhou Pingjian, "A Review and Outlook on the Forum on China-Africa Cooperation," in Zhang Hongming, ed., *Annual Report on Development in African (2014–2015)*, Social Sciences Academic Press, 2015, p. 243.

After the 4th Ministerial Conference, people-to-people exchanges began to receive special attention. During this period, China officially launched the China-African joint research and exchange plan and held the China–Africa People's Forum, the FOCAC Think Tank Forum, and the China–Africa Youth Leaders Forum. The 5th Ministerial Conference in 2012 pushed the people-to-people exchange between China and Africa to new heights, a focus that was closely related to China's attempt at reforming the FOCAC mechanism in order to strengthen its soft power in Africa.[23] After the 5th Ministerial Conference, the following achievements related to people-to-people exchanges were recorded: the program of China–Africa mutual visits between cultural personnel, the China–Africa Radio and Television Cooperation Forum, the launch of the China–Africa people-to-people friendship action, the China–Africa People's Forum, the FOCAC Think Tank Forum, and the institutionalization of the China–Africa Youth Leaders Forum. Furthermore, the China-African joint research and exchange plan was initialized, and academic institutions and scholars from the both sides were sponsored in order to carry out academic research and mutual visits.

The third phase is from the 2015 Johannesburg Summit to the present. The economic cooperation between China and Africa has entered a new period of industrial capacity cooperation, industrial partnering, and jointly advancing the Belt and Road Initiative. At the same time, peace and security cooperation has become an important area of cooperation as well as experience sharing on the issue of state governance. Peace and security cooperation was part of FOCAC from its establishment. The 1st Ministerial Conference in 2000 pledged to cooperate to prevent and combat the problem of illicit proliferation, circulation, and the trafficking of small arms and light weapons, while the 2nd Ministerial Conference in 2003 stressed the importance of cooperation on non-traditional security issues and counterterrorism, and the 5th Ministerial Conference in 2012 proposed launching the Initiative on China–Africa Cooperative Partnership for Peace and Security. Before 2015, the overall progress of peace and security cooperation between China and Africa was very slow, and no formal security cooperative partnership was established,[24] however,

[23] Interview with Lu Shano on October 22, 2018.

[24] Zhang Chun, "Global Public Goods with Chinese Characteristics: A Case Study of FOCAC," *Foreign Affairs Review*, Vol. 36, No. 3, 2019, p. 9. For the study on China-Africa security cooperation by 2015, see David Shinn, "FOCAC: The Evolving

this changed at the Johannesburg Summit. In his address at the opening ceremony of the Johannesburg Summit, President Xi Jinping identified mutual assistance in security as one of the "five major pillars" to forging China–Africa comprehensive strategic and cooperative partnership. Peace and security cooperation was therefore included in the ten cooperations plans and China agreed to "provide US$6 million for grants to support the building and operation of the African Standby Force and the African Capacity for the Immediate Response to Crisis. China will continue to participate in UN peacekeeping missions in Africa and support African countries' capacity building in areas such as defense, counterterrorism, riot prevention, customs and immigration control."[25]

The importance of peace and security cooperation was further highlighted at the 2018 Beijing Summit. In his keynote speech, President Xi Jinping said, "let us build a China-Africa community with a shared future that enjoys common security."[26] Peace and security initiatives were included in the "eight initiatives." In light of this, China will take the following measures: setting up the China–Africa Peace and Security Fund; continuing to provide military aid to the AU; supporting countries in the Sahel region and those bordering the Gulf of Aden and the Gulf of Guinea in upholding security and combating terrorism in their regions; establishing the China–Africa Peace and Security Forum and the China–Africa Law Enforcement and Security Forum as platforms in order to conduct more exchanges in peace and security cooperation; and launching fifty security assistance programs to advance China–Africa cooperation under the Belt and Road Initiative and in areas of law and order. Areas of cooperation also included UN peacekeeping missions, piracy, and combating terrorism. These measures were quickly implemented following the 2018 Beijing Summit. In July 2019, the 1st China–Africa Peace and Security Forum was held in Beijing. The China–Africa Peace

China-Africa Security Relationship," *The China Monitor* (Special Edition: FOCAC VI: African Initiatives toward a Sustainable Chinese Relationsship), 2015, pp. 6–13.

[25] Xi Jinping, "Open a New Era of China-Africa Win–Win Cooperation and Common Development," http://www.china.org.cn/chinese/2015-12/07/content_37255657.htm?t=wdrnmh.

[26] Xi Jinping, "Work Together for Common Development and a Shared Future-Full text of Chinese President Xi Jinping's speech at opening ceremony of 2018 FOCAC Beijing Summit," http://www.chinadaily.com.cn/a/201809/04/WS5b8d5c25a310add14f389592.html.

and Security Fund has subsequently been put in place, and the first 200 million RMB of aid will support the development of the African Standby Force and the African Capacity for the Immediate Response to Crisis program.

A highlight of the 2018 Beijing Summit was undoubtedly the shared experience of state governance.[27] The exchange of governance experience has been mentioned many times since the 2006 Beijing Summit, but it has only been a subheading of "High-Level Visits and Dialogue" or "Exchanges between Political Parties," and has lacked systematic measures that rendered it far from capable of becoming the focus of China–Africa cooperation. The 2018 Beijing Summit promoted the importance of experience sharing on state governance in China–Africa cooperation. According to the Forum on China–Africa Cooperation Beijing Action Plan (2019–2021), the capacity-building initiative was included in the eight major initiatives, the experience sharing on state governance was upgraded to the level of "political cooperation," and its aim was identified as supporting "African countries in independently exploring development paths suited to their national conditions." In order to achieve this aim, China made some innovations at the institutional level, including organizing the China–Africa State Government Forum, promoting high-level exchanges among political parties, and increasing the frequency of high-level exchanges among political parties. It also included raising the level of cooperation it invested in training cadres and stepping up anti-corruption cooperation by holding a training session every year from 2019 to 2021 in order to train 100 African anti-corruption officials.[28]

3 Characteristics of FOCAC

In the 20 years it has been developing, the FOCAC has formed a set of unique characteristics, which are embodied in the principle of "extensive consultation, joint contributions, and shared benefits." This is based on

[27] On the historical changes of experience sharing on state governance between China and Africa, see Luo Jianbo, "Exchanges of Governance Experiences Between China and Developing Countries: History, Theory and World Significance," *West Asia and Africa*, No. 4, 2019, pp. 3–23.

[28] Forum on China-Africa Cooperation Beijing Action Plan (2019–2021), the website of FOCAC, September 5, 2018, https://www.focac.org/eng/zywx_1/zywj/t1594297.htm.

the philosophy of including the development needs of African countries and a cooperative attitude with solid implementation. These three characteristics are not only important symbols of how FOCAC is different from other international multilateral cooperation mechanisms with Africa, but also act as an important guarantee that the FOCAC can continue to develop and achieve further results.

"Extensive Consultation, Joint Contributions, and Shared Benefits"

Since its establishment, the nature of the roles played by China and Africa in the FOCAC has been the focus of numerous academic debates. While China has always insisted on the equality of both sides and identified the FOCAC as an important platform for collective dialogue between China and Africa, some Western scholars have questioned this assertion. For example, Ian Taylor argued that "China is very much in control of the whole process and it is Beijing that sets the agenda and the declarations and outcomes. Africa plays a bit-part at best."[29]

Indeed, as Lu Shaye has pointed out, China has objectively played a leading and coordinating role in the FOCAC, but "China has never thought of controlling it. No one can control it, we only accept the right suggestions to make decisions."[30] Therefore, contrary to Ian Taylor's argument that China is in control of the FOCAC, China has always adhered to and firmly implemented the principle of "extensive consultation, joint contributions, and shared benefits," which is the most important reason for its steady development.

As mentioned above, in the process of the FOCAC's establishment and institutionalization, China has committed to the spirit of extensive consultation and joint contribution and has also given full respect to Africa's agency and ownership. The spirit of extensive consultation and joint contribution has long been a factor for China's dealings with Africa, especially when deciding which cooperative projects would be chosen in each Ministerial Conference. Before each Ministerial Conference, China would, on the basis of full respect for Africa, ask African countries to

[29] Ian Taylor, *The Forum on China-Africa Cooperation (FOCAC)*, Routledge, 2011, p. 110; Ian Taylor, "The Institutional Framework of Sino-Africa Relations," in Arkebe Oqubay and Justin Yifu Lin, eds. *China-Africa and Economic Transformation*, Oxford University Press, 2019, p. 118.

[30] Interview with Lu Shaye on October 22, 2018.

submit two or three alternative proposals. Thereafter, the outcome documents would be finalized only after discussing those proposals between the two sides.[31]

The spirit of shared benefits, however, is more important than the spirit of extensive consultation and joint contribution. China's cooperation with Africa has never been exclusively for its own private interests, or simply to obtain Africa's resources and markets to promote its own development. China hopes to achieve mutual benefits and common development with African countries through cooperation under the framework of the FOCAC. After President Xi Jinping put forward the principles of sincerity, real results, affinity, good faith, and upholding justice while pursuing shared interests, China has further pursued common interests and put friendship first in its pursuit of China–Africa cooperation. In this way it "follows the principle of giving more and taking less, giving before taking, and giving without asking for return" in its cooperation with Africa.[32] It can be said that China has elevated the spirit of shared benefits to a new height.

It should also be pointed out that, since the founding of the People's Republic of China, many of China's most important diplomatic thoughts and concepts were first directed at Africa. After developing these diplomatic thoughts on Africa, they became the guiding principles for China's relations with developing countries and even became the guiding principles for China's overall diplomacy. The same is true of the principle of "extensive consultation, joint contributions, and shared benefits," which has gone beyond the scope of China–Africa cooperation and become an important guiding principle for China's multilateral diplomacy, especially for jointly promoting the Belt and Road Initiative.[33] Wang Yi, State Councilor and Foreign Minister of China, even pointed out that the main reason for the joint promotion of the Belt and Road Initiative is

[31] Li Anshan and Liu Haifang, "On the Operational Mechanism of the Forum on China-Africa Cooperation and Its Relationship with African Integration," *Teaching and Research*, No. 6, 2012, p. 60.

[32] Xi Jinping, "Work Together for Common Development and a Shared Future-Full text of Chinese President Xi Jinping's Speech at Opening Ceremony of 2018 FOCAC Beijing Summit," http://www.chinadaily.com.cn/a/201809/04/WS5b8d5c25a310add14f389592.html.

[33] Zhang Chun, "On Theory and Practice of China's Diplomatic Relations with Developing Countries in the New Era," *Pacific Journal*, No. 7, 2018, p. 11.

based on following the sound principle of "extensive consultation, joint contributions, and shared benefits."[34]

Meeting the Needs of Africa

The mechanism of the FOCAC remains in a state of development, but fitting itself to African needs is the concept that China always followed. In an interview in 2010, Lin Songtian, China's former Ambassador to South Africa, argued that when carrying out cooperation with Africa, "the first thing for China to think is Africa's needs, the second is whether China can provide assistance. If they don't need your help, or if you cannot meet their needs, it would be meaningless."[35] China is the world's largest developing country, and Africa is the continent with the largest number of developing countries. Compared with Western countries, China is better positioned to understand the needs of Africa, and China will not "attach strings" to its aid or interfere in their internal affairs when responding to their needs.[36]

At the turn of the century, African countries were deep in an economic recession, which largely manifested in the decline of foreign direct investment and increased debt. In this context, in addition to promoting economic and trade cooperation between China and Africa, China also made debt relief and cancellation a top priority at the 1st Ministerial Conference and promised to reduce or cancel debts amounting to 10 billion RMB yuan owed by the most heavily indebted and least developed African countries in the coming year. Of course, there was also a deeper consideration for China to reduce or cancel the debts owed by African

[34] "Wang Yi Talks About Belt and Road Initiative: It Is a Sunshine Initiative, and the Principle of Eextensive Consultation, Joint Contribution and Shared Benefits' Is a Golden Principle," https://www.focac.org/chn/zywx/zywj/t584788.html.

[35] "Lin Songtian: The Establish of the FOCAC to Strengthen China-Africa Cooperation in the New Century," http://www.china.com.cn/news/2010-02/12/content_19417254.htm.

[36] During a lecture on March 29, 2012, Johnny Moloto, the Deputy Chief of Mission at the South African Embassy in the United States, compared the difference between the FOCAC with the African Growth and Opportunity Act, noting that non-conditionality is one of the biggest differences between them. "Lecture on the Strategic Implications of China's Growing Trading Partnership with Africa, by Mr. J Moloto, Deputy Chief of Mission at the South African Embassy in Washington DC, March 29, 2012," http://saembassy.ogt11.com/?p=1280.

countries. Ambassador Liu Guijin argued in an interview that, "The debt issue was very serious for African countries at that time, and they have been calling Western countries for debt relief. As a developing country, China is also very poor. But if China can take the lead in debt relief for African countries, it will be a great support for them to pull their debt burdens down, and it will also drive Western countries to reduce their debt."[37] It was for this reason that China has reduced or canceled debt for some African countries in each ministerial conference.

After entering the second decade of the twenty-first century, Africa has changed from "a hopeless continent" to an "Africa rising" on the Economist's cover. The economic growth rates of Ethiopia, Kenya, Rwanda, and other countries rank among the highest in the world, the level of regional integration is continuously improving, and the continent's industrialization, agro-modernization, and urbanization are constantly moving forward. However, Africa still faces three major developmental bottlenecks: poor infrastructure, a shortage of professional personnel and funding, and a lack of the capabilities needed for self-driven development. Solving these problems is the biggest demand of African countries. For this purpose, since the Johannesburg Summit, China has focused its cooperation with Africa on pursuing complementarity with the AU Agenda 2063, the UN 2030 Agenda for Sustainable Development, and the development programs of African countries. It has promoted the "ten cooperation plans" and "eight major initiatives" and placed great emphasis on the joint promotion of the Belt and Road Initiative with Africa. Through those measures, China hopes to help African countries accelerate their industrialization and agro-modernization, enhance their capabilities for self-driven development, and achieve independent and sustainable development and economic independence.

Besides its economic development, Africa is currently facing a series of problems in the areas of state governance, peace and security, and the livelihood of its people. It therefore has many requirements for outside stakeholders. In order to meet these demands, China has continuously strengthened its cooperation with Africa in select fields since 2012, and finally proposed working together with Africa to build a China–Africa community with a shared future that assumes joint responsibility, win–win cooperation, happiness for all, cultural prosperity, common security,

[37] Interview with Liu Guijin on March 30, 2018.

and harmony between humanity and nature. It can be said that meeting African needs and promoting African development have been internalized as the spirit of China–Africa cooperation.

Cooperation in Implementation

Due to a lack of restraint, the implementation of the outcomes of international multilateral cooperation mechanisms with Africa is always far from satisfactory. For example, at the 2nd Turkey–Africa Summit in 2014, Nkosazana Dlamini Zuma, the former Chairperson of the AU Commission, reminded Turkey that many of the commitments of the first summit still had yet to be implemented.[38] After its establishment, the FOCAC has always prioritized follow-up actions and achieved tangible achievements. For example, the 10 billion yuan debt relief promised by the first forum was completed by June 2002, with a total of 156 cases of debt relief paid out to 31 African countries and amounting to 10.5 billion yuan. In November 2009, then Premier Wen Jiabao put forward eight new measures to strengthen China–Africa cooperation. The progress of the follow-up actions was reported to him twice in May and November from 2010 to 2012, which effectively promoted the implementation of these measures.[39] At the 5th Ministerial Conference in 2012, China promised to provide a US$20 billion credit line to Africa. It was delivered when Premier Li Keqiang visited Ethiopia and Nigeria in May 2014, with China subsequently adding another US$10 billion credit line and raising $2 billion for the China–Africa Development Fund. At the 2015 Johannesburg Summit, China pledged to provide US$60 billion of funding support to Africa, about 92 percent of which has either been delivered or been arranged to be delivered by March 2018.[40] The reasons that China can effectively implement the outcomes of the FOCAC are not only closely related to the Chinese virtues of "giving more and taking less,

[38] Sinem Cengiz, "Will Erdogan's 'Walk with Africa' Policy Prove Successful?" *Arab News*, March 2, 2018, http://www.arabnews.com/node/125799.

[39] "An Interview with Shu Zhan, the Former Chinese Ambassador to Eritrea," *African Studies*, Vol. 1, 2010, p. 370.

[40] Dai Bing, "The New Circumstances of Africa, the New Development of China-Africa Relations and the New Prospect of China-Africa Cooperation," in Zhang Hongming, ed., Annual Report on Development in African (2017–2018), Social Sciences Academic Press, 2018, p. 153.

giving before taking, and giving without asking for return," but more importantly, provides an institutional guarantee and adopts the policy "based on bilateral and implemented country-by-country."

As for the mechanisms themselves, the Chinese Follow-Up Committee of FOCAC—comprising the MFA, Ministry of Commerce, Ministry of Finance, and other ministries—was established in November 2000 and was responsible for coordinating the implementation of the outcomes of the FOCAC. Secondly, the FOCAC follow-up mechanisms were established and built at three levels: the Ministerial Conference, the Senior Officials Follow-up Meeting and the Senior Officials Preparatory Meeting for the Ministerial Conference, and the consultations between the African Diplomatic Corps in China and the Secretariat of the Chinese Follow-up Committee. Finally, various sub-forums were established, including the China–Africa People's Forum, the China–Africa Young Leaders Forum, the Forum on China–Africa Media Cooperation, the China–Africa Think Tanks Forum, and the China–Africa Peace and Security Forum. The establishment of these mechanisms has effectively ensured the implementation of the outcomes of the FOCAC.

In terms of concrete measures, due to the large differences among African countries in terms of their individual social and economic development levels, resources endowment, infrastructure situation, and degree of economic openness, China has adopted policies "based on bilateral and country-by-country implementation." With the exception of exclusive projects such as grants and interest-free loans, most of the projects are implemented bilaterally. The 20 years of the development of FOCAC shows that bilateral implementation has its unique advantages: firstly, different projects can be implemented according to the needs of different countries; secondly, the speed of implementation can be accelerated and the effectiveness of implementation can be easily tracked. Finally, the interests of different countries, especially those of small and medium-sized countries, can be taken into account. Furthermore, after the Johannesburg Summit, China has adopted a policy of "from point to area, integrating point and sphere, and steadily advancing" its China–Africa industrial capacity cooperation and industrial partnering. Based on a comprehensive consideration of the advantages of location, resources, political situation, and friendship with China, it has selected

Ethiopia, Tanzania, Kenya, and the Republic of Congo to be pilot countries.[41] This policy has achieved remarkable results; for example, the Mombasa–Nairobi Railway has become a new benchmark for China–Africa cooperation.

4 Achievements of FOCAC

The FOCAC has had a fruitful 20 years. Driven by the principle of "extensive consultation, joint contributions, and shared benefits," as well as an attitude of solid implementation, it has enriched both the practice and theory of major country diplomacy with Chinese characteristics. It has therefore promoted the rapid and comprehensive development of China–Africa relations, and more importantly, improved the self-development capacity of African countries through the promotion of their economic development. Moreover, it has exemplified the cooperation between international communities and Africa.

Enriching Major Country Diplomacy

On October 18, 2017, President Xi Jinping noted in the report at the 19th National Congress of the Communist Party of China that China "[has] made all-round efforts in the pursuit of major country diplomacy with Chinese characteristics, thus advancing China's diplomatic agenda in a comprehensive, multilevel, multifaceted way and creating a favorable external environment for China's development." He also made clear that "major country diplomacy with Chinese characteristics aims to foster a new type of international relations and build a community with a shared future for mankind."[42] The issue of major country diplomacy with Chinese characteristics was put forward by President Xi at the central conference on work relating to foreign affairs on November 28, 2014. We are witnessing major changes unfolding in our world, unlike anything

[41] Lin Songtian, "Firmly Promote Implementation of FOCAC Summit Outcomes to Comprehensively Carry Forward China-Africa Win–Win Cooperation and Common Development", in Zhang Hongming, ed., *Annual Report on Development in African (2015–2016)*, Social Sciences Academic Press, 2016, p. 232.

[42] Xi Jinping, *Secure a Decisive Victory in Building a Moderately Prosperous Society in All Respects and Strive for the Great Success of Socialism with Chinese Characteristics for a New Era*, Delivered at the 19th National Congress of the Communist Party of China October 18, 2017.

seen in a century. In this context, major country diplomacy with Chinese characteristics will provide important theoretical guidance and strategic direction for China's diplomatic practices.

The concept of major country diplomacy with Chinese characteristics is a new strategy that has great potential for future growth. Developing China–Africa relations is an integral part of this diplomacy and the FOCAC has made important contributions to enriching its practice and theory, as well as acting as an important pillar of China–Africa relations.

The FOCAC is the earliest and most successful collective cooperation mechanism between China and developing countries. The China–Africa comprehensive strategic and cooperative partnership that it has promoted provides an example of the establishment of a new type of international relations fostered by the major country diplomacy with Chinese characteristics. More importantly, in his keynote speech at the opening ceremony of 2008 Beijing Summit on September 3, 2018, President Xi put forward the "five-no" approach to China's relations with Africa; that is, no interference in African countries' pursuit of development paths that fit their national conditions; no interference in African countries' internal affairs; no imposition of the Chinese will on African countries; no attachment of political strings to assistance to Africa; and no seeking of selfish political gains in investment and financing cooperation with Africa.[43] The "five-no" approach is not only a contribution to the establishment of a new form of international relations predicated on mutual respect, fairness, justice, and win–win cooperation, but its innovation and the further development of its practical applications can be used as a benchmark for measuring the relationship between China and other developing countries and regions.

In addition to the establishment of a new type of international relations, the policy framework of major country diplomacy with Chinese characteristics also includes building a community with a shared future for humankind and promoting the Belt and Road Initiative. These are the areas that the forum is currently working hard to put forward as themes of the current China–Africa cooperation situation. Given that jointly advancing the Belt and Road Initiative will be discussed below,

[43] Xi Jinping, "Work Together for Common Development and a Shared Future-Full text of Chinese President Xi Jinping's Speech at Opening Ceremony of 2018 FOCAC Beijing Summit," http://www.chinadaily.com.cn/a/201809/04/WS5b8d5c25a310add14f389592.html.

this paper briefly analyzes the China–Africa community with a shared future here. As early as his March 2013 visit to Tanzania, President Xi Jinping argued that "China and Africa have always been a community of shared destinies, and similar historical experiences, common development tasks, as well as shared strategic interests have bound the two sides together."[44] This means that building a China–Africa community with a shared future is one method of leading the way toward the construction of a community with a shared future. Under the theme of "China and Africa: Toward an Even Stronger Community with a Shared Future Through Win–Win Cooperation," the 2018 Beijing Summit put forward six pillars of building a China–Africa community with a shared future from the institutional level. Guaranteed by the cooperative attitude of solid implementation, China–Africa cooperation will become the first model for building a community with a shared future.

The Comprehensive Development of China–Africa Relations

The FOCAC has promoted the comprehensive and rapid development of China–Africa relations. As pointed out in *China's Second Africa Policy Paper*, "The forum has served as a platform for all-round pragmatic cooperation, pushing for leapfrog increase in China-Africa trade and mutual investment and promoting mutual benefit and common development. The forum has been a bridge for closer people-to-people exchanges and friendship between China and Africa, promoting bilateral exchanges in all areas, and consolidating and expanding the social and popular support for the friendship between China and African countries. It has helped enhance communication and collaboration between China and African countries in the international arena, facilitating them to work together in safeguarding the overall interests of the two sides and developing countries."[45]

Driven by the FOCAC mechanisms, China–Africa relations have achieved a historic height, from being a new type of partnership in 2000, to becoming a more strategic partnership in 2006, then on to a comprehensive strategic and cooperative partnership in 2015, and finally

[44] "Xi speaks on Africa Policy, Pledges Assistance," http://en.ce.cn/National/Government/201303/26/t20130326_1053274.shtml.

[45] *China's Second Africa Policy Paper*, http://www.china.org.cn/chinese/2015-12/07/content_37256882.htm.

to becoming a China–Africa community with a shared future in 2018. The relations between China and African countries have also improved. Up to now, China has established comprehensive strategic and cooperative partnerships with Mozambique, the Republic of Congo, Sierra Leone, Senegal, Namibia, and Zimbabwe, comprehensive strategic partnerships with Algeria, Egypt, and South Africa, and strategic partnerships with Angola, Morocco, and Nigeria. Driven by FOCAC mechanisms, the members of the China–Africa family have also been expanded. When the FOCAC was established in 2000, there were still eight African countries which had not established or resumed diplomatic relations with China. With the resumption of diplomatic relations with Burkina Faso in May 2018, only Swaziland has held out on establishing diplomatic relations with China.

As an important part of China–Africa relations, summit diplomacy promotes top-level design and the strategic planning of China–Africa relations, while also laying a solid foundation for further China–Africa relations.[46] With the comprehensive and rapid development of China–Africa relations, high-level exchange visits have been frequent, especially after the 18th National Congress of the Communist Party of China.[47] President Xi Jinping chose Africa for his first international visit after he was elected and re-elected in both 2013 and 2018. He has, in total, visited the African continent four times, visiting eight African countries including South Africa, Tanzania, and Zimbabwe. Meanwhile, nearly 100 African heads of state or government officials have visited China. The FOCAC has also become an important stage for summit diplomacy. At the 2018 Beijing Summit, President Xi Jinping held talks and meetings with all African leaders present. At the time of the most intensive activities, he held meetings with 11 African presidents within the space of ten hours.

As the most important economic cooperation initiative proposed by China to the international community, the Belt and Road Initiative has been fruitful ever since it was introduced in 2013. Africa, being part of the historical and natural extension of the Belt and Road, has been an important participant in this initiative. The 2018 Beijing summit opened

[46] Li Anshan, "China-Africa Relations in the Discourse on China's Rise," *World Economics and Politics*, No. 11, 2006, pp. 7–14.

[47] Zhang Ying, *China-Africa Relations from the Perspective of Summit Diplomacy*, Current Affairs Press, 2017, pp. 50–57.

a new era of China–Africa cooperation under the Belt and Road Initiative. As of November 2019, 44 African countries and the AU commission have signed Belt and Road cooperation agreements with China, accounting for one-third of the total number of countries that have signed BRI cooperation agreements with China. Africa has become one of the most active and determined regions to participate in the joint advancement of the Belt and Road Initiative.

Improving the Capacity for Self-Development of African Countries

The lack of self-development capacity has been a serious problem for African countries since their independence, causing them not only to rely excessively on international aid for a long time, but also forcing them to adhere to the political conditions of Western countries in order to gain support for their economic development plans, such as the New Partnership for Africa's Development.[48] In the twenty-first century, Africa became the second fastest-growing region of the world after Asia. Rapid economic development has helped African countries to enhance their self-development capacity but poor infrastructure and a shortage of both personnel and funding still restrict their development. In the case of infrastructure, according to the *African Economic Outlook 2018* released by the African Development Bank, it is estimated that African investment needs are \$130–\$170 billion a year. However, in 2016 commitments from all sources were only \$62.5 billion, which means that there is a financing gap in the range of \$67.6–107.5 billion.[49] China's support is therefore essential for African countries to improve their capacity for self-development.

Under the framework of the FOCAC, China has been enhancing African countries' self-development capacity by promoting their economic development and human resources training. Economically, China has become an important driver of Africa's development. This was seen in 2009, after China became Africa's largest trading partner and bilateral trade volume continued to grow steadily, reaching \$221.67 billion in 2014: a record high. Although it declined from 2014 because of negative

[48] Ian Taylor, "NEPAD: Towards Africa's Development or Another False Start?" *African Affairs*, Vol. 107, No. 426, 2008, pp. 126–128.

[49] *African Development Bank, African Economic Outlook 2018*, p. 82.

international environment, it reached $204.1 billion in 2018 and $208.7 billion in 2019, an increase of 19.7% and 2.2%, respectively. Secondly, China has become one of the most important sources of investment for Africa. China's net foreign direct investment increased from $200 million in 2000 to $5.3 billion in 2018, and its stock of foreign direct investment increased from less than $500 million to $46.1 billion over the same period.[50] Finally, China has become the largest contributor to infrastructure construction in Africa. According to Deloitte's *Africa Construction Trends Report 2018*, infrastructure financing provided by China to Africa averaged $11.5 billion between 2012 and 2016. In 2018, 18.9% of Africa's projects were funded by China and 33.2% were built by China.[51] By 2019, China financed and built over 10,000 km of roads, 6,000 kms of railways, 150 plus stadiums, conference centers, and parliament buildings, 200 schools, 80 power plants, 80% of communications infrastructure, and many airports and ports.[52] Altogether, this contributed 2% of the GDP growth achieved by African countries.[53]

In the field of human resources, since the establishment of the FOCAC, China has trained a large number of Africans through the African Human Resources Development Fund, government scholarships, training and capacity-building seminars for African youth and officials, Luban Workshops, and tailor-made programs. The number of human resources training scholarships and government scholarships promised by China to African countries increased from 7000 at the 1st Conference to 50,000 at the 2018 Beijing Summit, and the number of the latter has increased from 6,000 at the 2nd Conference to 50,000 at the 2018 Beijing Summit. Up to this point, the combined total number of scholarships are 172,000 and 132,500, respectively (see Table 1). After finishing their studies or training, former scholarship holders will undoubtedly go on to greatly improve their countries' self-development capacity.

For most African countries, in addition to accelerating their economic development and strengthening their human resources training, they also

[50] China National Bureau of Statistics data, http://data.stats.gov.cn/easyquery.htm? cn = C01.

[51] Deloitte, *Africa Construction Trends Report 2018*, pp. 12–52.

[52] Lin Songtian, "Tampering with the Concept of Colonialism Is Ill-intentioned," *Pretoria News*, September 12, 2019.

[53] Ma Xinmin, "Join Hands to Make FOCAC Thrive," *Sudan Vision*, September 18, 2019.

Table 1 Number of human resources training and government scholarships promised by China to African countries

Year	*Conference*	*Human resources training*	*Government scholarships*
2000	First	7,000	–
2003	Second	10,000	6,000
2006	Third	15,000	12,000
2009	Fourth	20,000	16,500
2012	Fifth	30,000	18,000
2015	Sixth	40,000	30,000
2018	Seventh	50,000	50,000
Total		172,000	132,500

Source Data was collected by the author according to the website of the FOCAC (https://www.focac.org/chn/)

need to improve their state governance capacity. This will be improved by peaceful domestic and regional environments and increased regional integration in order to truly improve their capacity for self-development. In this regard, peace and security cooperation, experience sharing on state governance, and transnational and cross-regional infrastructure construction are also directly or indirectly responsible for improving the self-development capacity of African countries.

Improved Cooperation Between International Communities and African Countries

Since the colonial era, the unfair and unreasonable international political and economic order has placed Africa in a difficult position in terms of international cooperation, especially with regard to Western countries. Furthermore, the end of the Cold War left many African countries as "cold war orphans." After its establishment, the Forum actively promoted the principles of equality, win–win progress, and common development, implemented a model of all-round cooperation in the political, economic, social, and cultural fields led by economic cooperation, and achieved tangible achievements in its promotion of African economic development and enhancing African countries' agency and ownership. These principles, cooperation models, and achievements have had a positive impact on international communities' cooperation with Africa and have even improved the cooperation between international communities and African countries to some extent.

As a benchmark for South–South cooperation and for cross-regional cooperation, the Forum has provided a set of cooperation mechanisms for other countries to learn from and has highlighted what's possible with global summit diplomacy with Africa. Having been influenced by the Forum, emerging countries such as South Korea, Brazil, India, Turkey, Iran, Indonesia, and Russia have all held similar forums or summits. For example, the 2nd India–Africa Summit, which was held in 2011, has great similarities with the FOCAC. It defined "Enhancing Partnership: Shared Vision" as its theme and discussed economic and trade cooperation, scientific and technological cooperation, energy cooperation, international affairs cooperation, and marine cooperation as its cooperation areas. It also aimed to establish a development-oriented cooperation mechanism.[54] More importantly, the principles of noninterference in internal affairs, non-political conditionalities, and development priorities that China upheld in its cooperation with Africa are also being accepted by these countries.[55] Western countries have also begun to transform and innovate their own cooperation mechanisms with Africa. For example, at the request of African countries, the Tokyo International Conference on African Development, which was established in 1993 and held every five years in Japan, was forced to change to an every-three-years schedule that would be held in Japan and Africa alternatingly. Furthermore, much like FOCAC, the first US–Africa Leaders Summit, the largest ever held by the US with African countries, wasn't even held until 2014. The European Union-African Union summit, which emerged after some initial development difficulties, is on track after the third summit in 2010.

It should also be noted that, since China is at the forefront of international cooperation with Africa, it has played the role of "game changer" on the continent. These innovations in cooperation are of great importance to Africa and the international community. On the one hand, China–Africa cooperation has promoted the rapid socio-economic development of Africa and improved African countries' agency and ability

[54] Ma Libing, Guo Dongyan, "India-Africa Forum Summit and the Impact of India-Africa Cooperation Mechanism," *International Forum*, Vol. 20, No. 5, 2018, pp. 15–22.

[55] Office of the Special Adviser on Africa, "Africa's Cooperation with New and Emerging Development Partners: Options for Africa's Development," United Nations, 2010, pp. 37–28.

to independently choose their partners. On the other hand, China–Africa cooperation has improved Africa's overall international status, forced the outside world—especially the Western countries—to reconsider Africa, and therefore greatly increased African countries' weight in the international game and maximized their benefits from it.

5 Conclusion

At the second session of the 13th National People's Congress, Wang Yi, State Councilor and Foreign Minister said that, "after decades of diligent nurturing, the sapling of China-Africa cooperation has grown into a towering tree that no force can topple."[56] Through this process, the FOCAC undoubtedly played a vital role because in the 20 years since its establishment, it has come to represent the growth of China–Africa cooperation. In these 20 years, by promoting China–Africa relations and accelerating the socio-economic development of Africa, the Forum has gradually matured and even entered its best years.

Nevertheless, there are still some challenges it needs to overcome. First, the Forum is facing some problems in institutional innovation, regulation of the scale of cooperation, the expansion of cooperation activities, the implementation of cooperation outcomes, and the evaluation of cooperative effectiveness. Second, as the international community, especially Western countries, continue to strengthen their game with China in Africa, the Forum will not only face further competition from them but also face pressure from the international norms and standards that they dominate and control. Finally, as African countries are increasing their consciousness of home-grown solutions and self-development, improving their ability to independently choose partners, and accelerating the process of regional integration, their ability to develop common positions on specific topics has also improved. This may shrink the space for China–Africa cooperation and reduce the autonomy and flexibility of China's policy.[57]

[56] "State Councilor and Foreign Minister Wang Yi Meets the Press," *Xinhuanet*, March 8, 2019, http://www.xinhuanet.com/politics/2019lh/zb/wzjzh61366/wzsl.htm, visit time: November 13, 2019.

[57] Zhou Yuyuan, "Fifteen Years of Forum on China-Africa Cooperation: Achievements, Challenges and Prospects," *West Asia and Africa*, No. 1, 2016, p. 14; Zhang Chun, "On African Efforts and Achievements in Developing Diplomatic Autonomy," in Zhang

In order to better overcome these challenges, realize the ongoing sustainable development of the FOCAC and continuous promotion of friendship between China and Africa, China can take measures at both the strategic and policy levels. At the strategic level, China should make at least three interventions. First, China should further improve Africa's position in its major country diplomacy with Chinese characteristics, comprehensively consolidate the foundation of the new type of China-African international relations, and maximize the convergence of the common interests that exist between China and Africa. Second, China should effectively manage the role of major powers in China–Africa relations and the role of Africa in its relationship between the major powers, as well as strive to continue to take the lead in international cooperation with Africa. Third, China should further innovate the FOCAC's working mechanisms and build it into an international organization co-dominated by China and Africa, thus forging a new path ahead for China to be able to participate in global governance.

At the policy level, there are four suggested measures for addressing this. First, China should further strengthen its supervision of feasibility studies, follow-up activities, and evaluations of the Forum's projects. For some projects that cannot be completed within three years, their time limits should be appropriately relaxed with effectiveness, rather than speed, as the sole criterion for evaluation. Second, China should continue to strengthen its cooperation on peace and security with African countries to help them build a more secure environment for social and economic development, and at the same time, maximize the interests and security of China and Chinese people in Africa. Third, China should innovate its people-to-people exchanges with Africa. For example, it can make full use of the overseas Chinese already in Africa and their organizations, as well as organize various kinds of people-to-people exchanges together with the local communities.[58] Fourth, China should rationalize and strengthen the publicity works regarding the Forum and China–Africa friendly relations, telling good stories about Chinese development and China–Africa friendly

Hongming, ed., *Annual Report on Development in Africa (2017–2018)*, Social Sciences Academic Press, 2018, p. 104.

[58] The "Dream Show" organized by overseas Chinese Zhao Ke in Zimbabwe is a good example. Since its establishment in 2014, it has become an important platform for people-to-people exchanges between China and Tianjin. See Shen Xiaolei, *Friendly Cooperation Between China and Zimbabwe*, China Social Sciences Press, 2019, pp.103–106.

cooperation so as to minimize misunderstandings and resistance from both home and abroad. In conclusion, by strengthening joint efforts, relying on successful experiences, and seeking greater innovation, the FOCAC will have broader and brighter prospects going forward into the future.

CHAPTER 19

China–Africa Joint Industrial Parks: History, Challenges, and Solutions

Hongyi Wang

With their vigorous economic cooperation in recent years, the China–Africa cooperation model has recently undergone profound changes. Production capacity is one of the key areas where this change has occurred. In order to promote cooperation in this area, China and Africa have jointly planned, built, and operated a number of industrial parks, aiming to effectively utilize both sides' comparative advantages. This is also in line with the spirit of China's Belt and Road Initiative (BRI), the strategic goals of the African Union's Agenda 2063, and the common interests of both China and Africa.

H. Wang (✉)
Institute of West-Asian and African Studies, Chinese Academy of Social Science, Beijing, China
e-mail: wxy03@ssap.cn

X. Li and C. An (eds.), *The Changing World and Africa*,
https://doi.org/10.1007/978-981-16-4983-7_19

1 The Development History of China–Africa Joint Industrial Parks

China–Africa joint industrial parks emerged around 2000. After the 2006 Beijing summit of the Forum on China–Africa Cooperation (FOCAC), large-scale enterprises began to join in on the construction. Recently, since the 2018 FOCAC Beijing Summit, we have seen the introduction of a new period characterized by the Special Economic Zone (SEZ).

Initial Period

The China–Africa joint industrial park is a new cooperation model originally created by trading companies. With the implementation of the country's "Go Global" strategy around 2000, many Chinese firms flooded into Africa. Because of fierce competition, they witnessed falling profits and were forced to change their investment strategies. Many of them have since chosen to build factories locally, mainly for market-based considerations. While many Chinese manufacturers face difficulties back home due to Western quota restrictions, particularly in electronic products, textiles, and the light industry, Africa enjoys quota-free access and preferential tariff treatment in EU and US markets. This has attracted more Chinese companies to operate in Africa. For example, after trading with West African countries for several years, in 2002 Henan Guoji Construction Group decided to build an industrial park directly in Sierra Leone, transforming an abandoned railway station into a productive assembly plant. Plenty of Chinese firms have been introduced since then, producing construction materials, electrical components, plastic products, coatings, and more. This represents a new trend of Chinese investment in Africa for that period, although the park itself didn't survive the Sierra Leone civil war.

Another type of park has been formed on the basis of trade promotion centers. In the 1980s and 90 s, China's Ministry of Commerce (MOFCOM) established trade promotion centers in eleven African countries, relying on provincial commerce departments that gradually became the bridge for enterprises in these provinces to invest in Africa. Firms from the same provinces tended to go out together and build factories near each other, making an industrial park the ideal choice. This kind of park or processing zone has strong "characteristics of the province of origin," and can be seen in Nigeria, Uganda, Cameroon, Tanzania, and Ethiopia.

The third kind of park is dominated by large Chinese enterprises. These enterprises came to Africa in search of resources during the early years of Chinese expansion into Africa. Later, upstream and downstream firms came, too, especially as resource exploitation required a significant degree of support that was unavailable in Africa at that time. Industrial parks were thus formed spontaneously. In 1998, the China Nonferrous Metal Mining Group (CNMC) obtained land development rights on the surface of the former Chambishi copper mine in Zambia. In 2003, CNMC started to establish China's first overseas nonferrous metal industrial park there. Textile parks also appeared early on. In 2004, the Yuemei Group of Zhuji, Zhejiang built a textile industrial park in Nigeria, followed by the Esquel Group of Guangdong, which set up another one in Mauritius in 2005.

Some other parks have been started by Chinese enterprises that were specifically invited to Africa by African governments themselves. Since the late 1990s, many African leaders have proposed building their own "Shenzhens" or "Suzhous," ideas that were welcomed by the Chinese government. In 1994, then Egyptian President Hosni Mubarak visited the Tianjin Economic-Technological Development Area (TEDA) and proposed the construction of industrial parks jointly with TEDA. In 2003, Taida Group from TEDA purchased one square kilometer of land in Egypt to independently build the Suez Industrial Park.

Institutionalization Period

The 2006 Beijing summit of FOCAC led China–Africa economic and trade cooperation into a new stage, focused on the goal of "promoting Africa's industrial development and enhancing Africa's production and export capacity."[1] The Beijing Action Plan adopted at the summit declared that "China is ready to encourage, in the next three years, well-established Chinese companies to set up three to five overseas economic and trade cooperation zones in African countries where conditions permit"[2] the expansion of future Chinese investment in Africa.

[1] "Forum on China-Africa Cooperation Beijing Action Plan (2007-2009)," Forum on China–Africa Cooperation, https://www.focac.org/eng/zywx_1/zywj/t280369.htm.

[2] Ibid.

During the summit, China and Africa decided to facilitate the negotiation, conclusion, and implementation of the Agreement on Bilateral Facilitation and Protection of Investment and the Agreement on Avoidance of Double Taxation to create an enabling environment for investment cooperation and to protect the lawful rights and interests of investors from both sides, thus providing policy guarantees for China–Africa joint industrial parks. Meanwhile, the two sides pledged whatever facilitation was necessary in terms of administrative approval procedures, customs clearance, and border entry and exit for company investors on both sides. In addition, the China–Africa Development Fund was founded to encourage and support Chinese firms to invest in Africa, helping Africa achieve sustainable economic and social development.

Driven by FOCAC policies, seven existing industrial parks were selected to be part of MOFCOM's Overseas Economic and Trade Cooperation Zones (OETCZs). Some of the parks even become key industrial projects in the host countries. The Eastern Industrial Zone (EIZ) in Ethiopia, established by a private enterprise named Jiangsu Yongyuan Investment, was approved as MOFCOM's OETCZ in November 2007. In April 2015, it entered the list of OETCZs released by MOFCOM and the Ministry of Finance. The Ethiopian government also regarded EIZ as an important part of its Sustainable Development and Poverty Reduction Program, and as a priority of its national industrial plan.

After the 2006 FOCAC Beijing summit, building industrial parks in Africa became increasingly popular among Chinese investors. In early 2009, the Tiantang Industrial Park was registered in Uganda by Tian Tang Group, with over ten resident firms. At the end of that year, the Xinguang Industrial Group from Shandong built a textile park in South Africa and set up six factories that produced 30% of all blankets in Southern African countries. In 2010, PetroChina began construction of the Sunshine International Industrial Park in Chad, attracting more than ten upstream and downstream firms. Two years later, the Beira Economic and Trade Cooperation Zone (ETCZ) emerged in Beira, Mozambique, which was owned by the Anhui Foreign Economic Construction Group, while Jingrui Garments from Shaoxing, Zhejiang started the International Business Center in Togo. In 2013, China Merchants Holdings International and Qingdao Ruichang Cotton Industrial Co. followed closely, planning to build the Bagamoyo Special Economic Zone in Tanzania and the Agricultural Product Processing Park in Zambia, respectively. The agreement to build the largest steel city in Africa was signed in early 2014 between the Limpopo government in South Africa and more than ten Chinese firms, including the HBIS Group.[3] In the same year, a metallurgical SEZ and an industrial park for platinum, gold, lithium, and cesium were constructed by Qingdao Hengshun Zhongsheng and other companies in Zimbabwe. The following year saw the construction of three parks, namely, Huajian Group's International Light Industry City in Ethiopia's Addis Ababa, China Harbour Engineering Company (CHEC)'s high-tech industrial park in the economic capital of Côte d'Ivoire, and CGCOC's integrated industrial park in Senegal. More recently, the China Road and Bridge Corporation (CRBC) and BAIC Group also announced their own plans for development, with the former building a center of logistics, manufacturing, aviation, and capacity building in Port of Pointe-Noire, Congo,[4] and the latter a park in Kuha, South Africa with an investment of US$800 million.

SEZ Construction Period

The 2018 Beijing summit of FOCAC announced eight major collaborative initiatives with Africa for industrial promotion. Per these agreements, "China will step up support in the development of industries in Africa including processing and manufacturing and the development of special economic zones and industrial parks, and support Chinese private enterprises in setting up industrial parks in Africa and carrying out technology transfer, to help African countries build more diversified economies and stronger capabilities for self-driven development." Additionally, "China will encourage Chinese companies to increase investment in Africa, build and upgrade a number of economic and trade cooperation zones in Africa ... encourage Chinese companies to make at least US$10 billion of investment in Africa in the next three years."[5]

[3] "A Steel City Built in South Africa by Chinese," December 7, 2014, http://finance.ifeng.com/a/20141207/13335071_0.shtml.

[4] "Wang Yi Meets with Foreign Minister of the Republic of the Congo," Ministry of Foreign Affairs of PRC, January 11, 2017, https://www.mfa.gov.cn/web/zyxw/t1429524.shtml.

[5] "Elaboration on the Eight Major Initiatives of the FOCAC Beijing Summit," Ministry of Commerce of PRC, September 19, 2018, http://english.mofcom.gov.cn/article/policyrelease/Cocoon/201809/20180902788698.shtml.

In order to promote China–Africa production capacity cooperation, the summit also announced that the BRI will be closely in line with the UN 2030 Agenda for Sustainable Development, the African Union's Agenda 2063, and each individual country's own national development strategy. During the summit, 28 African countries signed memoranda of understanding (MOUs) with China on the BRI, rendering a total of 37 African signatories.

At the same time, facing the multipart problems of insufficient infrastructure, trade facilities, and public services, China and Africa decided to implement an infrastructure connectivity initiative, a trade facilitation initiative, and a capacity-building initiative. With regard to the financial difficulties of African countries, the Chinese government promised that it would be "willing to expand the use of local currencies in investment, funding and trade. China will give better scope to the China–Africa Development Fund, the China–Africa Fund for Industrial Cooperation and the Special Loan for the Development of African SMEs."[6] In his keynote speech, Chinese President Xi Jinping committed a further US$60 billion financial support to Africa.

Given all of these favorable conditions, the enthusiasm of Chinese enterprises has further increased. A series of manufacturing parks, logistics parks, port economic zones, and high-tech parks have been launched, which have not only enriched the diversity of entities, but also increased the scale and level of industrial parks themselves. For example, the 2013 Port of Pointe-Noire plan has since been upgraded into a national SEZ between China and the Congo. In July 2018, an industrial park constructed by the China National Building Materials Group (CNBM) opened up in Zambia. Four months later, CRRC established a joint R&D center in South Africa to promote the development of its rail transit equipment manufacturing base there. In late 2018, the Jimma Industrial Park in Ethiopia, built by China Communications Construction Company (CCCC), was completed, attracting textile, garment, and other enterprises and industries to settle in.

Meanwhile, the industrial parks built by private Chinese enterprises have further diversified into agriculture, tourism, and healthcare. In January 2018, based on the import quarantine agreement on cassava

[6] "Elaboration on the Eight Major Initiatives of the FOCAC Beijing Summit."

between China's General Administration of Quality Supervision, Inspection, and Quarantine and the Tanzanian government, a Chinese private company, TAEPZ, began the construction of a cassava processing zone in Tanzania. The planned investment is US$1 billion for more than ten productive processing plants.

2 The Main Types of China–Africa Joint Industrial Parks

Currently, there are more than 30 China–Africa joint industrial parks in operation and nearly 70 under construction. Due to the different industrial plans of each country, the functions and positioning of the parks and the incentives they receive vary widely. We can classify them according to the following five dimensions.

Investors

Most parks are designed, constructed, and operated by China–Africa joint investment firms. Though Chinese investors play a leading role and most enterprises in the parks are Chinese, their African counterparts are also important.

The industrial parks themselves can be divided into three types in terms of investors. The first type is those mainly invested in by local Chinese governments, central state-owned enterprises (SOEs), and local SOEs, such as the Suez ETCZ in Egypt and the Chambishi ETCZ in Zambia. The second type is led by private enterprises, like the Djibouti SEZ and the Huajian International Light Industry City. The third type of park, mainly invested in by commercial associations, can be seen in Cameroon and Uganda. Some Chinese firms own more than one park in Africa. For instance, the ETCZ in Zambia consists of two parts, the Chambishi copper mine and the industrial park in Lusaka, both of which are constructed and run by CNMC. Xinguang Industrial Group operates textile industrial parks in South Africa and Sudan. Companies like China State Construction Engineering Group (CSCEC), Hisense Group, and CNBM also have plans in several African countries. CCCC's planned industrial parks along roads, railways, and ports cover five African countries, while China Merchants Port's (CMPort) port economic zones involve six countries.

Ownership

Ownership structures vary across different parks, including some wholly owned by either the Chinese or the African side, but most of them are co-owned by a China–Africa joint venture. For those jointly owned, African governments usually share in the project by providing land and capital. A few joint parks are financed through a public–private partnership (PPP) model with local governments, while the Chinese partner does the planning, construction, investment promotion, and sometimes operation and management. China–Africa joint management committees have also been established in some parks, however. Parks wholly owned by China include TEDA's Suez Industrial Park and the Touchroad Group's park in Djibouti. In Suez, TEDA has land property and operational and management rights, while in Djibouti, Touchroad even possesses some administrative powers like the power of taxation and security. For parks wholly owned by the African side, some are also jointly planned, built, and promoted by both sides, like those in Senegal, Côte d'Ivoire, and Algeria.

Industrial Policies

Judging from policy applicability, China–Africa industrial parks fall into five types. First is the export-free zone. In these parks, firms enjoy preferential policies related to taxes, fees, and loans. Raw materials and semi-finished products can be imported duty-free, while local materials can be used with subsidies and concessions. The final products can be exported freely, but are restricted or completely prohibited from being sold locally. Most joint parks belong to this category, such as the EIZ in Ethiopia, TEDA's Suez Industrial Park in Egypt, and the integrated industrial park in Senegal. The second type is the free port or the free trade zone, such as the SEZs in Djibouti, Bagamoyo Port, and Pointe-Noire. In these zones, all manufacturers, processors, warehousing, trade, logistics, hotel, and other service providers can enjoy preferential policies. If the products are sold locally, import duties on the imported materials must be paid. Third is the free transit zone, such as the port industrial parks in Angola, Mozambique, and Tanzania. Products transiting from these zones to other countries are exempted from customs duties. Fourth is the economic and trade cooperation zone (ETCZ), such as the Zambia-China ETCZ, the Nigeria-Guangdong ETCZ, and Mauritius's

Jinfei ETCZ. Fifth is the high-tech industrial zone, which is mostly seen in South Africa. For this type of park, the necessary new technologies are imported from China, so as to raise the local industrial innovation capabilities and production levels. A large proportion of these parks are co-designed by Chinese and African governments and do not accept any backward production capacity.

Scale

China–Africa joint industrial parks can be found in most African countries that have diplomatic relations with China. In some countries, we can even find more than one, such as in South Africa, Egypt, Ethiopia, Nigeria, Algeria, Tanzania, Kenya, and Uganda. In terms of scale, there are three types of parks. First, the parks are listed in MOFCOM's OETCZs. These parks are usually large in terms of assets, number of resident firms, and economic profits, such as the EIZ in Ethiopia, the Chambishi ETCZ in Zambia, and the Suez ETCZ in Egypt. As landmark projects for China–Africa production capacity cooperation, they have played an important role in developing local industries, earning foreign exchange, increasing tax income, boosting employment, and improving the local fiscal situation. Second are the medium-sized parks. These parks are usually dominated by a single firm, such as PetroChina's industrial park in Chad and Haishan Group's park in Angola. By improving self-sufficiency and providing the import substitution of certain products, these parks also help promote related industries' development in the host country. Third, the small ones that are mainly built by private firms and local Chinese business associations, tend to promote economic diversity and development. The Tiantang Industrial Park in Uganda is an example of this third type.

Industrial Functions

Among China–Africa joint industrial parks, there are industrial parks, agro-processing zones, logistics parks, port economic zones, commercial zones, tourism parks, and other professional parks. Currently, the parks built by Chinese enterprises in Africa are mainly industrial parks, a category that can be further divided into raw material processing parks, export processing parks with supplied materials, and final product processing parks. Raw material processing parks use local raw materials like energy, minerals, and timber for rough processing before selling to international markets, as we can see in the Chambishi ETCZ. Deep processing is rare in this kind of park. For the second type, which is export-oriented, both the source of raw materials and the market itself are outside Africa, with the Huajian Light Industry City being a typical case. Most privately invested parks belong to the third type: the final product processing parks. They sell only locally, such as we see with the Tiantang Industrial Park and the Xinguang Textile Park in South Africa.

3 Opportunities for China–Africa Joint Industrial Parks

China–Africa joint industrial parks conform to the logic of historical and economic development, the interests of China and Africa, and the trend of China–Africa economic cooperation. Now, Africa enjoys the competitive advantages of developing industrial parks, which is in line with China's BRI and the African Union's Agenda 2063.

Conforming to the Trend of International Economic Cooperation

For a long time, Africa's role in the global economic chain has been that of a mere resource provider, not that of a major market. With the development of science and technology and the improvement of the global trading system, the world is more closely connected and integrated than ever. During the 40-year period of its reform and opening-up, China's economy has developed rapidly, contributing to the collective rise of developing countries worldwide. The relationship between Africa's and China's economic development has been continuously strengthened. Since 2000, Africa's economic growth curve has been roughly consistent with China's own development trend.[7]

With increasing international cooperation and close bilateral relations, the labor division between international actors has also become more sophisticated. The traditional role of Africa as a raw material provider no longer fits the needs of global production. Major manufacturers in the world, including China, are going through economic structural transformations, which may require Africa to begin providing primary industrial

[7] "Martin Davis: Close Connection between Chinese and African Economy," June 27, 2010, http://finance.sina.com.cn/hy/20100627/23208188441.shtml.

products. China–Africa joint industrial parks are in line with the logic of the historical development of international economic cooperation.

As for the law of global economic development, mainstream economists such as Raymond Vernon, W. Arthur Lewis, and Justin Lin believe that industrial transfer dominated by "economic rationality" plays an important role in any future development. Since the emergence of globalization, there have been four rounds of industrial transfer at the global scale, and this historic trend is still underway. As accelerated industrial upgrading has been taking place in China, labor-intensive industries will gradually transfer from China to other BRI countries, including Africa. Among these, Africa has its own unique advantages. Therefore, in the first two lists of MOFCOM's OETCZs, projects in Africa account for seven out of the nineteen. Moreover, CNBM's building material park in Zambia and China Overseas Ports Holding Company (COPHC)'s Djibouti Free Trade Zone are also expected to be included in this list.

The Trend of Cooperation is Conducive to the Construction of Industrial Parks

Since the BRI was proposed, the consensus of various departments in the Chinese government is for strengthening investment in Africa. China has been Africa's largest trading partner and third largest investor for many years.[8] In 2017, China's FDI stock in Africa exceeded US$100 billion, which is nearly 50 times the amount it was in 2010.[9] As more enterprises enter industrial parks, an industrial agglomeration effect has begun to emerge. The transformation of the China–Africa infrastructure cooperation model from EPC (engineering, procurement, and construction)[10] to

[8] "Structural Change of FDI Inflow in Africa," Ministry of Commerce of PRC, January 11, 2017, http://www.mofcom.gov.cn/article/i/jyjl/k/201701/20170102499833.shtml.

[9] "MOFCOM: China's Investment Stock in Africa Exceeds USD100 Billion," August 29, 2018, http://finance.sina.com.cn/china/gncj/2018-08-29/doc-ihikcahe9915490.shtml.

[10] Under the EPC contracting arrangement, the contractor performs the whole or part of the process from design, procurement, and construction to commissioning of a project. Conditioned by an overall budget, it is usually made responsible for the project's quality, security, expenditure and schedule.

BOOT (build, own, operate, and transfer)[11] has greatly improved Africa's infrastructure and the environment for further park development. Up until now, Chinese firms have built 5,756 km of railways, 4,335 km of roads, 9 ports, 14 airports, and 34 power plants, and more than 10 large hydropower stations and thousands of small ones for Africa. Private capital has also flooded into Africa since the implementation of the BRI under the guidance and promotion of sovereign funds like the China–Africa Development Fund, the China–Africa Fund for Industrial Cooperation, and the Silk Road Fund. Commercial financing has also become more popular among Chinese firms, which used to rely on sovereign guarantee financing.

Model of Industrial Park Needed in African Market

Considering the mixed risks and opportunities in Africa, relying on the collective advantages of industrial parks seems the way to invest in Africa, in order to deal with the challenges posed by an unfavorable market environment there. Most Chinese manufacturers choose this option. In Egypt, nearly 80% of Chinese enterprises are located in several joint industrial parks. Moreover, since the industrial production chain in Africa is not complete, manufacturers' production needs might only be met if upstream and downstream industries are highly spatially concentrated. With a closely connected industrial value chain, maximum production capacity can be achieved and promoted in order to achieve industrial polymerization. This has been verified as an effective model of industrialization in Europe, North America, and East Asia. The absorption effect is another advantage of industrial parks that helps expand the market, improve labor productivity, reduce industrial costs, and make preferential policies available. It's especially important in Africa, where the market is fragmented, infrastructure is relatively backward, and the costs of logistics, electricity, and administration are high.

[11] Under a BOOT model, infrastructure projects are built with financial support from private companies or international consortiums. They own and run the projects for several years according to the contract, and transfer the ownership to the government upon the contract's expiration.

Favorable Factors for the Construction of Industrial Parks in Africa

Africa is suitable for developing four kinds of industrial parks. First among these are raw material-oriented industrial parks, since Africa is rich in natural resources. Second, labor-intensive industrial parks, since Africa has the youngest population in the world. By 2050, 35% of the world's young people under 25 will be in Africa.[12] Third, market-oriented industrial parks for textiles, food, electronics, and other consumer goods, as Africa, with a population of 1.1 billion, is the world's largest market still to be developed. Fourth, export-oriented industrial parks, due to the preferable policies Africa enjoys among developed and emerging economies, including tariff exemption and deductibles.

Africa already has experience in building industrial parks. During World War I, Western colonists established processing industries in North, East, and Southern Africa. Although these industries didn't develop into modern industrial parks or economic development zones, they did leave Africa with an industrial base. In the 1970s, when African countries formulated their policies to achieve economic independence and industrialization, Egypt, Liberia, Senegal, and Mauritius, among others, built a series of industrial parks. By the 1990s, export-oriented industrial parks with preferential policies for taxation and labor had been established in more than 20 countries, including Kenya, Egypt, Mauritius, Nigeria, South Africa, and Zambia. In 2015, the number of industrial parks in Africa exceeded 200.

At present, it is the consensus of most African countries that they want to develop special economic zones. Driven by the African Union, the African Development Bank, and the United Nations Economic Commission for Africa, African countries have successively formulated implementation plans for the 2030 Agenda for Sustainable Development, Agenda 2063, the New Partnership for Africa's Development (NEPAD), and their respective national industrial plans. Some plans have already been completed, such as those of the four development zones in Côte d'Ivoire, the four SEZs in the Republic of the Congo, the Tangier SEZ in Morocco, and the Nkok SEZ in Gabon. Legislative work is also underway

[12] "Des Emplois pour les Jeunes en Afrique," Groupe de la Banque Africaine de Développement, 2018, https://www.afdb.org/fileadmin/uploads/afdb/Documents/Generic-Documents/Brochure_Job_Africa_Fr.pdf.

in some countries. For example, followed by Tanzania's promulgation of the Special Economic Zone Act in 2006, the South African Department of Trade and Industry formulated its own SEZ regulation in 2014, and Egypt enacted an SEZ investment law specifically for the Suez Canal region based on its 2002 and 2013 SEZ laws. Kenya passed its Special Economic Zone Act in 2015 and Zimbabwe in 2016. In 2017, the Zimbabwean government declared that the Indigenization and Economic Empowerment Act does not apply to investors in SEZs, which can have 100% share of total investment.

4 Challenges to China–Africa Joint Industrial Parks

From the African side, four challenges are obvious for China–Africa joint industrial parks—namely, security, financing, the investment environment, and economic policy. For investors and operators, homogeneous competition, lack of coordination, and service insufficiency stand out as the primary issues. For firms in the parks, the main issues comprise internationalization, localization, and legal compliance.

Problems of African Countries

Investment security is a common problem in Africa. It is the most important factor in determining an industrial park's development, as the parks usually attract strategic, long-term, and asset-intensive investment, which is vulnerable to major risks. In Africa, the forms of security risks most typically include regime change, tribal conflicts, military conflicts, and terrorist attacks.

The financial environment is also a constraint. Industrial parks are immovable and capital-intensive projects that face a high risk of asset devaluation and disruption in the capital chain. Due to Africa's small and fragmented market, weak and brittle fiscal structure, unstable exchange rates, and fragile financial order, financing costs in the African local market are high, with interest rates in some countries exceeding 15%, even 25% in Malawi.[13] The low credit rating of African countries is another problem

[13] "High Interest Rates Ruin Malawi's 2015," Ministry of Commerce of PRC, December 30, 2015, http://www.mofcom.gov.cn/article/i/dxfw/gzzd/201512/20151201222695.shtml.

because it limits investors from using African assets as mortgages in international financing. In addition, during periods of financial turmoil, some African countries adopt foreign exchange control measures that can cause disruption to many firms' capital flow.

Third, the investment environment in Africa needs improvement. The development of industrial parks requires friendly social public services, as well as favorable soft and hard investment environments. However, in both its soft and hard aspects, Africa is relatively weak. This is the most urgent problem for foreign investors. For a soft environment, law and standard adaptability, trade facilitation, social security, tariff reductions, labor regulation, and human resource quality are the most critical items. For a hard environment, the main elements are the available supplies of electricity, transport facilities, construction materials, social services, and medical care.

Last but not least, policies in Africa are not always stable. Political stability and long-term planning are lacking in many African countries. When faced with conflicts between long-term and short-term interests, or between general and specific interests, plenty of African governments have proven to be short-sighted. They change policies frequently, abandon agreements signed with foreign firms, and sometimes even blackmail them. Moreover, as the political and economic lifeblood of many African countries is still controlled by the West, their autonomy is insufficient for formulating the domestic and foreign policies needed to stabilize their policy environments. With the implementation of the BRI and China–Africa production capacity cooperation, Western dominance in Africa is challenged. To avoid this, the West has used various obstructionist and destructive measures to constrain the equitable and win–win cooperation between China and Africa, hindering the development of China–Africa joint industrial parks.

Problems of Park Investors and Operators

Homogeneous competition is the biggest problem for park investors and operators. Most China–Africa industrial parks are built by enterprises spontaneously. These firms often rush into the same industry due to insufficient communication and coordination, resulting in cutthroat competition. Moreover, the parks are mostly manufacturing ones with the same development model. This problem can be witnessed at both the country and provincial levels.

A failure of coordination in terms of planning, construction, and operations ranks second in terms of problems. In the absence of a full understanding of the whole production chain and the difficulties of coordination, park investors usually face risks after the construction and investment promotion phase. What's more, the design of the factory, living, and commercial areas in the park often disappoint enterprises, since the constructors lack professional knowledge of the manufacturing industry. In addition, most parks are invested in by a single enterprise, which places more emphasis on construction than operation. Due to a lack of professional experts in the early stages of planning and development, the enterprise may find itself in a predicament later.

A clear understanding of the local business environment also matters. Park investors tend to focus only on individual factors within the production chain, ignoring other key elements such as safety, the market environment, finance, commerce, and logistics. There are actual cases of huge losses that were caused by an incomplete study of local social security, judicial fairness, financial freedom, investment and commercial convenience, logistical costs, and labor productivity, among other factors.

Fourth, management and service capabilities. Most of the operators of China–Africa joint parks are contractors with little park management experience. When running the parks, these operators directly pull from their experiences back home, drawing on everything from design drawings to management regulations. Some even still regard themselves as administrators rather than servers. However, in Africa, most park management committees cannot enjoy the administrative powers bestowed by local governments.

Problems Faced by Enterprises in the Parks

Internationalization. The history of Chinese companies in Africa is not particularly long. Many firms have little experience in dealing with problems caused by the differences in business environments between China and Africa. Additionally, most Chinese firms that came to Africa in the earliest wave of industrialization to engage in trade, energy, and mineral development, and project contracting, did so with inadequate experience in manufacturing. Even though "single window" or "one-stop service" measures can be helpful with many administrative issues, many problems specific to Africa remain difficult to settle.

Localization. Chinese firms face the dual challenge of management localization and employee localization. Management localization is far from being achieved, as Chinese SOEs tend to adopt centralized management structures while private companies are run by families. Breaking promises and sacrificing principles for profits have even been witnessed in a small number of Chinese enterprises. With regard to employing local labor, although it is required by African governments and people and is, in fact, written into many countries' labor laws in the form of a minimum percentage of local employees, some Chinese firms still ignore it. The reputation of China–Africa joint industrial parks is affected by cases in which Chinese firms violate host countries' visa regulations by using Chinese labor illegally.

Legal compliance. Due to some African countries' low governance capacity and the Chinese government's weak supervision, some Chinese firms in Africa disrespect the law. A few companies only seek short-term benefits without caring about brand reputation. They get involved in problems like evading taxes, forging contracts, using poor-quality materials, and producing counterfeit and inferior commodities, among other legal issues.

5 Solutions to the Challenges Facing China–Africa Joint Industrial Parks

Nowadays, China–Africa cooperation on industrial parks has entered a period of rapid progress. On one hand, this has provided unprecedented opportunities for economic development in both China and Africa. On the other hand, it also brings challenges, such as risks to the security of investment and homogeneous competition. In this regard, the Chinese government and the firms concerned should plan proactively and respond appropriately, in order to push industrial park cooperation to a higher level.

For the Chinese government itself, there are four pieces of advice:

First, strengthening the planning of and guidance for further industrial park development. At present, the parks fight each other in terms of industrial positioning and investment promotion. To avoid such competition, related government agencies should formulate an overall plan for overseas industrial parks that makes it more relevant to the strategic needs of China–Africa cooperation and the real conditions of each African country. Key projects should be identified and arranged in advance, according to the basic development logic of industrial parks and SEZs. More policy support is also necessary.

Second, the government should focus on harmonizing the management of parks. Currently, government departments at all levels seem to have some sort of say in managing overseas industrial parks. The government could upgrade its production capacity cooperation to the national strategic level and establish unified coordination processes for all relevant departments' efforts.

Third, better coordination with African countries is a must. The ideas, measures, and mechanisms of government administration in Africa are underdeveloped. Improving Africa's administrative governance is beyond individual investors' capacity. Therefore, the Chinese government should take the initiative to help African countries rationalize their development ideas and improve their management and services.

The fourth suggestion involves helping firms with financial difficulties. Grants should be allocated carefully, focusing on the planning of industrial parks, training of local employees, and improvement of the overall investment environment. The two preferential facilities from the Export-Import Bank of China—namely the Chinese Government Concessional Loan and the Preferential Export Buyer's Credit—need to be reasonably used to support key projects. Financial innovation is worth trying, so new models could be explored, such as offshore guarantees for onshore loans. Banks, insurance companies, and other domestic financial institutions ought to be prompted to enter overseas industrial parks.

For park operators and enterprises in the parks, we also have four pieces of advice:

The first is to choose industrial positioning rationally. The choice of industries should be decided on the basis of local resources and market demand, so that resources, capital, technology, and labor can all be integrated to improve productivity and the level of economic development in the local community. Park operators must be careful in choosing partners in the planning stage, constructors in the construction phase, and enterprises in the operation stage.

Second, highlighting capacity building. To give enterprises in the parks a better business environment requires park operators to improve their management and services. Resident firms should also enrich their knowledge of the local environment so that, together with park operators, risks and problems encountered locally can be resolved. To maintain the upward trend of industrial development, technological upgrades as well as skill training and retraining for workers are vital.

Third, synergizing with local governments' efforts. As the bridges between local governments and resident enterprises, park operators can inform enterprises of local policies, while at the same time reporting enterprises' needs to local governments. Only by knowing local industrial policies can the firms comply with local laws and meet local needs.

Fourth and finally is to uphold justice while pursuing shared interests. The balance between individual interests and other people's interests and between short-term and long-term interests should be kept. Relations with host governments, local people, business partners, and employees should also be handled appropriately.

CHAPTER 20

China and Africa: Toward a Stronger Community with a Shared Future Through a Shared Knowledge System

Hongwu Liu

In recent years, China has proposed a global community with a shared future and a new model of international relations. At the opening ceremony of the 2018 Beijing Summit of the Forum of China–Africa Cooperation (FOCAC), Chinese President Xi Jinping proposed in his keynote speech to work together for the common development and a shared future with a stronger China–Africa community that assumes joint

This article is supported by the National Social Science Fund of China (15ZDA066) and Provincial Humanities and Social Sciences Research Project for Universities in Zhejiang Province, China (2013GH00).

The original version of this chapter was revised: Affiliation of the chapter author has been updated. The correction to this chapter is available at https://doi.org/10.1007/978-981-16-4983-7_23

H. Liu (✉)
Institute of African Studies, Zhejiang Normal University, Jinhua, China
e-mail: wxy03@ssap.cn

X. Li and C. An (eds.), *The Changing World and Africa*,
https://doi.org/10.1007/978-981-16-4983-7_20

responsibility, pursues win–win cooperation, delivers happiness for all of us, enjoys cultural prosperity, enjoys common security, and promotes harmony between humanity and nature. In this context, it is necessary to build a shared knowledge system to support, explain, and serve this stronger China–Africa community with a shared future and a new model of international relations.

1 Creating Shared Knowledge: Inspiring Shared Action and Building up Shared Interests

A shared knowledge system develops from its own sources. It cannot be replaced by the knowledge of any single region, any single nation, or any single culture. Since the end of middle ages, the West has been developing first and then expanding their own knowledge all over the world. Among the types of Western knowledge, some can be considered knowledge creations of universal value, especially in the field of natural science and technology. However, much of Western knowledge is regional or native knowledge that developed from experiences and environments specific to the West, especially in the fields of human and social sciences. It's worth noting that in the past decades, lots of Western regional knowledge is regarded as universal knowledge and has been promoted as such to the rest of the world, resulting in many contradictions and conflicts for humans' mutual relations and knowledge creation.

At present, it is only through the construction of international and regional studies around the globe on the basis of developing and carrying forward the traditional knowledge and modern wisdom of various countries, regions, and nations; by linking, fusing, and comprehensively innovating humans' knowledge, thoughts, and cultures of the world through equal dialogue and research cooperation, that we can create a truly universal shared knowledge and culture that surpasses the limits of the regional and overcome all the challenges that we face today. In this sense, the concept of knowledge production has a special theoretical and practical significance for building a China–Africa community of development, with shared interests and a shared future. By boosting the disciplinary construction of African Studies in China and Chinese Studies in Africa, we can create, accumulate, and spread a shared knowledge system for Africa and China.

The Chinese emphasize the significance of a "right time," "right place," and "right people." This traditional thought has special meaning for the building of the new international regional studies and a China–Africa shared knowledge system. The "right time" refers to the universal spirit of civilization and principle of rules, the shared laws of nature and society of humankind (which the ancient Chinese called "the way of Heaven *(tian dao)*," "the rule of law (*fa ze*)," or "the principle of Heaven (*tian li*)"). The "right place" means that the diversity of geographic space and the eco-environment of human existence are the causes of various characteristics and traits belonging to different regions, nations, and cultures, which are themselves the foundation for the development of international and regional studies. And the "right people" refers to how people connect the knowledge systems and wisdom from various countries and regions through their own endeavors in mutual communications, equal dialogue, and learning from each other. Like mountain ranges interconnecting or rivers flowing into the sea, the "right persons" can make the river of regional knowledge and culture flow into the sea of human knowledge, thereby forming a world of universal harmony that integrates "right time" with the "right place" and "right people." Only in this way can the universal value of humankind promote the building of a harmonious world where everyone can talk to each other, understand each other, and share with each other.

The construction of African Studies, a special discipline within area studies, should assume this kind of responsibility. Its innovative development plays a vital role in the creation of mutual China–Africa shared knowledge, in improving the China–Africa academic system, and in promoting the construction of a China–Africa community with a shared future.[1] We say that academics are both the producers of and products of their own times. By tracing history, we can see how both Africa and China have been through twists and turns in their own processes of national rejuvenation. In order to rejuvenate themselves, they need to inherit and carry forward their cultural and spiritual traditions by self-esteem, confidence, and self-reliance. At the same time, they should open their mind to embrace cultures from different countries and nations, and promote different civilizations to talk and cooperate in a more balanced, diverse, and equal way. It is in this way that new international relations and a China–Africa community with a shared future can be built.

On a global scale, when regional studies—such as African Studies, East Asian Studies, South Asian Studies, Middle Eastern Studies, European

[1] Shan Min, "Prof. Liu Hongwu Attended the International Conference of Regional Studies Held by the Ministry of Education and Delivered a Speech on Constructing Shared Human Knowledge". http://www.zjnu.edu.cn/2018/0728/c3997a266155/page.htm, August 16, 2018.

Studies, American Studies, and Australian Studies—and national studies—such as Chinese Studies, Indian Studies, Nigerian Studies, Brazilian Studies, US studies, and Japanese Studies—are being developed, there is naturally an increasing richness in the disciplines and ideology of research across regions, countries, and nations. If these research disciplines can communicate frequently, talk equally, and learn from each other, we will finally be able to create a shared knowledge system for humankind, which would provide the necessary action, wisdom, and intellectual support necessary for the construction of a community with a shared future, new international relations, and the promotion of a new globalization.

The challenges and opportunities facing humankind today are so complicated that no countries, nations, or regions can solve all these problems alone. The world must join hands and work together. The present world is at a critical period of development and reformation as it welcomes unprecedented changes. From the perspective of China itself, after more than a century's hard work, the country is developing and improving at a historical speed. China's focus has shifted from the pursuit of its own rejuvenation and development to an outlook more focused on initiating a community with a shared future and the common development of the world; from playing catch-up in terms of its development through reform, opening up, following, and learning to the development of strategic partnerships by exchanges, mutual learning, cooperation, and win–win methodology; from older forms of development that relied on resources, technologies, human resources, and finances to new development based on knowledge, thoughts, systems, and cultures. Against this background, China needs to develop the independent and strategic vision to pursue its own knowledge and ideas and to review and summarize Chinese academic tradition. 5000 years of evolving Chinese civilization, 3000 years of national governance, 200 years of rejuvenation, and 40 years of reform and opening-up have been the foundation for China to return to its origins. This is also the basis upon which it should and carry its traditions forward to look at the height of mutual relations between the Chinese nation and global development in the future, to confirm and understand once again China's position in the world, China's identity, role, and aspirations in the new era, and then to contribute Chinese wisdom to the promotion of global governance and development. Today, China–Africa cooperative relations are developing at a higher speed than ever, which requires China to construct its own type of African Studies.[2] This discipline has two main fields: one is African Civilization Studies, which focus on the historical processes of the African continent; another one is African Development Studies, which focus on the problems of the African continent.[3] Generally speaking, the former centers on history, which is primarily made up of background and foundational studies, and the latter stresses the contemporary world, especially practical and applied research. As a matter of fact, these two parts are two sides of the same coin; they depend on each other and serve as the foundation for each other.

2 The Times Call for Building a Universal African Knowledge System

Whether a discipline can be established or not depends on the following prerequisites: (1) whether or not it has its own research subject and object; (2) whether or not it has its unique research value and significance; (3) whether or not it has its own research methods and technical measures; (4) whether or not it represents the wants of society and application space.[4] The urgency of development issues on the African continent today and the abundant practices and current necessities of China–Africa developmental cooperation provide these foundations and conditions, which can facilitate us forming the appropriate research methods, approaches, and knowledge system necessary for recognizing, understanding, and explaining Africa. This will blaze a trail for the construction of a relatively uniform African Studies. Therefore, the building and developing of African Studies is compelled by the objective situation, and it is of great significance for China to expand its knowledge outside its region to improve the contemporary Chinese academic system, as well as to accumulate shared China–Africa knowledge to promote

[2] Liu Hongwu, "A Primary Thought on Constructing the Distinctive 'Chinese Africanology'" [J] *West Asia and Africa*, No. 1, 2010, pp. 5–10.

[3] Liu Hongwu, "The Chinese School of African Studies: Approaches" [J] *West Asia and Africa*, No. 5, 2016, pp. 3–31.

[4] Liu Hongwu, *Introduction to Humanities*. [M] China Social Sciences Press, 2002, p. 148.

the construction of a China–Africa community with a shared future. However, tremendous efforts must be made to move this cause forward.[5]

First, African studies can be a discipline focusing on the fact that the African continent, whether from the perspective of geography or civilization, the past or the present, has some common pan-continental and pan-regional properties. There are some common connections and characteristics across various nations in terms of nature, geography, history, society, culture, etc., which make it possible for us to know and understand this continent integrally, connectively, and macro-graphically. In the past century, many African historians, elites, thinkers, and political leaders have proposed that Africa should be treated as a united region because it shares historical experiences and a modern destiny. They always emphasize and think highly of "the historical offshoots of a set of peoples and societies united by bonds centuries old" among "the various civilizations of the African continent".[6] The existence of the history of these connections and ties reminds us that we need to view Africa with an eye toward wholeness and connection, and form professional knowledge and academic related to Africa accordingly.

Second, in contemporary world order, the African continent faces several common developmental issues and holds some of the same targets. They share not only historical encounters, but also a practical destiny; therefore, these common problems and goals can be studied as common academic issues. Modern African thinkers and politicians all know that the "African continent is a community of history, culture and realistic destiny," as Kwame Nkrumah, the first President of Ghana and one of the founding members of the Organization of African Unity (the present African Union), once said. "Divided we are weak; united, Africa could become one of the greatest forces for good in the world." Therefore, in the past hundred years, the pan-Africa integration movement has continued from the "Pan-Africanism movement" during the national liberation movement to the establishment of the Organization of Africa Unity after independence to the "African Union" in twenty-first century, which currently plays a significant role. And today, a unified Africa Free

[5] Liu Hongwu, "Creating Shared Knowledge for Humankind to Promote a Global Community of a Shared Future". http://www.cssn.cn/zx/bwyc/201807/t20180728_4512627.shtml, April 8, 2018.

[6] Amadou-Mahtar M'Bow, "Preface". J.Ki-Zerbo, (ed.) *General History of Africa I: Methodology and African Prehistory*, University of California Press, 1981. p. xviii.

Trade Area is being promoted to establish the ideal of a "United States of Africa," something supported by many African politicians. The existence of "Africanology" is because of the existence of "African issues." The research activity, exploration, reflection, and accumulated knowledge and thoughts on "African Issues" constitute the basis for the formation and development of "Africanology."

Third, compared with the rest of the world, the existing knowledge formation on Africa already has a long tradition of its own, and today there is greater demand for it than ever. The Africa continent, as a community with a shared historical destiny, requires us to have a holistic and macroscopic grasp of the complex issues outlining various countries and regions, i.e., we need recognize and understand Africa as a discrete entity because the African continent is both a natural and human realm with its own specific historical attributes and connections. In this way, just like how "a view from the peak of a mountain dwarfs the world around, a grasp of the outline provides understanding into the general situation," we can take a panoramic view toward issues on the African continent. On this basis, we can grasp the initial outline of and do further research on the various and diverse issues shared across all countries, regions, and special topics of the 54 African countries and subregions. For example, this might include our research on Francophone African countries today alongside an in-depth study of special topics, which allows us to "see the forest and the trees" and then achieve a thorough understanding.[7] That is to say, Africanology is an area of learning that is inclusive of area studies, field studies, country studies, and special topic studies.[8]

Fourth, in the realistic process of China's academic development, African studies is becoming a relatively independent field of knowledge and thought, whose significance and role are increasingly gaining recognition and attention. In contemporary China, comparatively speaking, a convergent and linked academic group of African studies is gradually forming. There are a growing number of academic institutions, think tanks, and cooperation mechanisms related to African studies. The significance of this relatively new, independent African studies is, to some extent, being recognized by the relevant government policy makers. For example,

[7] Liu Hongwu, *African Culture and Contemporary Development* [M], People's Publishing House, 2014, p. 9.

[8] Liu Hongwu, "the Significance of the Study of African National History" [J] *African Studies*, Vol. 1, 2016, p. 250.

I was appointed as Changjiang Scholar for the Ministry of Education in China in the post of "African Studies" in 2013,[9] which indicates that the Chinese government increasingly values research on Africa. Another example is that there is a column called the "interdisciplinary channel" on the official website of China Social Sciences, under which there are inter-disciplines like "African Studies," "Frontier Studies," "Dunhuang Studies," "Tibetan Studies," "Jiangnan Studies," "Huixue" (Anhui Studies), etc.[10] The "African Studies" column collected and reprinted many articles on African studies, and readers can find some of my papers and academic writings there on constructing China's Africanology. It is thus obvious that academia in China and abroad increasingly views "Africanology" as a field of professional knowledge, and Africanology is being recognized and valued correspondingly.[11]

3 Disciplinary Development in China: the Situation, Defects, and Solutions

There have long been many obvious defects in the disciplinary construction, teaching, and research in the humanities and social sciences in Chinese universities, which restricts modern China's ability to understand the world and or itself.

The first defect is that our recognition of world history is not comprehensive. Most of the time, the history of the US and Europe is regarded as the history of the world and the history of non-Western areas, countries, and nations is neglected. For a long time now, when teaching world history in Chinese universities, courses like Western history or European and American history have always occupied an absolutely dominant position, while the courses devoted to non-Western history, such as Asian history, African history, and the history of Latin America, are marginalized. Among hundreds of Chinese universities today, only a few of their history departments even offer history courses for Africa, Latin America,

[9] Ministry of Education: A Notice of the Ministry of Education on the List of Scholars awarding Distinguished Professors of Changjiang Scholars in 2013 and 2014, http://old.moe.gov.cn/publicfiles/business/htmlfiles/moe/s8132/201502/xxgk_183693.html, 2015-01-20.

[10] See: http://indi.cssn.cn/kxk, March 30, 2018.

[11] Liu Hongwu, "The Chinese School of African Studies: Approaches", http://indi.cssn.cn/kxk/fzx/201709/t20170918_3644181.shtml, February 12, 2018.

the Middle East and other regions and countries. Although China has promoted the Belt and Road Initiative for several years, courses related to the histories of the One Belt and One Road countries are relatively absent from both teaching and research at the history department of many Chinese universities. Despite it being necessary to conduct comprehensive cooperation with these counties, it is very difficult to get proper intellectual support. Comparatively, almost all the history departments of Chinese universities have courses like Western Cultural History, History of Western Political Systems, History of Western Thought, or country-specific courses like History of America, History of France, etc. And most of these courses are regarded as elementary courses or professional courses for the history major curriculum. What's more, a large majority of papers, articles, and dissertations published in China on world history are really only about European and American history. As a matter of fact, under the influence of the Chinese and Western dualist view of history, the so-called comparison of Chinese and foreign history is actually the comparison of Chinese and Western history, while Foreign Cultural Studies is actually Western Cultural Studies, where all the conversations are only about Greece or Europe and America. This understanding of world history is bad for Chinese students to be able to understand the world history with any scale or scope or to be able to form their own view comprehensively.

The second is that the recognition of Chinese history is itself incomprehensive; the History of Han Nationality is usually regarded as the History of China, which fails to pay enough attention to Chinese Frontier History and the History of Ethnic Minorities in various regions. Under the influence of an orthodox historical viewpoint that has long approached Central Han Nationality as central to the historical curriculum, the teaching of Chinese History is done by teaching the History of Han Nationality or the History of the Central Plain. China is a country of 56 ethnic groups, 55 of which are ethnic minorities. However, when talking about Chinese Studies, the main attention is on Chinese Han nationality, while little attention is paid to the Chinese ethnic minorities. That is, to a large extent, the so-called Chinese Studies is really the Studies of Chinese Han Nationality or Confucianism. Generally, the curriculum system of any given history department in Chinese universities mainly teaches the orthodox historical transformation of Chinese dynasties, which is also called "official history" (*zheng shi*). This "official history" has occupied an absolutely dominant position, while having little to do with the history of China's ethnic minorities or the history of local ethnic dynasties in the vast frontier areas. As a matter of fact, the history departments in Chinese universities that offer any courses at all in frontier history and the history of ethnic minorities can be numbered. The key universities in China should have had the overall vision and comprehensive layout of the whole country in mind when they formed; however, in the teaching system and curriculum system of Chinese history being offered there, the number of professors who can teach Chinese Frontier History, Chinese Regional History (such as the History of Xinjiang, the History of Tibet, the History of Mongolia, the History of the Dai Nationality) is very rare. It seems to have been an unwritten practice for a long time that only in a few Universities of Nationalities can there be specialized teachers who can offer a series of courses on the frontier area or ethnic minorities. As a result, two systems have come into being when considering the teaching of history in Chinese universities: many key universities do not teach about the frontier history, and only the Minzu (ethnic) universities, both central and regional, offer these courses. However, these courses in the MINZU universities have become a separate part of the national education system and do not fit into the teaching and curriculum system of general Chinese history. Under the influence of this teaching system, most history graduates in China know little about the history of the Chinese frontier and ethnic minorities, and the non-history graduates know even less. Obviously, this is not good for the young generation, who should have a sprawling knowledge of China's history, culture, and national conditions, and is also bad for the long-term interests of the "big family" of the Chinese nation.[12]

Third, the history pedagogy system in Chinese universities lacks any attention to local chronicles and indigenous knowledge, which means that youth have less access to their own local culture. History is a way of life and an emotional world through which human beings are able to remember the past and his or her hometown. That is, history itself should be endowed with a strong sense of family, country, and homeland. However, the function that shapes the spirit is severely lacking today in the way China's teaching of history is set up, which is caused by a profound neglect of indigenous knowledge and local culture. As a result of the country's rapid modernization and urbanization, many Chinese have been displaced spiritually. Those indigenous forms of knowledge,

[12] Liu Hongwu, "Variety of National Culture and the Long-term Interest of Chinese Nation" [J] *Thinking*, Vol. 5, 1997, p. 4.

folk life, and the local culture that soothes our soul and comforts our homesickness are disappearing in the process of the rapid modernization. Universities in China today can be divided into two categories: central universities and local universities. The former includes the "double top" universities (earlier known as the 985 project/211 project universities) under the direct administration of the Ministry of China; the latter includes the non- "double top" universities, like provincial universities, municipal universities, etc. In fact, what kind of university it is is dependent upon its location, such as Beijing, Tianjin, or Shanghai. Any given university has its own "soil and water" that feeds its growth. But nowadays these universities neglect the teaching of indigenous knowledge and fail to impart the local culture of that place. For example, most "double top" universities in Beijing seldom offer courses on the local history and culture of Beijing; it is the same in "double top" universities in Tianjin, Nanjing, and Shanghai. Therefore, even if a student studies for a few years in a university in Beijing, he or she might not acquire a certain local knowledge of Beijing, even if he or she majors in history.

A sense of homeland, a sense of family and country, and a heart aimed outwards toward the world are the most fundamental elements necessary for humans to form a complete spiritual world for themselves. The knowledge system and mental structure of a well-educated person should have a global vision, a sincere patriotism, a tender notion of the local, and a warm and respectful gratitude for that piece of land and the bunch of fellows with which he or she is surrounded. It will provide the youth with a multidimensional, rich, and comprehensive education that supports the development of individual character and a sound mind, which is the basic requirement for a history department to train qualified historical workers. However, the country's historical teaching and curriculum as they stand now are not good for the younger generation to be able to recognize the panorama of China and the world; that is to say, it is difficult, if not impossible, to form a complete "perception of China" or "perception of the world" that mirrors the reality around them. A lack of education and recognition of local culture and history will not help the younger generation to form a sound global vision and sense of family or country.

Modern China is evolving, fitting into, and even leading this changing world. Thousands of years' worth of cultural traditions and ways of life are being transformed at a rapid pace. Whether we can move forward more steadily and better, strategic thinking and foresight are needed. Under this background, for the construction of the humanities and social sciences in

China, we need to take positive actions to overcome all the educational system's existing defects.

First of all, it is necessary to bring together Chinese history and world history. Chinese history should be taught in the context of world history and world history should be taught from the perspective of Chinese history. Clarifying the relationship between Chinese history and world history can help the youth develop the ambition to integrate their homeland and the external world, as well as connect the ancient with the modern.

Second, courses on the history of non-Western world, i.e., the history of Asia, Latin America, and Africa, and especially the history of Belt and Road countries, should be added to the curriculum of world history, which can then be adapted to China's development strategy. This can then be used to train the youth to have a more holistic view of world history and a global vision.

Third, courses on the history of the Chinese frontier and ethnic minorities should be added to the teaching and curriculum system of Chinese history, which can cultivate the holistic view and awareness of the Chinese nation on the part of the nation's youth.

Fourth, courses about the history and culture of the city where the universities are located should be offered in the system of history and liberal education in Chinese universities. Attention should be paid to the teaching of local knowledge, which is a way to cultivate a sense of family and country for the students.

Fifth, the internationalization of world history should be promoted by offering foreign languages courses. Meanwhile, more chances should be offered to world history majors encouraging such an international outlook, such as internships abroad and/or field work, which can facilitate the students' understanding of existing and developing on a global scale. It can also improve their adaptive multicultural experience, by which they would know how to live together with people from different nations and countries.

Sixth, positive actions should be taken to supplement the existing teaching of the history of African nationalities and the history of Chinese ethnic minorities, which have historically been marginalized in the teaching of Chinese history. It is worth trying to connect these two distant worlds that may seem unrelated and pursue comparative research. On the one hand, we can view the nations, cultures, and histories of Africa from the perspective of the histories and cultures of the Chinese ethnic minorities; on the other hand, we can also observe the histories and cultures of ethnic minorities on China's frontier area from the outside view of Africa. This kind of academic endeavor, "viewing the margin from the margin, and observing the frontier from the frontier," as a brand-new form of cognitive perspective and research efforts, has a special significance for promoting the development of the modern Chinese academic system.

4 Responding to the Call of the Times: Promoting the Innovation of Disciplinary Development in China

In order to understand the question of disciplinary development, it is necessary to elucidate and analyze the notion of what we mean by "discipline" in the first place.

If we study the evolution of humanity's knowledge systems, we can say that "discipline" is a process in which humanity improves its cognitive ability and ability to apply knowledge.[13] Concepts like "politics," "economics," "physics," "biology," etc. are fields created by humans that will continue developing in the future. From this perspective, the "regional disciplines" (like African Studies, Middle Eastern Studies, Asian Studies, etc.) are constructed around regional research activities, which combine the local knowledge and scholarship together and then focus on some basic issues and common concerns in this one specific area. African studies used to be a marginal academic field in China. But today, it is being transformed into a popular discipline and research field that is good for innovation and attracts many researchers' attention.[14]

"African Studies" is a discipline within regional studies that centers on the African continent. Regional Studies stresses the geographical adaptability and spatial–temporal correlation of the discipline. That is, if we do African studies, such research would focus on the specific time and spatial structure of the African continent, the construction of its own knowledge systems, and would form a group of associated regional study disciplines

[13] Liu Hongwu, *Introductory to Humanities*, China Social Sciences Press, 2002, p. 20.

[14] Liu Hongwu, "Breaking the Boundary Between the Existing Discipline is the Way to the Development of Humanities," http://www.cssn.cn/gd/gd_rwhd/gd_mzgz_1653/201406/t20140624_1225205.shtml?COLLCC=3837000743, August 22, 2016.

that can illuminate and address African issues. From the perspective of discipline construction and systems in China, the construction and development of African studies, an emerging interdisciplinary field, can balance and complement the construction of international and regional studies. International and region studies tend to feature field studies in China, which can help us understand the diversity and complexity of our world.

Traditionally, the disciplinary setting in Chinese universities is based on the associated professional field, such as politics, economics, sociology, law, history, literature, etc. The basic premise and hypothesis of the field study are that there are clear differences in our lives based on politics, the economy, society, law, and so on, and these differences can be categorized and studied in a specialized way. This is how the various disciplinary knowledge systems come into being. This is similar to how Western medicine cures patients by dividing the hospital into different departments to treat specific ailments, such as internal medicine, surgery, ophthalmology, dermatology, neurology, etc. As a result of the influence of the Western knowledge system, this kind of categorization is necessary. However, this kind of category is not the natural continuation of all the traditions available in Chinese scholarship.

There is a consciousness of this variety of "field study" in traditional Chinese scholarship and thoughts, such as the category of the "Classics (*Jing*), History (*Shi*), Philosophy (*Zi*), Belles-lettres (*Ji*)," or the category of "*the Book of Songs (Shi), the Book of History (Shu), the Book of Rites (Li), the Book of Changes (Yi), the Book of Music (Yue).*" Du You from the Tang Dynasty classified all things in the world into different categories: "food and goods (*Shi Huo*)," "selection and election (*Xuan Ju*)," "office and position (*Guan Zhi*)," "rites (*Li*)," "music (*Yue*)," "army (*Bing*)," "punishment and law (*Xing Fa*)," etc. These categories roughly correspond to today's economics, politics, administration management, sociology, art, military science, law, and so on. Ma Duanlin from the Song dynasty further divided these areas into an additional 24 categories. However, traditional Chinese scholarship not only pays attention to the specialization of different fields, but also to the comprehensive mastery of various objects, such as the time and space relation between human society and national governance, the comprehensive understanding of integrality, otherness, and the diversity of human culture in place and time. If area studies is like the specialized cure in West medicine, then "regional study" is more like a comprehensive cure. The former depends on specialization and analysis and views the world through individual disciplines, while traditional Chinese medicine relies on comprehensiveness and relationality and looks at the world as a whole.

The culture and system of humankind are generated and have evolved in specific geographical spaces and eco-environments. Different regions are like all different points of the historical stage, and the question of what type of show to play on the stage is at the core of civilization studies. That is, it is worth stressing the environmental elements of history and civilization, time relations, developmental foundations, and the motivations behind cultural evolution when studying cultures. Doing African studies with this belief and from this perspective requires that we immerse ourselves in the African continent to be able to do research. We should try to avoid playing abstract games or considering abstract concepts. This is characteristic of the regional studies, like African Studies.

From the background of contemporary China's international regional studies, we can understand that African Studies is a special case of regional study in China. It is necessary to discuss the academic tradition and modern significance in the field of international and regional studies in China and how they relate to African Studies. In recent years, the issue of the disciplinary construction of regional studies has been gradually attracting people's attention. If we observe the domestic academic world in China, we find that there is a view that "regional studies" is a Western import that can be dated back to America's area studies after WWII. There is a lot for us to learn from area studies in the case of contemporary US and Europe. But it is not true that China is short on academic traditions and the knowledge of regional studies. As a matter of fact, from ancient times to the present, China has been a great civilized country with great cultural diversity within the region, as well as a complex ethnic and national diversity alongside its neighboring countries. For China to govern its country requires the basic recognition of the diverse cultural patterns, social systems, and economic life inside and outside of the region.

When considering the academic traditions of China, we see that when Chinese people view the world, they call for a vision and mindset defined by an "awareness of time and space for civilization's development." Traditional Chinese scholarship always values the background time and motivation of history. The best way for us to understand different regions and countries differs depending on the various histories and traditions being considered. Each country has its own ecological environment; we need to remember that "when in Rome, do as the Romans do" and "sing different songs on different mountains." This means adjusting measures to local conditions. In this way, we can make the best of the "right time," the "right place," and the "right man," and then make the right policy to make our country prosperous and people live in harmony. The disciplinary construction of international regional studies in contemporary China needs to carry forward this outstanding tradition that is so central to the Chinese nation.

5 Making a Fresh Start: Returning to Chinese Classical Works

It is necessary to elucidate how we might carry forward the traditional wisdom of nation governance alongside the academic tradition of regional study in China.

The Book of Songs (*Shijing*), compiled by Confucius 2000 years ago, could be seen as the first work of regional studies in China, which embodies the rational enlightenment and wisdom of knowing the world.[15] By the Han Dynasty, the Chinese intellectual and ideological system was becoming more mature. Ban Gu, the great scholar of the Han Dynasty, wrote a special chapter on the "geographical chronicle (*Dilizhi*)" in his book, *The History of the Han Dynasty* (*Han Shu*), which carried forward the tradition of "the Book of Songs." The *Han Shu* chronicled the evolution of national boundaries, the setting of territory, and various governing policies in ancient times, like those of the Xia Dynasty and Shang Dynasty, and offered a detailed account of the historical traditions and cultural characteristics of various regions both inside and outside of China, as well as the products and social customs that provided the knowledge foundation for their governance. This tradition of "*Han Shu, Dilizhi*" has been carried forward through all subsequent generations, which facilitated the forming of a "territorial geography" or "administrative geography" in the ancient history of China that could advise policymakers. The governing elites and scholars of all dynasties dedicated their lives to writing history from the perspective of nation governance and economic development by deeply analyzing the interrelation of different natural geography and human geography in and around

[15] Liu Hongwu, "Crossing the Borders: from China to the African Continent: Action and Thinking on Cross-Cultural Regional Studies", World Affairs Press, 2017, p. 52.

China. As a result, under the structure of Chinese scholarship, all kinds of works on regional and national governances came into being and endured along with the times.

Under the influence of this tradition, all dynasties in China paid a great deal of attention to the different climates, products, ethnic groups, and natural landscapes of specific times and specific places within their territories. They gave a detailed account of the mental state and psychological characteristics of various ethnic groups and the evolution and interrelation of the political system, which accumulated the national intellectual and knowledge-based systems of governance in ancient China. For example, Zhao Ye, from Shanying county during the Eastern Han Dynasty, recorded the story of successes and failures of contenders for the throne, Wu and Yue's, battle for king at the end of Chunqiu Dynasty, providing abundant historical materials and chronicles in his work, "Wu Yue in the Spring and Autumn Period (*Wu Yue Chun Qiu*)." Su Xun, from the Northern Song Dynasty, eloquently talked about China's state of good governance and bad governance and the causes for each in his article "On Six States (*Liu Guo Lun*)," which was famous for its style, choice of words, and vision. Just as it is fundamental to do field work on the African continent by observing the local customs and cultures if we are to understand Africa and do African Studies, to learn the specific situations in different countries, regions, and the whole continent, and to follow the general trend of African development and do solid research, we must resist putting on abstract labels to recognize the colorful living world. Only in this way can we know and understand Africa.

The Chinese nation typically takes global trends and human nature into consideration. This kind of fine tradition and historical wisdom can be carried forward when we are learning about other regions and nations outside China, including the regional situation of the African continent, the national conditions of African countries, and the situation of African people and societies. Therefore, we have always emphasized that African studies should be based on the African continent, various African countries, and regions, instead of copying concepts from books and making deductions from the literatures.

Chinese traditional scholarship stresses practicality and the necessary involvement of academic thought, which has significance for contemporary China in terms of building a discipline around regional studies. Generally speaking, political thinkers in Chinese history seldom write or propound upon ideas simply for the sake of writing, instead propounding upon ideas in order to realize their ambition for politics. They are statesmen first, engaged in the governance of a province or district. If their political aspiration and ideals have been practiced in the process of governing, it will not matter anymore whether to write or not. As scholar Qian Mu said, "We should know that behind all the systems formulated and implemented in the past dynasties of China there is a set of thoughts and theories. If it has been implemented well, they would not entrust it in words. Over half of Chinese scholars since Qin Dynasty have been working as civil servants for the traditional government and they are the real statesmen. Therefore, most of their thoughts and theories have been practiced in the society, which is obvious that there is no need putting down these theories as a book. It seems that there is no specialized works on political thought and there are no professional statesmen in the history of Chinese academic thought. However, our history of political systems is a good material for the history of political thought. This needs our attention."[16] This is similar to what Wang Yangming, a great philosopher of the Ming Dynasty, advocated when he said that knowledge and action should be one, because one is inseparable from itself. He claimed, "knowledge is the beginning of action, and action is the result of knowledge" because there is knowledge in action and vice versa. Action embeds knowledge, and knowledge is bound to be put into practice. Only when knowledge and action are unified as one can there be genuine knowledge and a state of harmony. Then and only then can we can seek truth and achieve great things.[17]

Scholarship in traditional China has always been combined with the development of the nation and the improvement of people's livelihoods. It seeks truth from facts, learns so that it can be applied in practice, and is used for governing the country. All these spiritual characters and cultural traditions are the sources of an academic spirit that is worth exploring in order for us to construct the discipline of African Studies and to promote the international cooperation of China's regional studies. Although there are many differences between the past and present situations, some basic principles remain consistent. Because of the historical

[16] Qian Mu, *Research Method of Chinese History*, SDX Joint Publishing Company, 2001, p. 29.

[17] Wang Yangmin, proofread by Ye Shengtao, *Instructions for Practical Living and Other Neo-Confucian Writings*, People's Publishing House, 2014, p. 8.

awareness of diverse regions and different time–space structures, the traditional central regime paid great attention to the diversity, flexibility, and adaptability of governance and diplomacy in its relations with all kinds of nationalities inside and outside China, all local governments, and some other countries. Therefore, the idea of governing a state using Chinese wisdom came into being, such as with the Jimi System and Tusi System. This is how the central regime governs its ethnic minorities on the frontier, and can be viewed as the traditional wisdom of China's governance across vast regions. It can help us understand the historical background of China's system of regional national autonomy by analyzing the governing practices of these traditional pieces of political wisdom.

As a matter of fact, these governing experiences can be identified as "seeking truth from facts" and "stressing the importance of investigation and research," a process which has been carried forward creatively by the Chinese Communist Party. *The Report on the Hunan Peasant Movement,* written by Mao Zedong, established the Chinese Communists' tradition of being against book worship and dogmatism, which has had a profound significance on Chinese academia ever since. Later, when he was in Yan'an, he stated the importance of conducting in-depth regional studies in his *Improving Our Study*. He said, "For a political party like us, although there have been certain research achievements on the current situation of both home and abroad, our understanding and knowledge are still fragmentary, and our work is far from systematic. Generally speaking, for twenty years, we haven't done any research systematically and thoroughly. We lack the spirit of objectivity." He then pointed out that if China's revolution is to succeed, "it is necessary to proceed from the reality at difference levels—the world, the country, the province, the county, or even the district. We should find out the intrinsic law from facts other than imaginations. That is, seeking the internal relation among the things and changes, and using it as the guideline for our action."[18] This tradition has been carried out for several generations by the leaders of the Chinese Communist Party. Comrade Deng Xiaoping governed the country based on accurately grasping the varieties of situations and then dealing with internal affairs and foreign affairs. His thought was pragmatic and approachable. The chief distinguishing

[18] Mao Zedong, *Improving Our Study, Selected Works of MAO Zedong*, People's Publishing House, 1982, p. 253.

feature of regional studies is that both the guiding principle and individual policies should be approachable at the same time that they care for the public mood. Current President Xi Jinping, himself, worked in Zhengding county, Hebei Province, in his earlier years. There, he learned about the basic condition of the county through field work a put forward a theory of regional development, the "twenty-word principle" of "living on a mountain, living off the mountain; living by the water, living off the water; living near a city, living off the city," which is "hitting on what it likes, giving it what it needs, taking advantage of its strengths, making up its weaknesses, and changing it with time."[19] This "twenty-word principle" is an example of China developing itself based on its realities, a practice that has a theoretical significance even today. At the same time that Xi Jinping was developing his governing principles, the well-known economist Yu Guangyuan proposed to establish a "Rural Institute," which studies the new socialist rural construction with Chinese characteristics and solves China's developmental issues. These are examples of how to carry forward the traditional Chinese regional studies.

China has formed its own principle of external relations based on traditional wisdom during its contact with other nations and peoples from Asia, Africa, and Latin America. After the People's Republic of China was founded, the Chinese government put forward the following principles of external relations: "seeking common ground while reserving differences," the "Five Principles" for Asia—Africa cooperation, the "Ten Principles" for the Bandung Conference, the "Eight Principles" for China's aid to Africa, and so on, all of which were based on China's understanding and respect of itself, others, global civilizations, and the cultural diversity of human beings. This is the most fundamental reason why China–Africa cooperation relations can be such a special area in China's diplomatic field and among the world's top ranks. If we want to do African studies today, especially African politics studies, we need to go deep into practical China–Africa cooperation and African society, endeavor to understand the process of making and implementing China's strategy and policy for Africa, observe the operations of China's enterprises and companies in Africa, combine fieldwork with theoretical thinking, draw conclusion from history, conclude truths from facts, and connect theory with practice. Only in this way can we grasp Africa and China–Africa relations correctly. Thus the guiding theory can be effective, available, and workable.

[19] Zhao Derun, the Turning Over Story of Zhengding County, *People's Daily*, June 17, 1984.

6 African Studies in China: National and Epochal Characteristics

In contemporary China, "understanding China–Africa relations" or "understanding China's strategy or policy for Africa" has been an important part of "understanding contemporary China." Under this background, "African Studies" has been a part of the field of knowledge and thought most closely related to "New Chinese Studies" in contemporary China, and is also a special area for thought creation and the expansion of knowledge.

Generally speaking, the rapid development of cooperative China–Africa development relations with the internal force of African countries is what is propelling the Africa continent to move on from "the burden of international aid" to a "chance for global development." It is also what is responsible for the core connotation of human modernity moving from a "Western context" to a "global context" and for China's national identity to shift from "seeking China's development" to "leading the development of Asia and Africa," among other knowledge-based shifts. The new concept of China–Africa cooperation in dealing with African issues has manifested itself insofar as China–Africa cooperation has a special significance for the moral construction of the field of international communication. It poses a special opportunity for China scholarship to realize its own self-innovation by systematically concluding the practices of contemporary China–Africa developmental cooperation, by promoting the construction of standardized discursive patterns for African studies with Chinese characteristics and spirit.[20]

Chinese scholarship has a long history. China's understanding of other nations and ethnicities, those of Africa included, can be traced back for many years. However, as a discipline with discrete meaning, African studies is a new field with new types of knowledge in the contemporary China's academic world. One basic feature of the appearance and development of this discipline is that it has been promoted alongside China's modern communicative cooperative relations with Africa and expanded with modern China's understanding of Africa, thus becoming a mature discipline. Therefore, this discipline has been characterized by two basic

[20] Liu Hongwu, "Sino-African Development Cooperation: Identity Transformation and System Reconstruction" [J] *Journal of Shanghai Normal University* (Philosophy & Social Sciences Edition), Vol. 40, No. 6, 2011.

features from its beginning: one is that it is intended to meet the needs of contemporary China's development and also to meet the needs of China–Africa cooperation relations. It is issue-oriented and practical in order to best serve the needs of China-African development. The other characteristic is that it is the result of the modern Chinese endeavor to understand the external world. It is not imported from foreign countries. Although we have learnt diligently from many Western academic works in this process, it is still China's creation based on the ancient tradition of Chinese scholarship, and it is a manifestation and product of the traditional Chinese scholarship as it has evolved in and understood the external world. Therefore, it has the basic spirit and character of China's scholarship.

The epochal character and Chinese character of Chinese African scholarship mean that African studies in contemporary China must meet the issues of China and Africa in China–Africa cooperation, construct its own foundation, cultivate its own character, expand its own perspective, and extract its own discourse. All these cannot be achieved without Chinese scholars being rooted in Africa, walking on African soil, observing Africa, and studying Africa for a long time.[21] Generally speaking, the basic spirit of African studies in China is centered on China's position, Africa's feelings, and a global vision linking the two.

Compared with other traditional disciplines, present-day China's understanding and study of Africa is still in the early phase of material accumulation and experiential exploration. The basic situation is that African studies in China have not generated any in-depth and professional work in even the basic respects, such as academic problems, systemic problems, or methodological problems.[22] However, since the country's reform and opening-up policy in 1978, Chinese society and the relationship between China and the external world have been changing tremendously. Among these changes, the construction of a China–Africa win–win cooperative relationship and the influence of China's development experience on Africa have been of great symbolic significance to

[21] Zhang Ming, "Vice Foreign Minister Zhang Ming put forward new requirements for Chinese Research Institutions to Strengthen Their Original Research on Africa," http://www.fmprc.gov.cn/wb/wjbxw_673019/t1492905.shtml, August 15, 2018.

[22] Zhang Hongming, "A Summary of the Development of African Studies in China" [J] *West Asia and Africa*, Vol. 5. pp. 3–13.

the changing times. For 40 years, the development of China–Africa relations has accumulated a plethora of rich experiences and feelings for the promotion of China–Africa bilateral development, all of which provide the foundation for the creation of related theories and knowledge. As the saying goes, "all roads lead to Rome." This new academic world reflects itself in diversity and inclusiveness.

In recent years, thanks to the appreciation of the Chinese central government and the needs of society for such a field, many institutes of African studies have opened in China. If these institutes are to make a difference, they need to focus on the following points: (1) strategic thinking and planning in accordance with the spiritual tradition of "learn extensively, inquire thoroughly, ponder prudently, discriminate clearly, and practice devotedly."[23] After developing their aims, developmental goals, and working paths, they must then follow up with diligent exploration and improvement of the practice. (2) These institutes must focus on developing a different and inclusive set of academic ideals and spiritual pursuits, forming a unique academic culture and research character, and manifesting it in every aspect of this cause; (3) a blueprint, together with a detailed plan, striving for excellence in practice, doing well in every detail and link, "many a little makes a mickle," avoiding being an armchair strategist; (4) gradually building a platform for disciplined construction and development, accumulating academic resources, and pooling various kinds of resources and conditions to provide space for further creative development; (5) an attitude of commitment to strict, and solid work, being "quick in action but slow in speech," exploratory in practice, examining and improving in practice; (6) cultivating a group of talented personnel with a common goal who love Africa and would like to root in Africa, who have an academic commitment to the continent and are willing to devote themselves to this cause.

Action, walking, and practice are both the starting points and the final destination for the theoretical problems in African studies, but this does not mean that the theory itself, the thought and spirit, are neglected. In fact, theoretical guidance is behind all practices and actions. Thinking is necessary and is already present while we are in the process of taking action. A doer, a person who thinks and observes in his or her actions, has to find the dynamic balance between reality and theory. The planning,

[23] *The Book of Rites: the Golden Mean*, No. 19, Chapter 20.

thinking, and conclusion of this practice include a great deal of content for both knowledge creation and the expansion of the mind. As the Chinese saying goes, "knowledge and action are one" or "action and knowledge can make it." Therefore, to construct the discipline of African studies is not a theoretical issue but a realistic one, which we must meet in practice.

The construction of African studies in China today should stick to the principle of seeking truths from facts, and explore forward in practice. According to the diversity and specificity of the research object, various kinds of research paths and methods should be taken to do research, that is, we cannot do this aimlessly. If an institute lacks the core theory and strategic goal and only follows the trend, pays no attention to the cultivation of the young researchers, there might be outstanding talents who can succeed in the academic world, it is hard to avoid the loose, disorderly, overstaffed and inactive situation.

Looking back at the construction and expansion of African studies in China, we can reach a few conclusions:

First of all, scholarship like this is the result of an accumulation over a long period of time. That is, both teaching and research cannot be done in one day, and are instead built gradually and steadily over time. Adequate time is necessary to carry forward professionally. African Studies in China must find its roots, and it needs several generations of adherence to this cause.

Second, as the saying goes, "you must learn more things beyond poems if you want to learn how to write them." If we are to succeed in something, we need to work hard not only on this one thing, but on the other things around it. It requires us to have an open mind, wisdom, and an ability that surpasses the thing itself. As for African Studies, besides engaging in the professional study of African politics, African economics, African education, African religion, etc., we must have more knowledge, a wider theoretical culture, and interests and hobbies. If we are to build a comprehensive institute of African Studies, we need a vision for how to arrange different subjects together. If the researchers want to make a difference, qualities like narrow-mindedness, a small field of focus, or a short-term horizon should be avoided. Knowledge of different subjects and fields, whether they are inside Africa or outside Africa (such as politics, they economy, culture, education, history, religion, nations, society, nature, science, technology, geography, and so on), would be helpful for researchers engaged in African studies. If a scholar wants to do African studies well, it is also necessary for him or her to understand not only China or America, but also Asia, Latin America, and Europe.

Third, taking root and holding fast to this field. To comment on whether a scholar is genuine or not depends on whether he or she has tenaciously dug into the field. There are no shortcuts for most of the things in the world, only through time can anything be accomplished. In the academic world, the basic unit of time is ten years. Without ten or twenty years' hardworking, it is not easy to yield significant results. There is no special method by which to do African studies, except through diligence, concentrated attention, and a willingness to root in Africa and keeping on working. The construction of African studies in China is an open and ongoing process; perhaps there will never be a day that it will be fully completed. It will always be opening, self-constructing, and expanding its own way forward.[24]

[24] Wen Guowen, *Important Achievements have been Made in the Research on the Original Development of African Studies*, China Reading Weekly, September 9, 2017.

CHAPTER 21

African Students in China: Research, Reality, and Reflections

Anshan Li

With the rapid development of China–Africa relations, the African community in China has become an impressive phenomenon and is very attractive to international academia. Various African diasporic groups exist in China, including business people and merchants, students, artists, professionals, etc., yet it is the students who form the second largest

[1]See Wu, Bin. 2015. "Links between Chinese International Students and Overseas Chinese Communities: An Empirical Study in Nottingham, UK." *Overseas Chinese History Studies* 2: 1–11 and Wu, Bin. 2016. "Local Engagement of Chinese International Students in Host Societies: A Perspective of Diasporic Chinese Community Building." *The International Journal of Diasporic Chinese Studies* 8 (2): 13–30.

This is a revised and updated version of the article "African Students in China: Research, Reality, and Reflection" published in *African Studies Quarterly* (Volume 17, Issue 4, February 2018, p. 44) of University of Florida. I would like to thank the journal for all the generosity and kindness for the publication.

A. Li (✉)
Center for African Studies, Peking University, Beijing, China
e-mail: wxy03@ssap.cn

X. Li and C. An (eds.), *The Changing World and Africa*,
https://doi.org/10.1007/978-981-16-4983-7_21

group among the diasporic African community. Although there have been debates about whether international students can be counted as immigrants, typically they are identified as such.[1] Studying African students in the context of China–Africa relations, this article is divided into five parts. The first provides a survey on the current research related to the topic, followed by an overview of the history of African students in China. The third part is an analysis of China's policy toward African students. Why do the African students want to come to China? What attracts them? What are their purposes? The fourth part will discuss the favorable conditions for their coming to China and their motivations. The last part will describe their contributions to both African and Chinese societies. I will argue that African students' existence in China and their interactions with the Chinese people have contributed to both China and Africa alike.

1 Research, Debates, and Views

The international community currently evinces great interest in China–Africa relations. Different issues, contradictory views, and various debates have been presented for some time.[2] The African diaspora has also become a fashionable topic among academia. It is presumed that the African community appeared in China in recent years, yet there has actually been a long history of diaspora and contact between China and Africa.[3] Owing to the rapid growth of bilateral trade volume, which increased from $10.8 billion in 2000 to more than $220 billion in 2014, it is undeniable that the size and impact of the African community have boomed in China during this same period.

Since China–Africa trade occupies a very important component in the countries' bilateral relations, businesspeople are by far the largest group among the African community in China. So far the studies have generally focused on African trading communities in China[4] or their economic

[2] See Li, Anshan. 2014. "Changing Discourse on China-Africa Relations since the 1990s." *World Economy and Politics* 2: 19–47.

[3] See Li, Anshan. 2015a. "African Diaspora in China: Reality, Research and Reflection." *The Journal of Pan African Studies* 7 (10): 10–43 and Li, Anshan. 2015b. "Contact between China and Africa before Vasco da Gama: Archeology, Document and Historiography." *World History Studies* 2 (1): 34–59.

[4] See Bertoncello, Brigite and Sylvie Bredeloup. 2006. "La migration chinoise en Afrique: accélérateur du développement ou 'sanglot de l'homme noir'." *Afrique Contemporaine* 218: 199–224; Bertoncello, Brigite and Sylvie Bredeloup. 2007a. "De Hong

activities in Guangzhou, the commodity hubs of China[5] and Yiwu, the

Kong à Guangzhou, de nouveaux 'comptoirs' africains s'organisent." *Perspectives chinoises* 98 (1): 98–110; Bertoncello, Brigite and Sylvie Bredeloup. 2007b. "The Emergence of New African 'Trading Posts' in Hong Kong and Guangzhou." *China Perspectives* 1: 94–105; Bertoncello, Brigite and Sylvie Bredeloup. 2009. "Chine-Afrique ou la valse des entrepreneurs- migrants." *Revue européenne des migrations internationales* 25 (1): 45–70; Bertoncello, Brigite, Sylvie Bredeloup, and Olivier Pliez. 2009. "Hong Kong, Guangzhou, Yiwu: de nouveaux comptoirs africains en Chine." *Critique international* 44: 105–121; Bodomo, A. 2007. "An Emerging African-Chinese Community in Hong Kong: The Case of Tsim Sha Tsui's Chungking Mansions." In Afro-Chinese Relations: Past, Present and Future, edited by Kwesi Kwaa Prah. *Cape Town: Centre for Advanced Studies in African Societies*: 367–389; Bodomo, A. 2012. *Africans in China: A Sociocultural Study and Its Implications on Africa-China Relations*. New York: Cambria Press; Dittgen, Romain. 2010. "L'Afrique en Chine: l'autre face des relations sino-africaines?" Economie, China Institute; and Cissé, Daouda. 2013. "South-South Migration and Trade: African Traders in China." *Policy Briefing*.

[5] See Bodomo, A, 2007. "An Emerging African-Chinese Community in Hong Kong: The Case of Tsim Sha Tsui's Chungking Mansions." In Afro-Chinese Relations: Past, Present and Future, edited by Kwesi Kwaa Prah. *Cape Town: Centre for Advanced Studies in African Societies*: 367–389; Bodomo, A. 2012. Africans in China: A Sociocultural Study and Its Implications on Africa-China Relations. New York: Cambria Press; Lyons, M., et al. 2008. "The 'Third Tier' of Globalization: African Traders in Guangzhou." *City* 12 (2): 196–206; Lyons, M., et al. 2012. "In the Dragon's Den: African Traders in Guangzhou." *Journal of Ethnic and Migration Studies* 38 (5): 869–888; Lyons, M., et al. 2013. "The China-Africa Value Chain: Can Africa's Small-Scale Entrepreneurs Engage Successfully in Global Trade?" *African Studies Review* 56 (3): 77–100; Li, Zhigang, et al. 2008. "The African Enclave of Guangzhou: A Case Study of Xiaobeilu." *ACTA Geographica Sinica* 63 (2): 207–218; Li, Zhigang, et al. 2009a. "The Local Response of Transnational Social Space under Globalization in Urban China: A Case Study of African Enclaves in Guangzhou." *Geographical Research* 28 (4): 920–932; Li, Zhigang, et al. 2009b. "An African Enclave in China: The Making of a New Transnational Urban Space." *Eurasian Geography and Economics* 50 (6): 699–719; Li, Zhigang, et al. 2012. "China's 'Chocolate City': An Ethnic Enclave in a Changing Landscape." African Diaspora 5: 51–72; Li, Zhigang and Du Feng. 2012a. "Production of China's New Social Space in City under 'Transnational Entrepreneurialism:' A Case Study on African Economic Zone in Guangzhou." *Urban Space Studies* 36 (8): 25–31; Li, Zhigang and Du Feng. 2012b. "The Transnational Making of 'Chocolate City' in Guangzhou." *Renwen Dili* 27 (6): 1–6; Bodomo, A. and Grace Ma. 2010. "From Guangzhou to Yiwu: Emerging Facets of the African Diaspora in China." *International Journal of African Renaissance Studies* 5 (2): 283–289; Diederich, Manon. 2010. "Manoeuvring through the Spaces of Everyday Life: Transnational Experiences of African Women in Guangzhou, China." Dissertation at the University of Cologne, Department of Geography; Müller, Angelo. 2011. "New Migration Processes in Contemporary China—The Constitution of African Trader Networks in Guangzhou." *Geographische Zeitschrif* 99 (2): 104–122; Haugen, H.

biggest commodity center in China.[6] (Le Bail 2009) Some have studied their business transactions and deals in Hong Kong[7] (Ho,2012) and Macau.[8] Other areas of research include the living conditions, social practices, or religious activities of the African diaspora (Li et al. 2011; Yang 2011),[9] barriers between Africans and Chinese, management of

Ø. 2012. "Nigerians in China: A Second State of Immobility." *International Migration* 50 (2): 65–80; and Bredeloup, S. 2012. "African Trading Posts in Guangzhou: Emergent or Recurrent Commercial Form?" *African Diaspora* 5 (1): 27–50.

[6] See Bertoncello, Brigite, Sylvie Bredeloup, and Olivier Pliez. 2009. "Hong Kong, Guangzhou, Yiwu: de nouveaux comptoirs africains en Chine." *Critique international* 44: 105–121; Osnos, Evan. 2009. "The Promised Land—Guangzhou's Canaan Market and the Rise of an African Merchant Class." *The New Yorker*, February 9–16: 50–56; Pliez, O. 2010. "Toutes les routes de la soie mènent in Yiwu (Chine). Entrepreneurs et migrants musulmans dans un comptoir économique chinois." *Espace Géographique* 2: 132–145; Bodomo, A. and Grace Ma. 2010. "From Guangzhou to Yiwu: Emerging Facets of the African Diaspora in China." *International Journal of African Renaissance Studies* 5 (2): 283–289; Bodomo, A. and Grace Ma. 2012. "We Are What We Eat: Food in the Process of Community Formation and Identity Shaping among African Traders in Guangzhou and Yiwu." *African Diaspora* 5 (1): 1–26; and Ma, Enyu. 2012. "Yiwu Mode and Sino-African Relations." *Journal of Cambridge Studies* 7 (3): 93–108.

[7] See Bodomo, A, 2007. "An Emerging African-Chinese Community in Hong Kong: The Case of Tsim Sha Tsui's Chungking Mansions." In Afro-Chinese Relations: Past, Present and Future, edited by Kwesi Kwaa Prah. *Cape Town: Centre for Advanced Studies in African Societies*: 367–389; Bodomo, A. 2009a. "Africa-China Relations in an Era of Globalization: the Role of African Trading Communities in China." *West Asia and Africa* 8:62–67; Bodomo, A. 2009b. "Africa-China Relations: Symmetry, Soft power, and South Africa." *The China Review: An Interdisciplinary Journal on Greater China* 9 (2):169–178; Bodomo, A. 2009c. "The African Presence in Contemporary China." China Monitor. University of Stellenbosch: South Africa; Bodomo, A. 2012. *Africans in China: A Sociocultural Study and Its Implications on Africa-China Relations*. New York: Cambria Press; Mathews, G. 2000. "Les traders africains a Kong Hong (Hong Kong) et en Chine." *Les Temps Modernes* 657 (Janvier-mars): 110–124; and Mathews, G. and Yang, Y. 2012. "How Africans Pursue Low-End Globalization in Hong Kong and Mainland China." *Journal of Current Chinese Affairs* 41 (2): 95–120.

[8] See Morais, I. 2009. "'China Wahala': The Tribulations of Nigerian 'Bushfallers' in a Chinese Territory." *Transtext(e)s Transcultures. Journal of Global Cultural Studies* 5: 1–22; Bodomo, A. 2012. *Africans in China: A Sociocultural Study and Its Implications on Africa-China Relations*. New York: Cambria Press; and Bodomo, A. and Roberval Silva. 2012. "Language Matters: the Role of Linguistic Identity in the Establishment of the Lusophone African Community in Macau." *African Studies* 71 (1): 71–90.

[9] See Li Zhigang, et al., 2008. "The African Enclave of Guangzhou: A Case Study of Xiaobeilu." *ACTA Geographica Sinica* 63 (2): 207–218; Li, Zhigang, et al. 2009a. "The Local Response of Transnational Social Space under Globalization in Urban China: A

African diaspora by Chinese authorities, or the reaction of Chinese citizens (Bredeloup 2007, Li and Du, 2009).[10] African entrepreneurs are also described in terms of how they transmitted their conception of China

Case Study of African Enclaves in Guangzhou." *Geographical Research* 28 (4): 920–932; Li, Zhigang, et al. 2009b. "An African Enclave in China: The Making of a New Transnational Urban Space." *Eurasian Geography and Economics* 50 (6): 699–719; Li, Zhigang and Du Feng. 2012a. "Production of China's New Social Space in City under 'Transnational Entrepreneurialism:' A Case Study on African Economic Zone in Guangzhou." *Urban Space Studies* 36 (8): 25–31; Li, Zhigang and Du Feng. 2012b. "The Transnational Making of 'Chocolate City' in Guangzhou." *Renwen Dili* 27 (6): 1–6; Li, Zhigang, et al. 2012. "China's 'Chocolate City': An Ethnic Enclave in a Changing Landscape." *African Diaspora* 5: 51–72; Bertoncello, Brigite and Sylvie Bredeloup. 2009. "Chine-Afrique ou la valse des entrepreneurs- migrants." *Revue européenne des migrations internationales* 25 (1): 45–70; Bodomo, A. 2009a. "Africa-China Relations in an Era of Globalization: the Role of African Trading Communities in China." West Asia and Africa 8:62–67; Bodomo, A. 2009b. "Africa-China Relations: Symmetry, Soft power, and South Africa." *The China Review: An Interdisciplinary Journal on Greater China* 9 (2):169–178; Bodomo, A. 2009c. "The African Presence in Contemporary China." *China Monitor*. University of Stellenbosch: South Africa; Bodomo, A. 2010. "The African Trading Community in Guangzhou: an Emerging Bridge for Africa-China Relations." *The China Quarterly* 203: 693–707; Xu, Tao. 2009a. "An Analysis on Africans' Social Relations and Interaction Logics in Guangzhou." *Youth Research* 5: 71–86; Xu, Tao. 2009b. "African's Social Support in Guangzhou: Weakening, Fracture and Reconstruction." *South China Population* 24 (4): 34–44; Müller, Angelo. 2011. "New Migration Processes in Contemporary China—The Constitution of African Trader Networks in Guangzhou." *Geographische Zeitschrif* 99 (2): 104–122; and Haugen, H. Ø. 2013b. "African Pentecostal Migrants in China: Marginalization and the Alternative Geography of a Mission Theology." *African Studies Review* 56 (1): 81–102.

[10] See Xu Tao. 2009a. "An Analysis on Africans' Social Relations and Interaction Logics in Guangzhou." *Youth Research* 5: 71–86; Xu, Tao. 2009b. "African's Social Support in Guangzhou: Weakening, Fracture and Reconstruction." *South China Population* 24 (4): 34–44; Osnos, Evan. 2009. "The Promised Land— Guangzhou's Canaan Market and the Rise of an African Merchant Class." *The New Yorker*, February 9–16: 50–56; Morais, I. 2009. "'China Wahala': The Tribulations of Nigerian 'Bushfallers' in a Chinese Territory." *Transtext(e)s Transcultures. Journal of Global Cultural Studies* 5: 1–22; Bodomo, A. 2010. "The African Trading Community in Guangzhou: an Emerging Bridge for Africa-China Relations." *The China Quarterly* 203: 693 – 707; and Haugen, H. Ø. 2012. "Nigerians in China: A Second State of Immobility." *International Migration* 50 (2): 65–80.

into their own countries, thus explaining the impact of Chinese development in a global context.[11] Now the general trend of such studies has gradually moved beyond Guangdong.[12]

However, African students form the second largest group of African diaspora in China. Yet far fewer scholars have been involved in the study of this topic. The earliest work was done by a Ghanaian student who had personal experience studying in China in the early 1960s. After their independence, fourteen African countries sent their students to China through the end of 1966, when China closed all universities because of the Cultural Revolution. Emmanuel Hevi was among the group. He wrote the first book about African students in China, complaining about racism and other unpleasant phenomena in China. He listed six causes of African student dissatisfactions, e.g., undesirable political indoctrination, language difficulty, poor educational standards, inadequate social life, outright hostility, and racial discrimination.[13] Although he indicated that many African students returned back home in 1961–1962, there was some disagreement about his statement.[14] Although the book raised attention from the West at the time of its publication, any further study of the subject almost entirely stopped once all the African students returned back to their countries after the Cultural Revolution. Until recently, scholars reviewed the historical context or explained the social background of China at the time as responsible for these dissatisfactory factors.[15]

China resumed educational cooperation with Africa in 1973, after which, trainees of railway technology arrived in China from Tanzania

[11] See Marfaing, Laurence and Alena Thiel, 2014. "Agents of Translation: West African Entrepreneurs in China as Vectors of Social Change", Working Paper, No.4, DFG Priority Program 1448, Accessed June 20, 2017 and Cissé, Daouda. 2013. "South-South Migration and Trade: African Traders in China." *Policy Briefing*.

[12] See Bodomo, A, 2016. *Africans in China: Guangdong and Beyond*. New York: Diasporic Africa Press.

[13] Hevi, Emmanuel, 1963. *An African Student in China*. Pall Mall.

[14] Larkin, Bruce D, 1971. *China and Africa 1949–1970: The Foreign Policy of People's Republic of China*. Berkeley: University of California Press.

[15] See Liu, P. H, 2013. "Petty Annoyances? Revisiting John Emmanuel Hevi's An African Student in China after 50 Years." *China: An International Journal* 11 (1): 131–145 and Cheng, Yinghong. 2014. "An African student's Impression of China of the 1960s." *Phoenix Weekly*, May 14, http://www.ifengweekly.com/detil.php?id=490l. Accessed: June 24, 2017.

and Zambia, followed by more formal students' enrolment. There are several works on the Tanzania-Zambia Railway (TAZARA), including documents, records, and paraphernalia where the African trainees are mentioned here and there.[16] As for a first systematic survey of African students in China, Sandra Gillespie's work is one of the few books that puts African student experiences in the context of South–South relations, emphasizing the knowledge transfer of China's educational exchange programs for Africa.[17] Several studies examine the conflicts between African students and Chinese students in the 1980s, criticizing Chinese racism.[18]

With the establishment of FOCAC, the number of African students greatly increased (See Table 2). As for the study of African students currently in China, there is a greater interest among Chinese scholars (or African scholars in China) than those outside China. The first study on the issue in China was by the Center for African Studies at Peking University. It is a general survey with a series of data based on the archives of the Ministry of Education, focusing on African students in China (China Africa Education Cooperation Group, 2005, thereafter CAECG). Current research about African students is generally focused on four subjects: cultural adaptation, China–Africa cooperation, educational management, and professional teaching such as language, mathematics, engineering, etc. Using the key words "African overseas students" as the search term, 47 articles can be found in the Chinese Journals Network (2003–2014)

[16] See Zhang Tieshan, 1999. *Friendship Road: A Report on the Construction of Tanzania-Zambia Railway*. Beijing: China Economic and Foreign Trade Press; Monson, J., 2009. *Africa's Freedom Railway: How a Chinese Development Project Changed Lives and Livelihoods in Tanzania*. Bloomington and Indianapolis: Indiana University Press; and Liu, Haifang and Jamie Monson. 2011. "Railway Time: Technology Transfer and the Role of Chinese Experts in the History of TAZARA." In *African Engagements: Africa Negotiating an Emerging Multipolar World*, edited by Ton Dietz, et al. Leiden and Boston: Brill: 226–251.

[17] Gillespie, Sandra, 2001. *South-South Transfer: A Study of Sino-African Exchange*. New York: Routledge.

[18] See Sautman, Barry, 1994, "Anti-Black Racism in Post-Mao China." *The China Quarterly* 138: 413–437; Seidelman, Raymond, 1989, "The Anti-African Protests; More Than Just Chinese Racism." *The Nation*. February 13; Sullivan, M.J. 1994, "The 1988–89 Nanjing Anti-African Protests: Racial Nationalism or National Racism?" *The China Quarterly* 138 (June): 438–457; and Li, Anshan and Liu Haifang, 2013. "The Evolution of the Chinese Policy of Funding African Students and an Evaluation of the Effectiveness." *Draft report for UNDP*.

containing journal articles and MA dissertations. Among them, five are on cultural adaptation, six on educational management, fourteen on teaching Chinese language, with the rest on China–Africa relations. Reports and memoirs have also provided information about the experiences of African students in China.[19] There is an important study regarding the evolution and effect of China's policy toward African students, for example, in which the author argues that the scholarship policy has contributed a great deal toward the success of China's international educational cooperation, especially with African students.[20]

Psychology is often applied in cross-culture study and two specific works in this area are well worth mentioning. One is an article based on the SASS (Study Abroad Stress Survey) of Africans and Western students in China. The SASS survey was carried out in 2003, with 200 forms sent out to foreign students at colleges in three cities in China. The thirty questionnaires are divided into four categories, e.g., interpersonal, individual, academic, and environment. 156 valid forms returned with 82 Africans (46 men, 36 women) and 74 Westerners (32 men, 42 women). It was intended to evaluate gender differences (women vs men) and cultural differences (Africa vs West) in the perception of stress. No group differences exist in the subtotal perception of the four stressors. Group variations exist only in their sub-divisional areas of stress. Cross-cultural orientation is suggested for foreign students. It indicates that academic and interpersonal sources of stress were the most common, with daily hassles being defined as high pressure and challenging among both male and female respondents.[21] Another psychological study is a MA dissertation based on an investigation of 181 pieces of feedback out of 210

[19] See Li Baoping, 2006. "On the Issues of China-Africa Educational Cooperation." http://www.docin.com/p-747065460.html, Accessed June 2, 2017; Li Jiangtao and Li Xiang. 2006. "China is my Second Hometown: African Students' Life in Beijing." http://news.xinhuanet.com/world/2006-10/21/content_5232813.htm; and China Africa Project. 2013. "Leading China Scholar Li Anshan Recalls his Experiences Teaching African Students." http://www.chinaafricaproject.com/leading-china-scholar-li-anshan-recalls-his-experiences-teaching-african-students-translation/, Accessed June 4, 2014.

[20] Liu Haifang, 2017, "A Study on the Evolution and Effects of China's Scholarship Policy towards African Students." In *Annual Review of African Studies in China 2015*, edited by Li Anshan and Liu Haifang: 141–192.

[21] See Hashim, Ismail Hussein, et al., 2003. "Cultural and Gender Differences in Perceiving Stressors: A Cross-Cultural Investigation of African and Western Students at Chinese Colleges." *Psychological Science* 26 (5): 795–799; and Hashim, Ismail Hussain,

forms, a rather high ratio for an investigation of this sort. The author is an African student, and the aim of the study was to get a real picture of the cultural shock and adaptation of African students in China. It found that all forms of cultural shock are common for African students in China and that the best way to ameliorate this is to increase social contact with local people. It is also found that although all African students experienced culture shock, the extent differs according to the grade and gender, e.g., it is more serious for undergrads than graduates, women than men.[22] Other studies are either on cultural adaptation,[23] cultural difference and its impact[24] or different concept of time and family.[25]

As for the role in China–Africa educational cooperation, Ketema suggested that Chinese universities played an important role in the bilateral cooperation (Ketema Meskela et al.,2009), King used African student in China as indicator of China's soft power in his chapter (King, 2012), Haugen analyzed China's policy for enrolment of African students and its effect and outcome,[26] and others argued that China's educational assistance formed an essential part of China–Africa cooperation and offered substantial support to Africa (Li & Luo, 2013).[27] Studies also

and Z. L. Yang. 2003. "Cultural and Gender Differences in Perceiving Stressors: a Cross-Cultural Investigation of African and Western Students in Chinese Colleges." *Stress and Health* 19 (4): 217–225.

[22] Disima. 2004. "Cultural Adaptation of Foreign Students in China." M.A. Psychology Thesis, Nanjing Normal University.

[23] See Yi Pei and Xiong Lijun, 2013. "An Empirical Study of Intercultural Adaptation of African Students in China." *Journal of Shenyang University (social science)* 15 (3): 364–368 and Gong, Sujuan. 2014. "A Study on African Students in China and their Cross-Cultural Adaptation." *Journal of Kaifeng Institute of Education* 34 (2): 127–130.

[24] Long Xia and Xiong Lijun, 2014. "The Influence of Sino-African Cultural Difference on the Education of African Students in China: Taking Angolan Students as the Example." *Journal of Chongqing University of Education* 27 (1): 133–136.

[25] Ye Shuai, 2011. "A Comparative Analysis of the Cross-Cultural Communication Based on the Somali Students and the Chinese Students on Time and Family Concepts." *Kexue Wenhui* 11: 30–31.

[26] Haugen, H. Ø. 2013a. "China's Recruitment of African University Students: Policy Efficacy and Unintended Outcomes." *Globalisation, Societies and Education* 11 (3): 315–344.

[27] See Li Baoping, 2006. "On the Issues of China-Africa Educational Cooperation." http://www.docin.com/p-747065460.html. Accessed June 2, 2017; Xu, Hui. 2007. "Sino-Africa Educational Cooperation under the FOCAC Framework." *Educational Development Research* 9: 1–7; He, Wenping. 2007. "A Summary Analysis of China-Africa

discussed the management of African students or graduates in China, either in universities or society (Cheng, 2012).[28] The fourth subject always involves teachers who are doing language teaching and looking for the best way to teach Chinese languages to Africans (Song, 2011).[29] African students themselves wrote about their experiences in China and emphasized African agency in their behavior in Chinese society.[30]

There are, however, criticisms to be had regarding the teaching methods and suggestions for improvement, including different views on the effect of China's educational policy toward the African students. One view holds that African scholarship holders are generally satisfied with their experiences in China, thus promoting a positive view of the potential for strengthening China–Africa friendship with African countries through educational programs (Dong & Chapman, 2008). Although there are shortcomings and room for improvement, China's policy is rather successful in promoting China–Africa relations, helping the African capacity building, and bettering China's image.[31] Yet Haugen asserts that

Educational Exchanges and Cooperation: Development Phases and Challenges." *West Asia and Africa* 3: 13–18; and Lou, Shizhou and Xu Hui. 2012. "The Development and Transition of China-Africa Educational Cooperation in the New Period." *Educational Research* 10: 28–33.

[28] See Zheng, Jianghua, 2012. "Research on Safety Management of African Students on University Campus." *Journal of Tianjin University of Technology and Education* 22 (4): 72–74; Zheng, Jianghua, et al. 2013a. "Construction of Community Management System for Foreign Students in Universities." *Vocational and Technical Education* 34 (23): 66–68; Zheng, Jianghua. 2013b. "Exploration of Compound Applied Talents Training Mode on African Students." *Journal of Tianjin University of Technology and Education* 23 (4): 64–70; and An, Ran, et al. 2007. "African Students' Educational Needs and the Recruitment Style." *High Education Exploration* 5: 110–113.

[29] See Lin Lunlun and Ren Mengya, 2010. "A Socioliguistic Study upon Chinese Language Learning Concept of African Overseas Students." *Journal of Hanshan Normal University* 31 (5): 32–37.

[30] See Amoah, Lloyd G. Adu, 2012. "Africa in China: Affirming African Agency in Africa-China Relations at the People-to-People Level." In *China-Africa Partnership: The Quest for a Win–Win Relationship*, edited by James Shikwati, 104–115. Nairobi: Inter Region Economic Network and Lokongo, Antoine Roger. 2012. "My Chinese Connection." CHINAFRICA 50, July.

[31] See Li Anshan and Liu Haifang, 2013. "The Evolution of the Chinese Policy of Funding African Students and an Evaluation of the Effectiveness." *Draft report for UNDP*; Niu, Changsong. 2016. "A Survey of African Students' Satisfaction of Chinese Government Scholarship." http://www.docin.com/p-1445264169.html. Accessed: June 25, 2017; and Liu, Haifang, 2017. "A Study on the Evolution and Effects of China's

China fails to reach its policy objectives because of African students' disappointment with the quality of the education they receive, noting that "Disappointment with the educational experience obstructs the promotion of Chinese values, thus obliterating the soft power potential of Sino-African educational exchanges" (Haugen, 2013).

The above-mentioned studies are characterized by similar features, e.g., cross-cultural theory with questionnaire as methodology and concrete suggestions provided. The studies are characterized as the collection of direct data with life experiences. Researchers sometime are African students themselves. The shortcoming is the studies are usually based on a case study of African students in a place (or a university) or from a country, thus limitations are inevitable. How to best apply theories in case studies is another issue. Bilateral migration provides both opportunity and challenge. There are similarities between Chinese culture and African cultures and mutual learning are always beneficial to both,[32] especially through people-to-people contact. Yet solid study on the topic is still lacking.

2 History, Reality, and Trends

The history of African students in China began in 1956, when four Egyptian students arrived in China. If we compare this figure with the 61,594 African students studying in China in 2016, the trend is both obvious and dramatic. The history of African students in China can be divided into four periods. The first lasted from 1956 to 1966, when the Cultural Revolution closed all the country's universities. The second mainly applies to the trainees for the Tanzania-Zambia Railway. The third runs from the 1970s to the year 1995. The fourth started from 2000, the year of the China–Africa Cooperation (FOCAC), to the present.[33]

Scholarship Policy towards African Students." In *Annual Review of African Studies in China 2015*, edited by Li Anshan and Liu Haifang: 141–192.

[32] Li Anshan, 2014. "Changing Discourse on China-Africa Relations since the 1990s." *World Economy and Politics* 2: 19–47.

[33] See Li Baoping, 2006. "On the Issues of China-Africa Educational Cooperation." http://www.docin.com/p-747065460.html. Accessed June 2, 2017 and He, Wenping. 2007. "A Summary Analysis of China-Africa Educational Exchanges and Cooperation: Development Phases and Challenges." *West Asia and Africa* 3: 13–18.

The period from 1956–1966 forms the core of the first contact of African students with China. In 1953, the Chinese youth delegation had broad contact with delegates from Egypt, Algeria, Tunisia, Morocco, Madagascar, and French West Africa during the International Conference of the Defense of Youth Rights, held in Geneva, and students of both sides established links at this early stage. Even before the establishment of China-Egypt diplomatic relations on May 30, 1956, the two countries had signed an agreement of cultural cooperation on April 15, 1956.[34] With the program of Egypt-China educational cooperation began the exchange of scholars and students from both countries. Four Egyptian students came to China in 1956, three of whom were there under the academic supervision of famous artist Prof. Li Keran to learn the skill of Chinese painting. They became well-known artists in Egypt after their studies at the China Central Academy of Fine Arts.[35]

In 1957, eleven African students came to study in China from Cameroon, Kenya, Uganda, and Malawi, which were not yet independent. During the 1950s, 24 African students came to China under the scholarship of the Chinese government. Many African countries won their independence in the 1960s and at that time China also initiated educational cooperation with those countries. African students or technicians came to Chinese universities for advanced study under various agreements or programs. In the 1960s, China sent various cultural delegations to Africa, learning different types of African dance while African countries sent young people to China for further study. In 1960, the number of African students in China increased to ninety-five. When the Cultural Revolution occurred in 1966, there were 164 students from 14 African countries. The African students had to go back home since all the universities were closed during this time (China Africa Education Cooperation Group, 2005, thereafter CAECG, 14–17).

Among the students from African countries, a Ghanaian student named Emmanuel Hevi wrote a book, complaining about racism and other unpleasant experiences in China.[36] His negative statement about China

[34] Jiang Chun and Guo Yingde, 2001. *History of China-Arab Relations*. Beijing: Economic Daily Press.

[35] See Li Baoping, 2006. "On the Issues of China-Africa Educational Cooperation." http://www.docin.com/p-747065460.html. Accessed June 2, 2017 and Jiang, Chun and Guo Yingde. 2001. *History of China-Arab Relations*. Beijing: Economic Daily Press: 530.

[36] Hevi, Emmanuel. 1963. *An African Student in China*. Pall Mall.

behind the iron curtain immediately brought about applause from the West. What is more, he was from Ghana, where President Nkrumah was strongly pro-socialist. Most importantly, the West was already looking for something negative about China and Hevi's book came out at just the right time. Hevi's complaint was understandable for several reasons. The first is the economic disaster in the early 1960s. Although African students, like all foreign students in China, enjoyed certain privileges and a higher living standard than ordinary Chinese citizens, China could not further improve these conditions as the early 1960s witnessed the worst economic period in China since the founding of the People's Republic of China. The second is the social conditions of the period. China's dogmatism, social taboos, and regulations set up a kind of "segregation" between African students and ordinary Chinese people, especially African men and Chinese women. Thirdly, to make things worse, the pervasive politics created a vacuum for social interaction that made life less interesting for foreign students, including Africans.[37] However, this was also a period in which young African students saw China with their own eyes and had their first personal contact with Chinese people.

The period of the 1970s is characterized by brotherly friendship, because many of the African students were related through TAZARA. During the 1960s and 1970s, two important events greatly improved the level of contact between China and Africa: the dispatch of Chinese medical teams and the building of the Tanzania Zambian Railway (TAZARA). After China sent its first medical team to Algeria in 1963, Chinese medical teams were dispatched to 47 African countries.[38] Supported by the Chinese government, TAZARA was built specifically to break the blockage of the white racial regimes of South Africa. TAZARA not only made a great contribution to the transportation of minerals from Zambia to the Port of Dar es Saalam, thus helping the frontier countries at the time, but also improved the livelihood of the local people. What's more, the process of building TAZARA provided an opportunity

[37] Cheng Yinghong, 2014. "An African student's Impression of China of the 1960s." *Phoenix Weekly*, May 14. http://www.ifengweekly.com/detil.php?id=490l. Accessed: June 24, 2017.

[38] Li Anshan, 2011. *Chinese Medical Cooperation in Africa: With Special Emphasis on the Medical Teams and Anti-Malaria Campaign*. Discussion Paper 52, Uppsala: Norkiska Afrikainstitutet.

for mutual contact, since more than 60 000 Chinese engineers, technicians, and workers joined the workforce in Africa, allowing Africans to better understand China and the Chinese people.[39] In order to help Tanzania and Zambia run TAZARA, China agreed to train engineers for the two countries from June 1972 on. Thus the trainees of railway technology came to China, followed by formal African students enrolled in Chinese universities in 1973.

This large group was trained for the future TAZARA project across various specialties. They started training courses at Beifang Jiaotong University (Northern University of Transportation). There, the 200 would-be engineers from Tanzania and Zambia enjoyed various life experiences. They took different basic courses in public transportation, then went on to train in different special fields, such as transportation, locomotive specialties, vehicle majors, communication majors, signal specialties, railway engineering specialties, and financial professionals. Among this group, 179 of them finally graduated in September 1975. In 1973, China resumed the enrolment of international students. At this time there were 37 formal African students, followed by 61 in 1974. However, there was a great increase of African students in 1975, reaching 113, probably because Chairman Mao Zedong met with Zambian President Kaunda in 1974 when Mao put forward the "Three Worlds Theory." In 1976, the number of African students increased to 144. Till the end of 1976, China had 355 students from 21 African countries all together and the number with Chinese scholarships increased as well (CAECG). After their return, they played a very important role in the field of transportation and other fields in their own countries.[40]

The period of 1978 to 1995 forms the third period that indicated increasing contact. Since the country's opening-up, China had resumed normal educational cooperation with African countries. However, the economic situation in China was inadequate for this task and the international students were rather limited as a result. In 1978, China enrolled

[39] Monson, J. 2009. "Africa's Freedom Railway: How a Chinese Development Project Changed Lives and Livelihoods in Tanzania". Bloomington and Indianapolis: Indiana University Press.

[40] Liu Haifang and Jamie Monson, 2011. "Railway Time: Technology Transfer and the Role of Chinese Experts in the History of TAZARA." In *African Engagements: Africa Negotiating an Emerging Multipolar World*, edited by Ton Dietz, et al. Leiden and Boston: Brill: 226–251.

Table 1 African students in China (1976–1995)

Year	*Scholarship*	*Self-financed*	*Total*
1976	144	0	144
1977	142	0	142
1978	121	0	121
1979	30	0	30
1980	43	0	43
1981	80	0	80
1982	154	0	154
1983	230	0	230
1984	247	0	247
1985	314	0	314
1986	297	0	297
1987	306	0	306
1988	325	0	325
1989	249	2	251
1990	252	6	258
1991	272	15	287
1992	267	20	287
1993	225	58	283
1994	220	246	466
1995	256	721	977
Total	4174	1068	5242

1236 new international students, with 95% enjoying a Chinese government scholarship (CGS thereafter). Among them, 121 were African students, about 10% of the total.[41] Together with the nearly 300 African students enrolled during 1976–1977, there were more than 400 African students in China, accounting for one quarter of foreign students then. However, only 30 Africans received CGS in 1979, 43 in 1980, and 80 in 1981 (CAEC G) (Table 1).

The statistics indicate that the number of African students continued increasing throughout the 1980s except for the year 1989, when the number dropped to 249 from 325 of the previous year. The number fluctuated between 200 and 300 in the following years, never surpassing 300. The situation might be explained by the clashes that took place between

[41] Chen Changgui and Xie Liangao, 2010. *Approaching Nationalization: Research on International Exchange and Cooperation of Education in China*. Guangzhou: Guangdong Educational Press.

Table 2 African students in China (1996–2011)

Year	*Scholarship*	*Self-funded*	*Total*
1996	922	118	1040
1997	991	224	1215
1998	1128	267	1395
1999	1136	248	1384
2000	1154	234	1388
2001	1224	302	1526
2002	1256	390	1646
2003	1244	549	1793
2004	1317	869	2186
2005	1367	1390	2757
2006	1861	1876	3737
2007	2733	3182	5915
2008	3735	5064	8799
2009	4824	7609	12,433
2010	5710	10,693	16,403
2011	6316	14,428	20,744
2012	6717	20,335	27,052
2013	7305	26,054	33,359
2014	7821	33,856	41,677
2015	8470	41,322	49,792
Total	67,231	169,010	236,241

Source China Education Yearbook, 2003–2015, Beijing: People's Education Press; *Brief Statistics of Foreign Students Studying in China, 2012–2015*, Department of International Cooperation and Exchanges, Ministry of Education of China

African students and Chinese students throughout the 1980s, especially with the clash at Hehai University in Nanjing in 1988.

With more African students enrolled, problems occurred, and racial tension reached a breaking point during the 1980s. For many Chinese people, it was the first time for them to see foreigners and they could not help pointing fingers at foreign students, especially African students. This became a very complicated issue, caused by various factors, such as African students' complaints about their economic or living conditions in China, political divergence between the US and Middle East, different social values, Chinese prejudice toward Africans, etc.[42]

[42] Li Anshan and Liu Haifang, 2013. "The Evolution of the Chinese Policy of Funding African Students and an Evaluation of the Effectiveness." *Draft report for UNDP*.

Complaints and grievances resulted in conflict and even demonstrations. Clashes between African and Chinese students occurred in Tianjin, Nanjing, Beijing, Shanghai, and other cities throughout the 1980s. African students voiced their grievances through different forms, such as demonstrations inside or outside of campus, boycotts of class, hunger strikes, petitions, etc. Occasionally, the Chinese students took part in these demonstrations, thus resulting in clashes. The incidents were described as "national racism" by some scholars.[43] Analyzing the situation from a contemporary perspective, differences in the social systems, values, and cultures between these groups might be the major cause. For a people with a rather conservative character, the Chinese were not used to seeing close relations between men and women in the public sphere, while African students held a more open attitude about the issue. Therefore, the trigger of conflicts was usually the close contact between male African students and Chinese girls, which was disliked by ordinary Chinese people. Of course, China was undergoing a dramatic social transformation of its own at the time. With six students in a room, the Chinese students were not happy with the better treatment received by the foreign students, who lived in a room with only two roommates. In addition, the foreign students enjoyed stipends and better conditions in other aspects of campus.

Therefore, it was natural for the Chinese students and ordinary citizens to complain about the special treatment given to foreign students. Combined with other inequalities or social dissatisfactions, this complaint gave vent to grievances, which then led to the conflicts.

Despite this, one phenomenon is most impressive, e.g., the fact that self-financed students from Africa were increasing during the first half of the 1990s. In 1990, there were six, a number that jumped to 15 in 1991. The number increased to 30, and then to 58 in 1993. The figure jumped to 246 in 1994 and 721 in 1995. More and more African youths wanted to come to China for further study and to benefit from its low fees and easy access to visas, which might explain the situation. After 1996, the history of African students in China entered a period of rapid

[43] See Sautman, Barry, 1994. "Anti-Black Racism in Post-Mao China." *The China Quarterly* 138: 413–437 and Sullivan, M.J. 1994. "The 1988–89 Nanjing Anti-African Protests: Racial Nationalism or National Racism?" *The China Quarterly* 138 (June): 438–457.

development. Readers may ask why 1996. What is the significance of the year?

In May 1996, Chinese President Jiang Zemin visited six African countries: Kenya, Ethiopia, Egypt, Mali, Namibia, and Zimbabwe. This was the first time a Chinese Head of State had visited any part of the sub-Saharan African continent. During the visit, Jiang put forward five proposals for China and Africa to build up a long-term stability and all-round cooperation for the twenty-first century, e.g., sincere friendship, equality, solidarity, and cooperation, alongside common development for the future. The visit and policy might have brought about a great increase of CGS to African students, which leapt from 256 in 1995 to 922 in 1996. With 118 self-financed students that year, the number of African students surpassed 1000 for the first time in China (CAECG, 16). That is the reason why 1996 was chosen as the division of the period (Fig. 1).

After the Forum on China–Africa Cooperation (FOCAC) was established in 2000, how to best promote China–Africa educational cooperation became an important issue. At the end of 2002, among 85,800 foreign students, 1,600 were Africans (*China Education Yearbook*, 2003, 343, thereafter CEY).In 2009, foreign students in China surpassed 230,000; African students reached 12,436 (CEY, 2010, 440). The figures

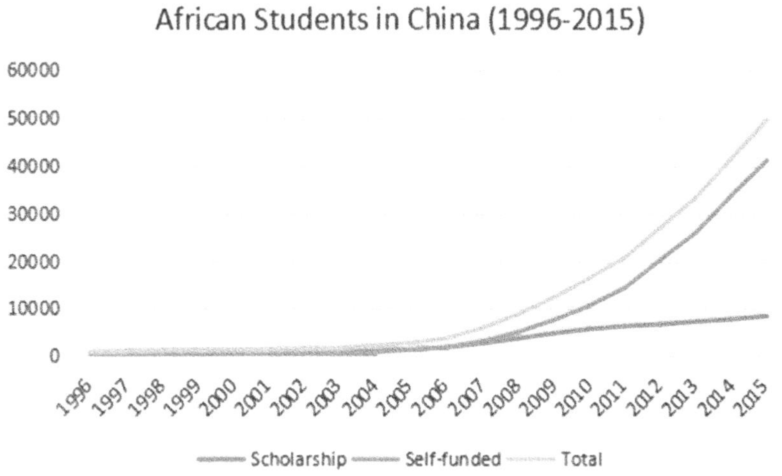

Fig. 1 African students in China (1996–2015)

indicate that the increase of African students is closely linked to that of international students in China overall. Between 1996 and 2011, there were 84,361 African students in China; 36,918 enjoyed CGS while 47,443 were self-funded.

The year 2005 was a turning point in which the number of self-funded students (1,390) from Africa outnumbered scholarship students (1,367) from Africa. This change may be due to the success of the scholarship programs and the Chinese Education Exhibition in Egypt and South Africa since 2003. However, this trend is in line with the situation of international students in China as a whole. In 2005, there were 133,869 self-funded students from 175 countries studying in China, about 94.88% of the total international students, and the growth rate was 28.56% more than 2004. In 2009, of the 238,184 foreign students from 610 Chinese universities and scientific research institutions, 219,939 were self-funded (CEY 2009, 440). The dramatic increase may be explained by Beijing Olympic Games. In 2011, the number of self-funded African students reached 14,428, doubled from African CGS holders (6,316). In 2015, there were 8,470 African CGS holders while the number of self-financed students reached 41,322, nearly five times of the CGS holders. Most of the African students in China, no matter whether they are CGS holders or self-financed students, are pursuing their degrees, and the number of degree-seeking students increased rapidly. In 2014, 84% of African students in China set up the degree as their goal, while only 16% chose non-degree courses.

These trends among the African students are impressive in three aspects. First, the increase is rapid, together with the increase of international students. Secondly, the self-financed African students increased faster than the CGS holders. Thirdly, the overwhelming majority of African students are pursuing their degrees.

3 Policy, Implementation, and Effect

China began accepting foreign students as early as the 1950s. However, until the end of the 1970s, the policy of educational cooperation between China and African countries was still ideologically oriented. The first stage of China's educational policy toward Africa was aimed at uniting African countries and opposed to the US-led capitalism. In the second stage, China opposed the two hegemonies of the US and the Soviet Union.

After the period of reform and opening-up, China started to emerge into the international arena of educational cooperation. As an international newcomer, the Chinese government had learned from its international partners, formulated its policy toward international students, and carried this policy out systematically. There is no specific law regarding any group of international students, such as US students or African students, yet the country's educational policy is itself the product of international relations. No doubt, China's policy of international educational cooperation is closely linked with or even decided by its international strategy. Since China's international education policy has been gradually formulated after the opening-up, this article will mainly focus on the development of this policy since the 1980s.

As early as 1978, the State Council issued a document requiring Chinese people to treat foreign students more amicably, including allowing foreign students to go shopping on the street and to marry Chinese people, from which they had previously been barred.[44] The 1980s witnessed the establishment of primary rules, regulations, and policies in the management of foreign students. In 1985, the State Council approved the "Measures of Administration of Foreign Students," issued by the State Education Commission, Ministry of Foreign Affairs, Ministry of Culture, Ministry of Public Security, and Ministry of Finance. The government realized that "the enrolment and training of foreign students is a strategic work in our diplomacy" and required ministries and different levels of administration to carry out the instruction. The document has eight chapters with 43 separate clauses, covering general principles, enrolment and status management, teaching, professional practice/field work, various types of management (such as ideological work and political activities), livelihoods, sociality, and organizational leadership.[45]

[44] I would like to thank Prof. Lin Fengmin, Dr. Xu Liang, Liu Qinglong and Li Zhen for their help.Ministry of Education, Ministry of Foreign Affairs, Ministry of Public Security, "Request for Instruction of better administration of social management of foreign students", April 29, 1978. See also Liu, 2017.

[45] State Council Document No.121[1985]. Circular of State Council's endorsement of "Methods of Management of Foreign Students" issued by State Education Commission, Ministry of Foreign Affairs, Ministry of Culture, Ministry of Public Security and Ministry of Finance, October 14, 1985. http://www.chinalawedu.com/news/1200/22598/22615/22822/2006/3/he99952431111836002357O-0.htm. (Accessed June 5, 2017).

It seems that the Chinese government regarded foreign students as an integral element of Chinese society and, as such, management was dogmatic and strict. The "Measures of Administration" covers a broad range of content, even the courses, Chinese language teaching, teaching materials, physical training, etc. An important document, it governed the management of international students for many years. On July 21, 1999, the Ministry of Education issued a document administering the enrolment of foreign students by primary and secondary schools.[46] In 2000, the "Provisions on the Administration of Foreign Students in Universities" was issued by the government, which contained eight chapters with 50 clauses. The provisions contained two additional chapters, e.g., "Scholarship" and "Entry-Exit and Residence Procedures," which make the implementation more applicable. The regulations are more systematic than the previous set of regulations.[47]

In March 2017, a new document was issued by three ministries regarding the enrolment and training of international students that was put into practice on July 1, 2017. The purpose of the decree was to standardize the enrolment, raising, and management of international students, provide them with better service in China, promote foreign exchange and cooperation with China's education, and in doing so raise the level of internationalization of China's education. It covers four levels of education, e.g., preschool, primary school, middle school, and university. Although the specific work is the responsibility of the local government, the educational administration under the State Council takes charge of the management of international students, including formulating the general policy of their enrolment and development and guiding and coordinating the concrete work of the local government. Meanwhile, the Ministries of Foreign Affairs and Public Security are in charge of their own respective management issues. Since this document covers every

[46] Decree No.4 [1999] "Provisional Measures of the Ministry of Education for the Administration on Enrolment of Foreign Students by Primary and Secondary Schools". July 21, 1999. http://www.pkulaw.cn/fulltext_form.aspx?Gid=23504. (Accessed June 5, 2017).

[47] Decree No.9 [2000] "Provisions on the Administration of Foreign Students in Universities" issued by Ministry of Education, Ministry of Foreign Affairs and Ministry of Public Security, January 31, 2000. http://www.moe.edu.cn/s78/A20/gjs_left/moe_861/tnull_8647.html. (Accessed June 5, 2017).

institution of education, the previous ones issued in 1999 and 2000 of the same type have been annulled.[48]

Although the related legal regulation and rule treat all foreign students as a whole, there are specific measures regarding African students, especially when some special events or unusual things occur. For example, during the troublesome days in the early 1980s, racial discrimination often occurred in Shanghai. Some Chinese verbally bullied the African students and there were clashes here and there. In one case, the Minister of Education of China had to call a meeting with the leader of African Diplomatic Corps in China and 15 African Ambassadors in February 1983 to explain the problems between Shanghai residents and African students. The African Ambassadors warned that the Chinese government should teach the policemen in addition to residents, since the African students were often harassed by the policemen. This often involved either being stopped by the local policemen without reason or being scolded, which became a nuisance to African students. If the situation went on like this, the African ambassadors warned, friendly Africa–China relations would be damaged. Therefore, different local governmental departments also promulgated various documents regarding specific issues related to African students.[49]

During the mid-1990s, after their graduation in China, some African students did not go back to their home countries. Instead, they took jobs in a country other than China or their motherland. This phenomenon did not agree with the original intention of the Chinese government to help develop capacity building for African countries. In 1996, the Ministry of Education of China issued a document that requested the institution of management to hand African graduates' return-ticket directly to the embassy of the students' country in Beijing at the end of the academic year, thus facilitating the direct return of the students back home. This has become a routine now. Recently, a report commented on China's policy that: "Due to Chinese visa rules, most international students cannot stay

[48] Decree No.42 [2017] "Measures of the Administration of Enrolment and training of International Students by the Educational Institution", issued by Ministry of Education, Ministry of Foreign Affairs and Ministry of Public Security, March 20, 2017. http://www.gov.cn/xinwen/2017-06/02/content_5199249.htm. (Accessed June 5, 2017).

[49] For specific cases, see Liu Haifang, 2017, 167–171 and Li, Anshan and Liu Haifang. 2013. "The Evolution of the Chinese Policy of Funding African Students and an Evaluation of the Effectiveness." *Draft report for UNDP*.

in China after their education is complete. This prevents brain-drain and means that China is educating a generation of African students who—unlike their counterparts in France, the US or UK—are more likely to return home and bring their new education and skills with them."[50]

In 2005, when Chinese President Hu Jintao participated in the High-Level Meeting on Financing for Development at the 60th Session of the United Nations, he promised that in the next three years, China would increase its assistance to developing countries, African countries in particular. He claimed that "China will train 30,000 personnel of various professions from the developing countries within the next three years so as to help them speed up their human resources development."[51] Since the CGS is closely related to China's international strategy, it is also a reflection of this focus of China's policy.

As we can see from the statistics, before 2005, the number of African students who received CGS was always smaller than that of European students. Yet the situation started to change in 2006, a year after Hu Jintao's declaration. Although in 2006 African students and European students accounted for equal proportions of scholars in residence, the actual number of African CGS holders (1,861) surpassed that of Europe (1,858) for the first time. From 2007 onward, the number of African students with CGS began to increase substantially.

There have also been continuous policy promotions. During the China-African Summit (as well the Third FOCAC) in 2006 in Beijing, the issue of setting aside CGS specifically for African students—from 2000 to 4000 annually—was raised. In the Fourth FOCAC in 2009, the CGS again increased to 5,500 every year. The number reached 5710 in 2010.[52] In 2012, the Fifth FOCAC announced that the number of scholarships would reach 20,000 for next three years. That's why the CGS for African students increased so rapidly. In 2011, the number was 6316 and, in

[50] The Conversation, 2017. "China Tops US and UK as Destination for Anglophone African Students." June 28. http://theconversation.com/china-tops-us-and-uk-as-destination-for-anglophone-african-students-78967. (Accessed: June 30, 2017).

[51] Hu Jintao, 2005. "Written Statement by President Hu Jintao of China at the High-Level Meeting on Financing for Development at the 60th Session of the United Nations." September 15, http://politics.people.com.cn/GB/1024/3696504.html. Accessed: June 12, 2017.

[52] Li Anshan, et al., 2012. *FOCAC Twelve Years Later: Achievements, Challenges and the Way Forward*. Uppsala: Nordic Africa Institute: pp. 58–60.

2012, 6717, followed by 7305 and 7821 in 2013 and 2014, respectively. In 2015, the number of African CGS holders reached 8,470 (CEY, 2010–2016).

In order to implement this policy, different strategies were planned by the agents, universities, municipalities, various departments, and even individuals.[53] Regarding CGS, there was no evaluation system until 1997, when the "Provisional Measures of Annual Review of Scholarship of Foreign Students" was issued by the State Education Commission. The measures made it clear for the first time that students had to undergo review according to certain standards predicated on "pass" and "no-pass" grades.[54] Within three years, 7,118 CGS holders took the review, 7008 passed, and 110 did not pass, accounting for 1.55% of total evaluations. In 2000, two documents were issued by the Ministry of Education regarding the CGS annual review system and method.[55] With more accurate standards and more autonomy for the institutions entitled to enroll foreign students, the receiving universities now had real authority to carry out their own review of foreign students. In the same year, 2342 CGS holders from 81 universities went through the review, 2314 (98.8%) passed, 28 did not pass. Among the unqualified students, 17 were Asian, seven European, two African, and two American (CEY, 2002). As for the universities, there are strict qualifications for those that have the authority to offer CGS. Usually only those universities that have top-quality levels of education and qualified professors who can offer courses in foreign languages and adequate educational facilities can take on the responsibility of enrolling international students. In 2015, only 279 designated Chinese universities under the CGS-Chinese University Program were entitled to accept individual scholarship applications.[56]

[53] Li Anshan and Liu Haifang, 2013. "The Evolution of the Chinese Policy of Funding African Students and an Evaluation of the Effectiveness." *Draft report for UNDP*.

[54] "Provisional Measures of Annual Review of Scholarship of Foreign Students", issued by the State Commission of Education, March 28, 1997. http://www.bjfao.gov.cn/affair/oversea/wglxsfg/23801.htm. (Accessed June 5 2017).

[55] Education (Jiaowailai) No.29[2000], "Notice of the Ministry of education on the implementation of the annual review system of the Chinese Government Scholarship", April 26, 2000. http://www.moe.edu.cn/s78/A20/gjs_left/moe_850/tnull_1183.html (Accessed June 5 2017).

[56] "Chinese Government Scholarship Application", August 12, 2015, China Scholarship Council, http://en.csc.edu.cn/laihua/newsdetailen.aspx?cid=66&id=3074.

The Chinese government wanted to become more actively engaged in international educational cooperation. Therefore, different ministries, provinces, municipalities, and even companies started to offer various types of scholarships. Owing to the complicated scholarship system and shortage of space, we would prefer to describe one specific example: the Shanghai Government Scholarship (Table 3).

At the end of the 1990s, a general framework for international student education had been established in China, which was compatible with Chinese culture and China's own education system. Since then, China has continued to improve its international educational cooperation as part of an effort to make itself one of the most popular destinations for foreign students in the world. In the meantime, as economic globalization accelerates, international demand also increases greatly for those young talents who can speak Chinese or who have a solid grasp of knowledge about China. As a result, the number of China's international students continues to grow rapidly. In 2016, the number of international students in China

Table 3 Shanghai government scholarship—Class A[57] (Unit/time: RMB¥/Annual)

Supporting categories	*Field of Study*	*Tuition*	*Accommodation*	*Stipend*	*Medical Insurance*	*Total*
Undergraduate Students	I	20,000	8400	30,000	800	59,200
	II	23,000	8400	30,000	800	62,200
	III	27,000	8400	30,000	800	66,200
Master's Students	I	25,000	8400	36,000	800	70,200
	II	29,000	8400	36,000	800	74,200
	III	34,000	8400	36,000	800	79,200
Doctoral Students	I	33,000	12,000	42,000	800	87,800
	II	38,000	12,000	42,000	800	92,800
	III	45,000	12,000	42,000	800	99,800

Note Full scholarships cover tuition waivers, accommodation, stipends, and comprehensive medical insurance.
Field of Study I includes philosophy, economics, legal studies, education, literature, history, and management; Field of Study II includes science, engineering, and agriculture; Field of Study III includes fine arts and medicine.

[57] "Shanghai Government Scholarship—Class A", China Scholarship Council, March 28, 2016. http://www.csc.edu.cn/Laihua/scholarshipdetailen.aspx?cid=105&id=1293.

increased to 442,773, 45,138 more than in 2015 and with a growth rate of 11.35%. The number of African students increased by 11,802, with the total number reaching 61,594 and a growth rate of 23.7%.[58] This fact demonstrates that the international education in China is becoming more popular than before.

The theory and practice of African development have long been dominated by Western countries. Since their independence, few African countries have developed properly under the guidance of the West; on the contrary, most of them have gotten into very difficult situations.[59] In recent years, the global economy has been volatile, and major changes have taken place in the international balance of power. On the one hand, the US financial crisis and the debt crisis in Europe have landed the Western economy in trouble; on the other hand, emerging economies have become the driving force of the world economy. Thus, "Look East" has become a prevalent tendency in some African countries.[60] Asia's experience of poverty alleviation and development has also become a lesson for Africans who want to find way forward for their own country, and China offers an alternative for the African government. Nigerian historian Femi Akomolafe explained in this way: "Now to the lessons Africa can learn from the world's new economic giant: The first and most profound is that: It is possible! Whichever way we throw it around, China's economic performance is nothing short of a miracle. It shows what a people with confidence, determination and vision can achieve."[61]

[58] "Statistics of Foreign students in China in 2016", March 1, 2017, Ministry of Education. http://www.moe.educn/jyb_xwfb/xw_fbh/moe_2069/xwfbh_2017n/xwfb_170301/170301_sjtj/201703/t20170301_297677.html.

[59] Professor of development studies at Johns Hopkins Deborah Brautigam says, "Where the West regularly changes its development advice, programs, and approach in Africa……China does not claim it knows what Africa must do to develop. China has argued that it was wrong to impose political and economic conditionality in exchange for aid, and that countries should be free to find their own pathway out of poverty. Mainstream economists in the West today are also questioning the value of many of the conditions imposed on aid over the past few decades" (Brautigam, 2009).

[60] ACET. 2009. "Looking East: A Guide to Engaging China for Africa's Policy-Makers." *Key Dimensions of Chinese Engagements in African Countries*, Vol. II, November, 2009.

[61] Akomolafe, Femi. 2006. "No One is Laughing at the Asians Anymore." *New African* 452 (June).

However, the African road to development should be determined by Africans themselves. The Chinese experience holds that, in order to pursue the development of its own economy, a country can only rely on the concerted efforts and determination of its own nation. Never, in all the history of humankind, did a nation rely entirely on foreigners for its own economy.[62] One of the ways for the African government to learn from China is to send their young people to China for further study. In 2005, the Rwanda Government signed an agreement with the Ministry of Education of China to train their undergraduates with a scholarship from the Rwandan government. In the same year, Tanzania's government also signed an agreement with the China Scholarship Council and agreed to train Tanzanian students in China's universities with Tanzanian scholarships (CEY, 2006).

4 Reasons, Motivation, and Purpose

Why do more and more Africans come to China for further study? There are various reasons, such as the favorable conditions provided by China, various motivations among young people, and pragmatic purposes aimed at personal development.[63] To know more about China and to learn more advanced technology from China are the main reasons for young Africans to come to China for further studies.

African media has been dominated by the West and, as a result, there are various untrue stories or even rumors about China. A typical one is that "China used prison labor in Africa," an accusation initially made by Ms. Roberta Cohen, former Deputy Assistant Secretary of State for Human Rights in the Carter Administration.[64] The BBC's irresponsible

[62] Li Anshan. 2013a. "African Countries Encouraged to 'Look East'." *Guangming Daily*, March 28.

[63] Liu Haifang, 2017. "A Study on the Evolution and Effects of China's Scholarship Policy towards African Students." In *Annual Review of African Studies in China 2015*, edited by Li Anshan and Liu Haifang: 141–192.

[64] In 1991, Ms. Roberta Cohen published an article in *New York Times*. She said, "I learned of the case of a Chinese construction company building a road in Benin using prison labour. 70 to 75 percent of the construction workers were known to be prisoners ...... The company was the Jiangsu Construction Company......The company was able to underbid all its competitors by a wide margin because its labor costs were so cheap" (Cohen, 1991). There was no source and no explanation. Since she is a former US top

report notes that "Angola's Chinese-built ghost town" is another case.[65] The residential area soon sold out after it opened for sale to the public.[66] Africans used to know very little about China and most of the young Africans come to China because they would like to see the real China with their own eyes. China's economic development and strong economy are the real attractions to young Africans. The Beijing Olympic Games in 2008 showcased China as never before to Africans, and they were surprised to find an impressive China on TV.[67] The event served as a strong draw to African youths, who soon wanted to know more about China's growth and its developmental experience with modern technologies. They want to understand why and how China is and can be Africa's largest trading partner for many consecutive years and is now the second largest economy in the world. It is China's growing presence in Africa, the commodities, television shows, Confucius Institutes, and Chinese people working in Africa that have aroused growing interest among African students. Maxwell Zeken, for example, is a 16-year-old Liberian who lives in rural Nimba County. Asked where he dreams of studying, he says: "I want to study engineering in China and come back to Liberia to build our roads and our cities. They say you must visit the Great Wall of China. I regret that my country didn't build something like that."[68]

China's readiness for educational cooperation has surely inspired the boom of African students coming to China. In recent years, the Chinese

official, the rumor of "China's prison labor" spread all over the world (Yan & Sautman, 2012, 2017).

[65] Redvers, Louise. 2012. "Angola's Chinese-Built Ghost Town." http://www.bbc.com/news/world-africa-18646243. July 2. Accessed: June 24, 2017.

[66] Situated in a spot about 30 km outside Angola's capital, Luanda, Nova Cidade de Kilamba is a newly built mixed residential project of 750 eight-story apartment buildings for half-million people with a dozen schools and more than 100 retail units, all sold out soon after its finish. To report an unfinished residential project as "no residents" and the area as "ghost town" before it opens for sale is really a biased view, if not a malicious slander. I visited the area in February 2016 and it was prosperous. At China–Africa Media Cooperation seminar held at Remin University on April 26, 2017, I exchanged ideas with Mr. Venancio Rodrigue, reporter of Angola, who verified that BBC's report is a twisted story.

[67] Several African students told me about their deep impression when they saw a different China in TV during the Olympic Game in 2008.

[68] Pilling, David. 2017. "Ports and Roads Mean China is 'Winning in Africa'." *Construction Review Online*, May 6. https://constructionreviewonline.com/2017/05/ports-and-roads-mean-china-is-winning-in-africa/. Accessed: June 25, 2017.

government has worked hard to strengthen its relations with African countries and adopted several measures to encourage African students to become familiar with China, such as the construction of multiple Confucius Institutes for young Africans to learn Chinese and the scholarship necessary to provide favorable conditions to attract African students.[69] Up until 2017, there were 48 Confucius Institutes and 27 Confucius Classrooms located in 33 African countries, each of which provide various levels of Chinese language instruction.

Many African students learned the Chinese language before they applied for Chinese scholarships or for enrolment in any Chinese university.[70] For example, Dr. Belhadj Imen first won the top prize in the Tunisian Chinese Bridge Competition, and the Chinese government subsequently offered her a scholarship to study in the Department of Chinese Language and Literature at Peking University. Since Peking University is the top university in China, many international students have to learn the language before applying for enrolment or a scholarship. It is the same case with other Chinese universities entitled to enroll international students. At the Shanghai Institute of Technology, about 130 African students are majoring in civil engineering and architecture. In their first year, they master Chinese and take a language proficiency test. This is a normal way for international students, including African students, to begin pursuing their degrees in China. Christian King, a student in philosophy and international trade at Renmin University, told Panview: "I started studying Chinese back home in Zimbabwe and it was very difficult at first. The tones and characters were challenging, but after several years in China I am almost fluent. I love and enjoy Chinese now."[71] (Table 4).

Scholarships also allow African students to come to China. With Africa's importance to China's international strategy, scholarships are

[69] Liu Haifang, 2008. "China-Africa Relations through the Prism of Culture—The Dynamics of China's Cultural Diplomacy with Africa." *Journal of Current Chinese Affairs (China aktuell)* 3: 9–44.

[70] Niu Changsong, 2016. "A Survey of African Students' Satisfaction of Chinese Government Scholarship." http://www.docin.com/p-1445264169.html. Accessed: June 25, 2017.

[71] "Africans learning Chinese can boost cooperation channels", March 23, 2015, *CCTV.com*.http://english.cntv.cn/2016/03/23/ARTIvEEYI0kItdGxV6F2JBK0160323.shtml. Accessed June 23, 2017.

Table 4 Confucius institutes and confucius classrooms in African countries (Febrauary 2017)

Country	*Confucius institutes*	*Confucius classrooms*
Angola	Agostinho Neto University	
Benin	University of Abomey-Calavi	1
Botswana	University of Botswana	
Burundi	University of Burundi	
Cameroon	University of Yaounde II	1
Cape Verde	University of Cape Verde	
Comoros		University of Comoros
The Republic of Congo	Marien Ngouabi University	
Côte d'Ivoire	University of Felix Houphouette Boigny	
Egypt	Cairo University, Suez Canal University	3 at the Nile Television of Egypt
Equatorial Guinea	National University of Equatorial Guinea	
Eritrea	National Board for Higher Education of Eritrea	
Ethiopia	Confucius Institute at TVET Institute of Ethiopia, Addis Ababa University	Mekelle University、Hawassa University (in total 5)
Ghana	University of Ghana, University of Cape Coast	
Kenya	University of Nairobi, Kenyatta University, Moi University	2 at CRI in Nairobi
Lesotho		Machabeng College International School
Liberia	University of Liberia	
Madagascar	Antananarivo University, University of Toamasina	1
Malawi	University of Malawi	
Mali		Lycee Askia Mohamed
Mauritius	University of Mauritius	
Morocco	University of Mohammed of V-Agdal, University Hassan II	
Mozambique	Eduardo Mondlane University	
Namibia	University of Namibia	

(continued)

Table 4 (continued)

Country	*Confucius institutes*	*Confucius classrooms*
Nigeria	University of Lagos, Nnambi Azikiwe University	1
Rwanda	College of Education, University of Rwanda	1
Senegal	Cheikh Anta Diop University, Dakar	
Seychelles	University of Seychelles	
Sierra Leone	University of Sierra Leone	
South Africa	University of Stellenbosch, University of Cape Town, Rhodes University, Durban University of Technology, University of Johannesburg	The Cape Academy of Mathematics, Science and Technology, Westerford High School, Chinese Culture and International Exchange Center (in total 5)
Sudan	University of Khartoum	
Tanzania	University of Dodoma, University of Dar es Salaam	Zanzibar Journalism and Mass Media College of Tanzania
Togo	University of Lome	
Tunisia		CRI in Sfax
Uganda	Makerere University	
Zambia	University of Zambia	2
Zimbabwe	University of Zimbabwe	
Total	48 Confucius Institutes	27 Confucius Classrooms

becoming more and more inclined to African students. As for CGS holders, Asian students are always at the top. It is natural both for geopolitical reasons and for the fact that many overseas Chinese in Asian countries are China's neighbors. Although Europe has much fewer CGS-receiving countries than Africa, it has long been in second place. Yet the situation has changed greatly since 2006, when Africa became the second place in terms of the CGS (Table 5).

As the above table shows, the percentage of CGS for Africa and Europe is the same in 2006, but the number of African students is a bit more. The number reached 2,733 in 2007, outnumbering Europe by 626. The CGS holders from Africa increased every year. Now, students from 51 African countries are eligible for Chinese government scholarships (for Europe, the number is 39). In 2010, CGS was extended to 22,390 beneficiaries, increased 4145, 22.72% more than 2009. 11,197 were offered

Table 5 Comparison of CGS holders between Africa and Europe (2003–2010)

Year	*Total scholarship*	*Africa*	*Percent (%)*	*Europe*	*Percent (%)*
2003	6153	1244	20.2	1442	23.4
2004	6715	1317	19.6	1880	23.5
2005	7218	1367	18.8	1761	24.4
2006	8484	1861	21.9	1858	21.9
2007	10,151	2733	26.9	2107	20.8
2008	13,516	3735	27.6	2628	19.4
2009	18,245	4824	26.44	3022	16.56
2010	22,390	5710	25.5	3283	14.66

to Asia (50.01%), 5710 to Africa (25.5%), 3283 to Europe (14.66%), 1761 to America (7.87%), and 439 to Oceania (1.96%) (CEY, 2011). Undoubtedly, Africa has become the second position in terms of CGS holders.

Besides the CGS, there are different types of scholarships that are offered to international students, such as provincial scholarships, ministerial scholarships, university scholarships, and various scholarships with specific purposes, such as those provided by companies, charity organizations, etc. CGS covers the waiver of various fees on campus, including tuition, the teaching material fee, research and survey fee, dissertation guidance fee, one-off resettlement fee, on-campus accommodations, medical insurance, one round-trip international airfare each year for home visits, and one-time round-trip international airfare for all the students. In addition, international students get a stipend monthly. With the country's economic development, the amount of the scholarship has been raised many times over the past years.[72] More and more African students enjoy CGS or other scholarships.

However, the number of self-funded African students has greatly surpassed the Chinese scholarship holders since 2005. In 2015, among 49,792 African students in China, only 8470 were CGS holders while 41,322 were self-funded. I once met a Zambian student in the *Shangdi* region in north Beijing where I live. He told me that he came to learn Chinese in a small language school in *Wudaocao*, an area well-known

[72] "China's government scholarship for international students raised", January 22, 2015. http://old.moe.gov.cn//publicfiles/business/htmlfiles/moe/s5147/201501/183255.html.

among foreign students. That surprised me because he looked very young and had come to China alone. He was living in a residential area far away from the city center and yet he still showed his determination to study Chinese language.

Then what are they up to? What is the purpose of African students in China?

There are different motivations. Some like the reputation of Chinese universities, some want to pursue specific fields.[73] China's experience of development with advanced technologies has inspired many young Africans. Chinese companies are building roads, bridges, hospitals, schools, dams, oil refineries, and modern railways in Africa. What is more, Huawei has been successful in the African IT industry and China is cooperating with Nigeria in the field of satellites. The localization of Chinese companies has attracted talented youngsters in many different countries. Young African students take it as an honor to work in Chinese companies like Huawei. I have also met various African students who are doing their master's programs, such as Serge Mundele at Beijing University of Science and Technology or Oodo Stephen Ogidi, a Nigerian student who worked as a postdoctoral fellow in electrical engineering at Dalian University of Technology. African students are also engaged in the study of social sciences, such as Erfiki Hicham, a Moroccan student who received his PhD in the School of International Studies at Peking University. Dr. Imen Belhadj, a Tunisian scholar, finished her MA in Chinese Language and Literature and her PhD in International Politics and postdoctoral studies in Arab Studies, all in Peking University.

All these phenomena make China an ideal country for young students to pursue further studies. In recent years, more African students have been engaged in professional studies.[74] According to a survey of 2000 samples in 2014, 84% of African students had earned a degree as their educational goal. Among them, 41.61% applied for medical science as their major, 21.56% chose engineering-related subject as a major, while 13.94%

[73] See King, K, 2013. China's Aid and Soft Power in Africa The Case of Education and Training. Suffolk: James Currey and Tsui, Chak-Pong Gordon. 2016. "African University Students in China's Hong Kong: Motivations, Aspirations, and Further Exchanges." In *Africans in China: Guangdong and Beyond*, edited by Adams Bodomo: 119–137.

[74] Bodomo, A, 2011. "African Students in China: A Case Study of Newly Arrived Students on FOCAC Funds at Chongqing University." PPT Outline, University of Hong Kong, p. 29.

went for business and management. The application for top five majors makes up 98.33% of all applicants.[75] A student from the Republic of Congo came to China in 2007. He once told me that after he saw several telecommunications products in the market were "made in China," he decided that he would go to China. With a dream of becoming the minister of telecommunications in his country, he is now a graduate student of telecommunications at the Beijing University of Posts and Telecommunications.

Obviously there are other practical reasons, too. In China, the tuition fee is lower and it is easier to get a visa than in the Western countries. Moreover, if you have grasped the Chinese language and have some understanding of the Chinese culture, it is more convenient for you to find a good job in good Chinese companies such as Huawei and ZTE in their own country. It's true that the Chinese people know very little about Africa and there is surely ignorance among the Chinese people regarding the black color of African students. However, the friendliness and warm feelings of Chinese people may also encourage young Africans to study in China.[76]

5 Role, Contribution, and Agency

Since African students are becoming a big group in China, it is important to ask, what role do they play? What contributions do they make to both China and Africa, or to bilateral relations?

The history of the human being is a history of (im)migration. Although international students are usually not classified as either immigrants or diaspora, their linkage with the diaspora is obvious. The first link is in their role as a bridge. As the second largest group among the

[75] "EOL & CUCAS jointly published 2014 Report of International Students in China", http://www.eol.cn/html/lhlx/content.html. Accessed: June 25, 2017.

[76] Regarding whether there is racism in China, views are different. A Ghanaian student talked about her experience in China, saying, "Others often ask me if I found Chinese to be racist, and whether their treatment of me as a spectacle—taking pictures, touching my hair, rubbing my skin, staring at me—does not indicate a racist attitude. I respond that I find them curious. Many of the experiences I had were borne of ignorance, not racism. Despite always being identified as 'black' and 'African', I never felt discriminated against or antagonized, but rather treated with warmth and friendliness. Because I spoke Mandarin, I could often understand what people said about me, and they were rarely disparaging or maligning" (Baitie, 2013).

African community in China, they constantly function as a bridge between African culture and Chinese culture. As soon as they enter into China and begin their social life on campus, they start the exchange of cultural ideas and play a role as the bridge between different cultures through conversations with their fellow students, contact with the authorities and ordinary Chinese, through courses and debates, social interaction, etc.

In a new situation, young Africans always face new challenges and will necessarily experience culture shock, which as we have seen is more serious for undergrads than graduates, women than men.[77] Cultural exchange or adaptation thus becomes important, since it can occur in daily life. A learning process and social contact bring about better relations in the host community.[78] Cultural adaptation becomes an active response to new conditions, a process of mutual learning. I personally supervised many African graduates, including three PhD students from Tunisia, Morocco, and the Democratic Republic of Congo. They told me various stories about their experiences. These experiences mix ignorance with biased views and friendliness with warm feelings.

African students become the bridge between Africa and China. Moses is a Nigerian student majoring in Chinese language teaching. He came to China in 2013 and has a typical Chinese name: "Wu Wenzhong." Having learned the Chinese language in Nigeria from his childhood on, he became addicted to Chinese culture during his stay in China. He has learned various kinds of Chinese arts and performances, including some superior arts such as *Xiangsheng* (Chinese cross-talk or comic dialogue) and lion dance, as well as attended various performances and art shows. With his profound interest in Chinese language and culture, Moses participated in the "Hebei Provincial Foreign Scholars' Chinese Talents' Show" in November 2014. He showed his kung fu, recited Chinese classic poems, and also performed self-composed Chinese cross-talk with a foreign partner. Thanks to his excellent performance and skill, he won "Best Creative Award," "Best Eloquence Award," and the "Silver Award of Recitation of a Classic Poem." He also received the "Best All-Around King" for his talented performance of Chinese culture. Owing to

[77] Disima, 2004. "Cultural Adaptation of Foreign Students in China." M.A. Psychology Thesis, Nanjing Normal University.

[78] Hashim, Ismail Hussein, et al., 2003, "Cultural and Gender Differences in Perceiving Stressors: A Cross-Cultural Investigation of African and Western Students at Chinese Colleges." *Psychological Science* 26 (5): 795–799.

his capability in Mandarin and understanding of Chinese culture, he was called a "China-hand" by his Chinese friends.[79]

African students' tuition has no doubt contributed to China's economy. Besides, their cultural knowledge about Africa has brought about multi-culturalism in China. Many cultural exchanges are going on between African and Chinese students. Africans are learning about Chinese language, culture, and work ethic.[80] At the same time, they are transmitting African culture, values, and skills.[81] Chinese students also have many different occasions to learn about African culture. There are various clubs devoted to African cultures in Chinese cities, including those devoted to topics such as African dance, African music, and African drums, which are all due to African students' contribution to society.[82] Francis Tchiégué, a Cameroonian student, came to China for further study many years ago. He originally received his PhD in Cameroon. In China, he was attracted by the similarity between Cameroonian culture and Chinese culture and thus started to learn Chinese art, skill, and Chinese cross-talk. Through various activities, he introduced African culture and even made a Chinese traditional costume using Cameroonian cloth. Francis was named "Envoy of Art Exchange between China and Africa." Now he is trying his best to introduce African films to China.[83]

There is an annual International Cultural Festival at Peking University and African students typically set up their stands to proudly introduce their own culture to a Chinese audience.[84] My African student, Antoine

[79] Yang Mengjie, et al., 2016, "China-Hand's Dream of Mandarin." *Chinanews*, November 28, 2016. China. http://www.chinanews.com/sh/2016/11-28/8076600.shtml. Accessed June 1, 2017.

[80] King, K, 2013, *China's Aid and Soft Power in Africa The Case of Education and Training*. Suffolk: James Currey.

[81] Amoah, Lloyd G. Adu, 2012, "Africa in China: Affirming African Agency in Africa-China Relations at the People-to-People Level." In *China-Africa Partnership: The Quest for a Win–Win Relationship*, edited by James Shikwati, pp. 104–115. Nairobi: Inter Region Economic Network.

[82] Shikwati, James, ed., 2012. *China-Africa Partnership: The Quest for a Win–Win Relationship. Nairobi: Inter Region Economic Network*: 93–97.

[83] "Cameroonian Tchiégué's life in China", http://tv.cntv.cn/video/C10616/3ce5c25b1bfc476095406544b5971b8a. Accessed: September 6, 2017.

[84] "African student's speech at International Festival in Peking University", October 29, 2013. http://www.fmprc.gov.cn/zflt/chn/zxxx/t1094003.htm. Accessed: June 23, 2017.

Lokongo, played the drum in the festival and many Chinese students tried to learn the skill. In order to introduce African culture to ordinary Chinese people, the Center for African Studies at Peking University and the *Half-Monthly Talks* co-run a special column entitled "Entry into African Culture." So far 15 articles on different topics about African culture have been published, covering African world heritage, languages, films, the role of chiefs, Léopold Sédar Senghor, Nobel Prize winner Wole Soyinka, Ibn Battuta, Ibn Khaldun, the civilization of Ethiopia, etc., some of which were written by African students (Li, 2015).[85]

The Chinese students in London serve as a bridge between Chinese culture and British culture, and between the Chinese diaspora in Britain and British society.[86] African students play the same role. They not only serve as a bridge between African culture and Chinese culture, but also a bridge between the African community in China and the Chinese people who are interested in China–Africa bilateral relations.[87] Thanks to the effort of African students, the Chinese people have begun to familiarize themselves with African values, ideas, dance, drums, pictures, sculptures, etc. There is another good example: my former student Wang Hanjie wrote her B.A. thesis on the theme of "The History and Spread of Djembe Drum in China." When asked why she chose this topic, she told me with a smile that she was a member of the Djembe Club at Peking University.[88] In Wuhan, an important metropolitan city in central China where advertisements projecting Western brands and tastes are rather popular, during an interview with some of the local ladies about their tastes for African cultural products in China, the young ladies claimed they found it *hen ku* (very cool). "Their choice showed that they were

[85] See Shikwati, James, ed., 2012. *China-Africa Partnership: The Quest for a Win–Win Relationship*. Nairobi: Inter Region Economic Network: 93–97.

[86] See Wu Bin, 2015. "Links between Chinese International Students and Overseas Chinese Communities: An Empirical Study in Nottingham, UK." *Overseas Chinese History Studies* 2: 1–11 and Wu, Bin. 2016. "Local Engagement of Chinese International Students in Host Societies: A Perspective of Diasporic Chinese Community Building." *The International Journal of Diasporic Chinese Studies* 8 (2): 13–30.

[87] Bredeloup, Sylvie, 2014. "West-African Students Turned Entrepreneurs in Asian Trading Posts: A New Facet of Globalization." *Urban Anthropology, Special Issue on African Global Migration* 43 (1/2/3): 17–56.

[88] Wang Hanjie, 2013. "The Spread and Distribution of African Drums in China" In *Annual Review of African Studies in China (2012)*, edited by Li Anshan and Liu Haifang. Beijing: Social Sciences Academic Press (China): 442–458.

avant-garde, cosmopolitan and even modern in their fashion tastes and preferences. This African cultural influence in Wuhan has been facilitated in no small measure by the annual Wuhan University Autumn International Cultural Festival."[89] In other universities in Beijing, such as Minzu University of China, there is even a special African Culture Day.

Although some African students choose to move to a third country after they finish their university in China, many of them decide to return to home after their graduation and make important contributions to their own country.[90] In addition to their participation in the work of different fields, some of them attain important positions in their countries and assume high posts in the government. As of 2005, eight former recipients of Chinese government scholarships held ministerial positions or above in their respective home countries, eight served as Ambassadors or Consuls to China, six worked as Secretaries to their countries' Presidents or Prime Ministers, and three as Secretary-generals of the Association for Friendship with China, not to mention many experts and elites in other fields (CAECG, 20–21). Taking Peking University as an example, its former student Mr. Mulatu Teshome Wirtu became Speaker of the Parliament in Ethiopia and now is the President of the country. After her education, Ms. Lucy Njeri Manegene worked for the Ministry of Foreign Affairs in Kenya. Ms. Rakotoarivony R.J.Manitra went back to Madagascar after her MA studies and now serves in the Madagascar Embassy in China. Ms. Mapulumo Lisebo Mosisili returned to Lesotho after being awarded her MA and is now the Principal Secretary of the Labor Department in Lesotho.[91,92]

Another important change instigated by African students related to their experiences in China is that they start their pan-African connections

[89] Amoah, Lloyd G. Adu, 2012. "Africa in China: Affirming African Agency in Africa-China Relations at the People-to-People Level." In *China-Africa Partnership: The Quest for a Win–Win Relationship*, edited by James Shikwati, 104–115. Nairobi: Inter Region Economic Network.

[90] See Bodomo, A, 2011. "African Students in China: A Case Study of Newly Arrived Students on FOCAC Funds at Chongqing University." PPT Outline, University of Hong Kong and Li, Anshan and Liu Haifang. 2013. "The Evolution of the Chinese Policy of Funding African Students and an Evaluation of the Effectiveness." *Draft report for UNDP*.

[91] Li, Anshan, 2013b. "My African Students". In *Chinese Ambassadors Telling African Stories*, edited by Cheng Tao and Lu Miaogeng. Beijing: World Affairs Press: 156–168.

[92] Lisebo Kikine to Li Anshan, Tue, 12 Feb 2013 00:18:24 + 0800 (CST).

on campus. When answering the BBC's question about why he came to China, Mikka Kabugo, a Ugandan student representing the African Students Association of Peking University, said he started to know China through a traditional Chinese medicine doctor in Uganda. When he came to China, he found that Beijing was a global village and he could exchange ideas about African affairs with fellow African students from other countries. He hoped that would definitely make them more broad-minded with a global outlook. With other African students in the same association, they look at African issues from a pan-African perspective and think about what they should do to contribute to the continent.[93] That makes the world very different for an African student. What's more, through classes, debates, and various seminars held jointly by the African Students Association and the Center for African Studies at Peking University, they have learned a great deal about world affairs and the African situation, problems, and solutions.

Following their fellow students at Peking University, the African students of Tsinghua University also formed the African Students Association on African Day, May 25, 2017. Students from 27 African countries joined the organization. During the opening ceremony, African students held discussions about various issues, such as the thoughts of President Julius Kambarage Nyerere and President Kwame Nkrumah, a presentation of the continent's contribution to knowledge development shared by Dr. Chabalala, a student from the School of Public Health at Tsinghua University, etc. What's more, Professor Tang Xiaoyang from the School of International Relations and the Carnegie–Tsinghua Center talked about the structural change underway with China–Africa relations and Professor John Akokpari from the Center of African Studies at Peking University opened up a discussion addressing opportunities for African students in the diaspora to be change-makers for the development of their countries. There are other African student associations as well, such as General Union of African Students in China (GUASC), General Union of African Students in Tianjin (GUAST), etc.[94]

[93] "Why are African students flocking to Chinese universities?" June 29, 2017, *BBC World Service Newsday*. http://www.bbc.co.uk/programmes/p0577s49?ocid=socialflow_facebook. Accessed: June 30, 2017.

[94] King, K, 2013, *China's Aid and Soft Power in Africa The Case of Education and Training*. Suffolk: James Currey.

Although most international students are not normally classified as immigrants, Bodomo correctly pointed out that the process of trade between Africa and China began with Africans who studied in China and remained there to do business,[95] and some African students in China have ended up trading business with China (Haugen, 2013). African students are the first to carry out business between their home country and China and Africa–China trading business began with those African students who remained in China and started their own businesses after graduation. Although they tend to have little capital to begin with, they have a solid social and linguistic background to their advantage. Gradually, they became major intermediaries of the trade between Africa and China, thus contributing to the economic activities between both sides. Dr. Abdul is a good example. Working as a veterinary official for the Niger Government, he received a Niger-China Friendship Scholarship from the Chinese government. After he finished his degree, he decided to change his profession by opening a new occupation unfamiliar to him but more profitable. Since 2000, Dr. Abdul has established himself in Guangzhou as an exporter of medicine and related products for veterinary use to Africa and Europe. He uses his knowledge to obtain those products directly from factories in northern China. After his success, he resumed his connections with the Niger government. Now fluent in Mandarin, Dr. Abdul serves as an honorary consul for Niger responsible for conveying the demands of Nigerian students with scholarships at Chinese universities to any Nigerian minister who visits Guangzhou. He describes his role as turning "brain drain" into "brain gain." According to Bredeloup, situations like this have resulted from two facts, e.g., the opportunities created by China's rapid economic development and the change of or even devaluation in the position of civil servants in Africa. There are quite a few examples like Abdul, such as Patrick from the Democratic Republic of Congo, Aziz from Mali, etc.[96] As indicated in other studies, some of the self-funding African students and even some CGS holders in Guangdong and Zhejiang entered trading activities for the first time in their lives and

[95] Bodomo, A, 2013, "African Diaspora Remittances are Better than Foreign Aid Funds." *World Economics* 14 (4): 21–28.

[96] Bredeloup, Sylvie, 2014. "West-African Students Turned Entrepreneurs in Asian Trading Posts: A New Facet of Globalization." *Urban Anthropology, Special Issue on African Global Migration* 43 (1/2/3): 17–56.

became settled businessmen while studying in China (Haugen, 2013).[97] However, many African students like Dr. Abdul now serve as a business and cultural bridge between Africa and China.

It is generally presumed that China has initiated every move in China–Africa relations while African agency in shaping and influencing her deepening relations with China is either paralyzed or non-existent. It is therefore interesting that some African students have started to research African agency in the making of Africa–China relations. Adu Amoah, a former Ghanaian government official, later became a student in China and married a Chinese woman. As the President of the African Students' Representative Committee of Wuhan University, he used his own observation and experiences to indicate that African students can be the masters of their own lives in China. He described a lively African migrant community emerging in Wuhan "which may potentially add to the makings of an African diaspora in contemporary China" and claimed that "this migrant African population is constituted fundamentally by students," "comprising a dynamic fashion of those pursuing their course of study and those who stay on after graduation." Taking Wuhan as an example, Amoah describes how the African presence in China influenced the reality of Chinese society in the form of fashion, inter-marriage and the exchange of language learning (African can explain their own worldview and experience in Mandarin while teaching English to Chinese students), along with the management of African SMEs, such as nightclubs run by Africans. "This is necessary to dispel the interpretation of Africa 'under the sign of crisis' (Quayson) in popular and academic discourses in general and specifically, the patronising idea of Africa as a clueless, pliant and suppliant partner in Africa-China relations."[98]

Of course, African students in China have not only immensely improved cooperation between Africa and China and contributed a great deal to cultural exchange, but have also promoted the internationalization

[97] See Bodomo, A, 2012. *Africans in China: A Sociocultural Study and Its Implications on Africa-China Relations*. New York: Cambria Press and Amoah, Lloyd G. Adu. 2012. "Africa in China: Affirming African Agency in Africa-China Relations at the People-to-People Level." In *China-Africa Partnership: The Quest for a Win–Win Relationship*, edited by James Shikwati, 104–115. Nairobi: Inter Region Economic Network.

[98] Amoah, Lloyd G. Adu, 2012. "Africa in China: Affirming African Agency in Africa-China Relations at the People-to-People Level." In *China-Africa Partnership: The Quest for a Win–Win Relationship*, edited by James Shikwati, 104–115. Nairobi: Inter Region Economic Network.

of China's universities.[99] For African students, there are definitely cultural shocks, homesickness, social adaptation, psychological stress and frustrations, difficulties, and problems in daily life. What's more, they experience misunderstanding and prejudice from Chinese students and international students or language barriers in the educational process. Chinese teachers' English proficiency is not always good enough, which makes African students' learning process more difficult.[100] In the future, there is much room for improvement.

In terms of the subjects treated in this article, international educational cooperation involves three parties: China, African countries, and African students. Here we can ask questions of the Chinese government, African countries, and African students themselves.

When addressing the Chinese side, it is important to keep in mind that Africa is not a totality but a continent of 54 countries which all have different conditions and needs.[101] African students, themselves, are not a totality either, but different individuals. Besides scholarship, does the Chinese government provide adequate living conditions for African students with various different religions, lifestyles, and cuisines in a society that is unfamiliar to them? Are Chinese teachers qualified to transmit their knowledge to African students? Are there good measures in place for African students to introduce their own culture to the Chinese society? Is there enough room for African students to exchange ideas and experiences with their Chinese partners? Better service should be provided for sure.

For the African countries, we must remember that the returning African students are those who love their own country and want to contribute the knowledge they learned in China to their motherland. Do African governments offer a good opportunity for African students to work at home after their graduation in China? Does the government

[99] Liu, Haifang, 2017. "A Study on the Evolution and Effects of China's Scholarship Policy towards African Students." In Annual Review of African Studies in China 2015, edited by Li Anshan and Liu Haifang: 141–192.

[100] Hashim, Ismail Hussein, et al. 2003. "Cultural and Gender Differences in Perceiving Stressors: A Cross-Cultural Investigation of African and Western Students at Chinese Colleges." *Psychological Science* 26 (5): 795–799.

[101] Apithy, Sedozan, 2013. "The Policy of Sino-African Educational Cooperation: What does Africa Expect for Sino-African Educational Cooperation?" *Annual Review of African Studies in China 2012*, Beijing: Social Sciences Academic Press (China): 326–329.

show enough concerns and care for their academics and life in China and create a better condition to facilitate their study and daily requirements? Does the embassy provide a suitable channel for communication with the African students in China and look after their interests and provide for their reasonable demands?[102] Better conditions should be prepared for those who would like to return home to serve.

For African students in China, we should remind them that they are pursuing a great dream, following the great expectation of their country and the great hopes of their family. We might ask them, do you make good use of the scholarship and provide your best efforts by studying hard to meet the challenge ahead, thus becoming fully prepared for future work? Do you take every opportunity to introduce African culture or the culture of your own country to ordinary Chinese people or to fellow students from other countries? Do you learn good experiences or developmental lessons from other countries, thus providing the opportunity to use them when you want to realize your dream after your return?

A report entitled, "China Tops US and UK as Destination for Anglophone African Students" appeared in *The Conversation* on June 28, 2017, noting that "According to the UNESCO Institute for Statistics, the US and UK host around 40,000 African students a year. China surpassed this number in 2014, making it the second most popular destination for African students studying abroad, after France which hosts just over 95,000 students."[103] It seems more and more African youth have come to China in recent decades and there is constant increase of African students in China. They serve as the bearers of African culture, mediators of bilateral trade and business, transmitters of social organization and ways of life, and bridge between Africa and China. "It's still too early to tell how these new dynamics might be shaping geopolitics on the continent."[104]

[102] I supervised several PhD students from Africa. Some of them came across difficulties in finding a decent job after return. An absurd case is that one of them was even asked to translate his PhD. Dissertation, which was written in Chinese, into French in order to prove his academic capability.First published in *African Studies Quarterly* (US),Volume 17, Issue 4 (February 2018) Special Issue "China-Africa Relations: Theoretical and Practical Perspectives on African 'Migrants' in China".Guest Editor: Agnes Ngoma Leslie. http://asq.africa.ufl.edu/files/v17i4.pdf.

[103] The Conversation, 2017, "China Tops US and UK as Destination for Anglophone African Students." June 28. http://theconversation.com/china-tops-us-and-uk-as-destination-for-anglophone-african-students-78967. Accessed: June 30, 2017.

[104] Ibid.

Yet, these African students are definitely creating a new world. To integrate into a host society does not mean to give up one's own culture. To build a bridge between two cultures and transform oneself from an "enclave" to "bridge" remains a difficult task. It is worth trying, and workable.

On November 11, 2017, the first "Amanbo Cup" of the Employment/Innovation Competition for African Students was held in Shenzhen, China. Jointly organized by the Center for African Studies of Peking University, the China–Africa International Business School of Zhejiang Normal University, and the Center for China–Africa Sustainable Development, the competition was sponsored by Shenzhen Right Net Tech Co. Ltd (Amanbo) and aimed at training and developing young African talents in the areas of innovation and entrepreneurship.[105] The six finalists were selected from more than 200 projects among African students in various universities in China. Drawn from African contestants across five universities, the projects involved health care (1st Award), biological pharmacy and ecological farms (2nd Award), water resources protection, the recycling economy, and preschool education (3rd Award). The competition was significant in three aspects. First of all, this was the first time for an African students' entrepreneurship competition in China and will definitely play an important role in the future of China–Africa cooperation. Secondly, it was initiated by a private enterprise and sponsorship while being presided over from the beginning to end by civil organizations. Thirdly, it was the first time a university and private institution combined their joint efforts to promote China–Africa cooperation with a focus on the development of young African talents.

Coincidentally, on November 15, 2017, the African Students' Job Fair was held in Beijing with the theme of "The Belt and Road Initiative and Prospects of African Youth Employment." It was organized by the Chinese People's Association for Friendship with Foreign Countries (CPAFFC) and the University of International Business and Economics. This fair was a recruitment drive by Chinese companies specifically aimed at African students in China. A total of 66 state-owned, local and private Chinese enterprises provided nearly 500 jobs on the African continent. African youth in both China and Africa showed great interest in the fair; nearly 400 African students in China came to the job fair in person, while

[105] http://www.chinafrica.cn/chinese/focus/201711/t20171114_800109951.html.

about 1000 African young people from 52 African countries submitted their resumes online beforehand.[106] I believe that with the opportunity for Chinese enterprises to enter Africa under the special opportunities provided by capacity cooperation, the road to China–African cooperation will continue to advance and grow.

The China–Africa Beijing Summit in 2018 marked a new beginning of bilateral cooperation in education in terms of capacity building. Among the "Eight Actions" of the Beijing Summit, the fifth is to implement capacity-building actions. China has decided to strengthen the exchange of development experience with Africa and support cooperation in economic and social development planning; set up 10 "Luban Workshops" in Africa to provide vocational skills training for African youth; support the establishment of China Africa innovation cooperation centers aimed at promoting youth innovation and entrepreneurial cooperation; implement the "Head Goose Project" to train 1000 elite talents for Africa; and provide 50,000 Chinese government scholarships to African students, 50,000 training positions for African youth, and invite 2000 African youth to China to promote mutual exchanges among young people.[107]

With such a vision in mind and concrete projects in action, China-African cooperation will definitely achieve greater progress.

References

ACET. 2009. "Looking East: A Guide to Engaging China for Africa's Policy-Makers." *Key Dimensions of Chinese Engagements in African Countries*, Vol. II, November 2009.

ACET. 2013a. "African Communities in China Hail Xi's Visit." *China Daily*, March 24.

ACET. 2009b. "African Community Needs more Attention." *China Daily*, November. 2.

ACET. 2013. "Africans Create Community in Guangzhou." *China Daily*, October 14.

[106] http://ge.cri.cn/20171116/c5c7ee5f-147e-a879-c315-8114acae8e4c.html.

[107] "Interpretation of the eight actions of the Beijing summit of China Africa Cooperation Forum", September 19, 2018, http://www.mofcom.gov.cn/article/ae/ai/201809/20180902788421.shtml.

African Union. 2005. "Report of the Meeting of Experts from the Members of the States on the Definition of African Diaspora." April 11–12, Addis Ababa, Ethiopia.

Akomolafe, Femi. 2006. "No One is Laughing at the Asians Anymore." *New African* 452 (June).

Amoah, Lloyd G. Adu. 2012. "Africa in China: Affirming African Agency in Africa-China Relations at the People-to-People Level." In *China-Africa Partnership: The Quest for a Win-Win Relationship*, edited by James Shikwati, 104–115. Nairobi: Inter Region Economic Network.

An, Ran, et al. 2007. "African Students' Educational Needs and the Recruitment Style." *High Education Exploration* 5: 110–113.

Apithy, Sedozan. 2013. "The Policy of Sino-African Educational Cooperation: What does Africa Expect for Sino-African Educational Cooperation?" *Annual Review of African Studies in China 2012*, Beijing: Social Sciences Academic Press (China): 326–329.

Baitie, Zahra. 2013. "On Being African in China." *The Atlantic*, August 28.

Bertoncello, Brigite and Sylvie Bredeloup. 2006. "La migration chinoise en Afrique: accélérateur du développement ou 'sanglot de l'homme noir'." *Afrique Contemporaine* 218: 199–224.

Bertoncello, Brigite and Sylvie Bredeloup. 2007a. "De Hong Kong à Guangzhou, de nouveaux 'comptoirs' africains s'organisent." *Perspectives chinoises* 98 (1): 98–110.

Bertoncello, Brigite and Sylvie Bredeloup. 2007b. "The Emergence of New African 'Trading posts' in Hong Kong and Guangzhou." *China Perspectives* 1: 94 –105.

Bertoncello, Brigite and Sylvie Bredeloup. 2009. "Chine-Afrique ou la valse des entrepreneurs- migrants." *Revue européenne des migrations internationales* 25 (1): 45–70.

Bertoncello, Brigite, Sylvie Bredeloup, and Olivier Pliez. 2009. "Hong Kong, Guangzhou, Yiwu: de nouveaux comptoirs africains en Chine. " *Critique international* 44: 105–121.

Bodomo, A. 2007. "An Emerging African-Chinese Community in Hong Kong: The Case of Tsim Sha Tsui's Chungking Mansions." In *Afro-Chinese Relations: Past, Present and Future,* edited by Kwesi Kwaa Prah. Cape Town: Centre for Advanced Studies in African Societies: 367–389.

Bodomo, A. 2009a. "Africa-China Relations in an Era of Globalization: the Role of African Trading Communities in China." *West Asia and Africa* 8: 62–67.

Bodomo, A. 2009b. "Africa-China Relations: Symmetry, Soft power, and South Africa." *The China Review: An Interdisciplinary Journal on Greater China* 9 (2):169–178.

Bodomo, A. 2009c. "The African Presence in Contemporary China." *China Monitor*. University of Stellenbosch: South Africa.

Bodomo, A. 2010. "The African Trading Community in Guangzhou: an Emerging Bridge for Africa-China Relations." *The China Quarterly* 203: 693–707.

Bodomo, A. 2011. "African Students in China: A Case Study of Newly Arrived Students on FOCAC Funds at Chongqing University." PPT Outline, University of Hong Kong.

Bodomo, A. 2012. *Africans in China: A Sociocultural Study and Its Implications on Africa-China Relations*. New York: Cambria Press.

Bodomo, A. 2013. "African Diaspora Remittances are Better than Foreign Aid Funds." *World Economics* 14 (4): 21–28.

Bodomo, A. 2014. "Africans in China: A Bibliographical Survey." In *Annual Review of African Studies in China 2013*, edited by Li Anshan and Lin Fengmin. Beijing: Social Sciences Academic Press (China): 109–121.

Bodomo, A. 2016. *Africans in China: Guangdong and Beyond*. New York: Diasporic Africa Press.

Bodomo, A. and Grace Ma. 2010. "From Guangzhou to Yiwu: Emerging Facets of the African Diaspora in China." *International Journal of African Renaissance Studies* 5 (2): 283–289.

Bodomo, A. and Grace Ma. 2012. "We Are What We Eat: Food in the Process of Community Formation and Identity Shaping among African Traders in Guangzhou and Yiwu." *African Diaspora* 5 (1): 1–26.

Bodomo, A. and Roberval Silva. 2012. "Language Matters: the Role of Linguistic Identity in the Establishment of the Lusophone African Community in Macau." *African Studies* 71 (1): 71–90.

Bork, T., et al. 2011. "Global Change, National Development Goals, Urbanization and International Migration in China: African Migrants in Guangzhou and Foshan." In *Megacities: Our Global Urban Future*, edited by F. Kraas, S. Aggarwal, M. Coy and G. Mertins. London: Springer.

Brautigam, Deborah. 2009. *Dragon's Gift: The Real Story of China in Africa*. Oxford U.P.

Bredeloup, S. 2012. "African Trading Posts in Guangzhou: Emergent or Recurrent Commercial Form?" *African Diaspora* 5 (1): 27 – 50.

Bredeloup, Sylvie. 2014. "West-African Students Turned Entrepreneurs in Asian Trading Posts: A New Facet of Globalization." *Urban Anthropology*, Special Issue on African Global Migration 43 (1/2/3): 17–56.

Brief Statistics of Foreign Students Studying in China, 2012–2015. Department of International Cooperation and Exchanges, Ministry of Education of China.

Chen, Changgui and Xie Liangao. 2010. *Approaching Nationalization: Research on International Exchange and Cooperation of Education in China*. Guangzhou: Guangdong Educational Press.

Cheng, Tao and Lu Miaogeng, eds. 2013. *Chinese Ambassadors Telling African Stories*. World Affairs Press.

Cheng, Yinghong. 2014. "An African student's Impression of China of the 1960s." *Phoenix Weekly*, May 14. http://www.ifengweekly.com/detil.php?id=490l. Accessed: June 24, 2017.

China Africa Education Cooperation Group. 2005. *China Africa Education Cooperation*. Beijing: Peking University Press.

China Africa Project. 2013. "Leading China Scholar Li Anshan Recalls his Experiences Teaching African Students. http://www.chinaafricaproject.com/leading-china-scholar-li-anshan-recalls-his-experiences-teaching-african-students-translation/ Accessed June 4, 2014.

China Education Yearbook, 2003–2015. Beijing: People's Education Press.

Cissé, Daouda. 2013. "South-South Migration and Trade: African Traders in China." *Policy Briefing*.

Center for Chinese studies n° 4 / 2013.

Cohen, Roberta. 1991. "China Has Used Prison Labor in Africa." *New York Times*, May 11.

Cui, Kuai, et al. 2009. "Study on the Trade Pattern of African Community in Guangzhou." *Economic Outlook the Bohai Sea* 5: 14–17.

Dang, Fangli. 2013. "British Media's Coverage of Immigrants and its Enlightenment to Chinese Media: Based on the Event of Africans' Protests in Guangzhou." *Tangdu Journal* 29 (5): 82–86.

Diederich, Manon. 2010. "Manoeuvring through the Spaces of Everyday Life: Transnational Experiences of African Women in Guangzhou, China." Dissertation at the University of Cologne, Department of Geography.

Disima. 2004. "Cultural Adaptation of Foreign Students in China." M.A. Psychology Thesis, Nanjing Normal University.

Dittgen, Romain.2010. "L'Afrique en Chine: l'autre face des relations sino-africaines?" Economie, China Institute.

Giese, Karsten and Laurence Marfaing, eds. 2015. "Entrepreneurs africains et chinois. Les impacts sociaux d'une rencontre particulière." Paris: Karthala.

Gillespie, Sandra. 2001. *South-South Transfer: A Study of Sino-African Exchange*. New York: Routledge.

Gong, Sujuan. 2014. "A Study on African Students in China and their Cross-Cultural Adaptation." *Journal of Kaifeng Institute of Education* 34 (2): 127–130.

Han, H. 2013. "Individual Grassroots Multilingualism in Africa Town in Guangzhou: The Role of States in Globalization." *International Multilingual Research Journal* 7 (1): 83–97.

Harris, Joseph E. ed. 1993. *Global Dimensions of the African Diaspora*, 2nd ed. Washington, DC: Howard University Press.

Hashim, Ismail Hussein, et al. 2003. "Cultural and Gender Differences in Perceiving Stressors: A Cross-Cultural Investigation of African and Western Students at Chinese Colleges." *Psychological Science* 26 (5): 795–799.

Hashim, Ismail Hussain, and Z. L. Yang. 2003. "Cultural and Gender Differences in Perceiving Stressors: a Cross-Cultural Investigation of African and Western Students in Chinese Colleges." *Stress and Health* 19 (4): 217–225.
Haugen, H. Ø. 2011. "Chinese Exports to Africa: Competition, Complementarity and Cooperation between Micro-Level Actors." *Forum for Development Studies* 38 (2): 157–176.
Haugen, H. Ø. 2012. "Nigerians in China: A Second State of Immobility." *International Migration* 50 (2): 65–80.
Haugen, H. Ø. 2013a. "China's Recruitment of African University Students: Policy Efficacy and Unintended Outcomes." *Globalisation, Societies and Education* 11 (3): 315–344.
Haugen, H. Ø. 2013b. "African Pentecostal Migrants in China: Marginalization and the Alternative Geography of a Mission Theology." *African Studies Review* 56 (1): 81–102.
He, Wenping. 2007. "A Summary Analysis of China-Africa Educational Exchanges and Cooperation: Development Phases and Challenges." *West Asia and Africa* 3: 13–18.
Hevi, Emmanuel. 1963. *An African Student in China*. Pall Mall.
Hu, Jintao. 2005. "Written Statement by President Hu Jintao of China at the High-Level Meeting on Financing for Development at the 60th Session of the United Nations." September15, http://politics.people.com.cn/GB/1024/3696504.html. Accessed June 12, 2017.
Jiang, Chun and Guo Yingde. 2001. *History of China-Arab Relations*. Beijing: Economic Daily Press.
King, K. 2010. "China's Cooperation in Education and Training with Kenya: A Different Model?" *International Journal of Educational Development* 30 (5): 488–96.
King, K. 2013. *China's Aid and Soft Power in Africa The Case of Education and Training*. Suffolk: James Currey.
Larkin, Bruce D. 1971. *China and Africa 1949–1970: The Foreign Policy of People's Republic of China*. Berkeley: University of California Press.
Li, Anshan. 2005. "African Studies in China in the Twentieth Century: A Historiographical Survey." *African Studies Review* 48 (1): 59–87.
Li, Anshan. 2011. *Chinese Medical Cooperation in Africa: With Special Emphasis on the Medical Teams and Anti-Malaria Campaign*. Discussion Paper 52, Uppsala: Norkiska Afrikainstitutet.
Li, Anshan. 2012. "China and Africa: Cultural Similarity and Mutual Learning." In *China-Africa Partnership: The Quest for a Win-Win Relationship*, edited by James Shikwati. Nairobi: Inter Region Economic Network: 93–97.
Li, Anshan. 2013a. "African Countries Encouraged to 'Look East'." *Guangming Daily*, March 28.

Li, Anshan. 2013b. "My African Students". In *Chinese Ambassadors Telling African Stories*, edited by Cheng Tao and Lu Miaogeng. Beijing: World Affairs Press: 156–168.

Li, Anshan. 2014. "Changing Discourse on China-Africa Relations since the 1990s." *World Economy and Politics* 2: 19–47.

Li, Anshan. 2015a. "African Diaspora in China: Reality, Research and Reflection." *The Journal of Pan African Studies* 7 (10): 10–43.

Li, Anshan. 2015b. "Contact between China and Africa before Vasco da Gama: Archeology, Document and Historiography." *World History Studies* 2 (1): 34–59.

Li, Anshan. 2015c. "A Long-Time Neglected Subject: China-Africa People-to-People Contact." In *FOCAC 2015: A New Beginning of China-Africa Relations*, edited by Garth Shelton, Funeka Yazini April and Li Anshan. Pretoria: Africa Institute of South Africa: 446–475.

Li, Anshan. 2016. "African Studies in China in the 21st Century: A Historiographical Survey." *Brazilian Journal of African Studies* 1 (2): 48–88.

Li, Anshan, et al. 2012. *FOCAC Twelve Years Later: Achievements, Challenges and the Way Forward*. Uppsala: Nordic Africa Institute.

Li, Anshan and Liu Haifang. 2013. "The Evolution of the Chinese Policy of Funding African Students and an Evaluation of the Effectiveness." Draft report for UNDP.

Li, Anshan and Liu Haifang, eds. 2017. *Annual Review of African Studies in China (2015)*. Beijing: Social Sciences Academic Press (China).

Li, Baoping. 2006. "On the Issues of China-Africa Educational Cooperation." http://www.docin.com/p-747065460.html. Accessed June 2, 2017.

Li, Jiangtao and Li Xiang. 2006. "China is my Second Hometown: African Students' Life in Beijing." http://news.xinhuanet.com/world/2006-10/21/content_5232813.htm.

Li, Jun. 2017. "Ideologies, Strategies and Higher Education Development: a Comparison of China's University Partnerships with the Soviet Union and Africa over Space and Time." *Comparative Education* 53 (2): 245–264.

Li, Zhigang, et al. 2008. "The African Enclave of Guangzhou: A Case Study of Xiaobeilu." *ACTA Geographica Sinica* 63 (2): 207–218.

Li, Zhigang, et al. 2009a. "The Local Response of Transnational Social Space under Globalization in Urban China: A Case Study of African Enclaves in Guangzhou." *Geographical Research* 28 (4): 920–932.

Li, Zhigang, et al. 2009b. "An African Enclave in China: The Making of a New Transnational Urban Space." *Eurasian Geography and Economics* 50 (6): 699 – 719.

Li, Zhigang and Du Feng. 2012a. "Production of China's New Social Space in City under 'Transnational Entrepreneurialism:' A Case Study on African Economic Zone in Guangzhou." *Urban Space Studies* 36 (8): 25–31.

Li, Zhigang and Du Feng. 2012b. "The Transnational Making of 'Chocolate City' in Guangzhou." *Renwen Dili* 27 (6): 1–6.

Li, Zhigang, et al. 2012. "China's 'Chocolate City': An Ethnic Enclave in a Changing Landscape." *African Diaspora* 5: 51–72.

Liu, Haifang. 2008. "China-Africa Relations through the Prism of Culture—The Dynamics of China's Cultural Diplomacy with Africa." *Journal of Current Chinese Affairs (China aktuell)* 3: 9–44.

Liu, Haifang, 2017. "A Study on the Evolution and Effects of China's Scholarship Policy towards African Students." In *Annual Review of African Studies in China 2015*, edited by Li Anshan and Liu Haifang: 141–192.

Liu, Haifang and Jamie Monson. 2011. "Railway Time: Technology Transfer and the Role of Chinese Experts in the History of TAZARA." In *African Engagements: Africa Negotiating an Emerging Multipolar World*, edited by Ton Dietz, et al. Leiden and Boston: Brill: 226–251.

Lin, Lunlun and Ren Mengya. 2010. "A Socioliguistic Study upon Chinese Language Learning Concept of African Overseas Students." *Journal of Hanshan Normal University* 31 (5): 32–37.

Liu, P. H. 2013. "Petty Annoyances? Revisiting John Emmanuel Hevi's *An African Student in China* after 50 Years." *China: An International Journal* 11 (1): 131–145.

Lokongo, Antoine Roger. 2012. "My Chinese Connection." *CHINAFRICA* 50, July.

Long, Xia and Xiong Lijun. 2014. "The Influence of Sino-African Cultural Difference on the Education of African Students in China: Taking Angolan Students as the Example." *Journal of Chongqing University of Education* 27 (1): 133–136.

Lou, Shizhou and Xu Hui. 2012. "The Development and Transition of China-Africa Educational Cooperation in the New Period." *Educational Research* 10: 28–33.

Lyons, M., et al. 2008. "The 'Third Tier' of Globalization: African Traders in Guangzhou." *City* 12 (2): 196–206.

Lyons, M., et al. 2012. "In the Dragon's Den: African Traders in Guangzhou." *Journal of Ethnic and Migration Studies* 38 (5): 869–888.

Lyons, M., et al. 2013. "The China-Africa Value Chain: Can Africa's Small-Scale Entrepreneurs Engage Successfully in Global Trade?" *African Studies Review* 56 (3): 77–100.

Ma, Enyu. 2010. "Walking into the Yiwu Muslim community." *China Religion* 6: 56–57.

Ma, Enyu. 2012. "Yiwu Mode and Sino-African Relations." *Journal of Cambridge Studies* 7 (3): 93–108.

Marfaing, Laurence and Alena Thiel. 2014. *"Agents of Translation": West African Entrepreneurs in China as Vectors of Social Change*, Working Paper, No.4, DFG Priority Program 1448, Accessed June 20, 2017.

Marsh, Jenni. 2014. "Afro-Chinese Marriages Boom in Guangzhou: but will it be 'til Death Do Us Part'?" June 1, http://www.scmp.com/magazines/post-magazine/article/1521076/afro-chinese-marriages-boom-guangzhou-will-it-be-til-death. Accessed June 24, 2014.

Mathews, G. 2000. "Les traders africains a Kong Hong (Hong Kong) et en Chine." *Les Temps Modernes* 657 (Janvier-mars): 110–124.

Mathews, G. and Yang, Y. 2012. "How Africans Pursue Low-End Globalization in Hong Kong and Mainland China." *Journal of Current Chinese Affairs* 41 (2): 95–120.

Monson, J., 2009. *Africa's Freedom Railway: How a Chinese Development Project Changed Lives and Livelihoods in Tanzania*. Bloomington and Indianapolis: Indiana University Press.

Morais, I. 2009. "'China Wahala': The Tribulations of Nigerian 'Bushfallers' in a Chinese Territory." *Transtext(e)s Transcultures. Journal of Global Cultural Studies* 5: 1–22.

Müller, A. and R. Wehrhahn. 2013. "Transnational Business Networks of African Intermediaries in China: Practices of Networking and the Role of Experiential Knowledge." *DIE ERDE–Journal of the Geographical Society of Berlin* 144 (1): 82–97.

Müller, Angelo. 2011. "New Migration Processes in Contemporary China—The Constitution of African Trader Networks in Guangzhou." *Geographische Zeitschrif* 99 (2): 104–122.

Niu, Changsong. 2016. "A Survey of African Students' Satisfaction of Chinese Government Scholarship." http://www.docin.com/p-1445264169.html. Accessed: June 25, 2017.

Osnos, Evan. 2009. "The Promised Land—Guangzhou's Canaan Market and the Rise of an African Merchant Class." *The New Yorker*, February 9–16: 50–56.

Pilling, David. 2017. "Ports and Roads Mean China is 'Winning in Africa'." *Construction Review Online*, May 6. https://constructionreviewonline.com/2017/05/ports-and-roads-mean-china-is-winning-in-africa/. Accessed June 25, 2017.

Pinto, Jeanette. 2006. "The African Native in Diaspora." *African and Asian Studies* 5 (3–4): 383–397

Pliez, O. 2010. "Toutes les routes de la soie mènent in Yiwu (Chine). Entrepreneurs et migrants musulmans dans un comptoir économique chinois." *Espace Géographique* 2: 132–145.

Redvers, Louise. 2012. "Angola's Chinese-Built Ghost Town." http://www.bbc.com/news/world-africa-18646243. July 2. Accessed, June 24, 2017.

Rennie, N. 2009. "The Lion and the Dragon: African Experiences in China." *Journal of African Media Studies* 1 (3): 379–414.

Sautman, Barry. 1994. "Anti-Black Racism in Post-Mao China." *The China Quarterly* 138: 413–437.

Seidelman, Raymond. 1989. "The Anti-African Protests; More Than Just Chinese Racism." *The Nation*. February 13.

Shelton, Garth, Funeka Yazini April, and Li Anshan, eds. 2015. *FOCAC 2015: A New Beginning of China-Africa Relations*. Pretoria: Africa Institute of South Africa.

Shikwati, James, ed., 2012. *China-Africa Partnership: The Quest for a Win-Win Relationship*. Nairobi: Inter Region Economic Network: 93–97.

Sullivan, M.J. 1994. "The 1988–89 Nanjing Anti-African Protests: Racial Nationalism or National Racism?" *The China Quarterly* 138 (June): 438–457.

The Conversation. 2017. "China Tops US and UK as Destination for Anglophone African Students." June 28. http://theconversation.com/china-tops-us-and-uk-as-destination-for-anglophone-african-students-78967. Accessed: June 30, 2017.

Tsui, Chak-Pong Gordon. 2016. "African University Students in China's Hong Kong: Motivations, Aspirations, and Further Exchanges." In *Africans in China: Guangdong and Beyond*, edited by Adams Bodomo: 119–137.

Wang, Hanjie. 2013. "The Spread and Distribution of African Drums in China" In *Annual Review of African Studies in China (2012)*, edited by Li Anshan and Liu Haifang. Beijing: Social Sciences Academic Press (China): 442–458.

Wang, Luxin. 2000. "Educational Exchange and Cooperation between China and African Countries." In *China and Africa*, edited by Lu Ting-en and Ma Ruimin. Beijing: Peking University Press.

Wu, Bin. 2015. "Links between Chinese International Students and Overseas Chinese Communities: An Empirical Study in Nottingham, UK." *Overseas Chinese History Studies* 2: 1–11.

Wu, Bin. 2016. "Local Engagement of Chinese International Students in Host Societies: A Perspective of Diasporic Chinese Community Building." *The International Journal of Diasporic Chinese Studies* 8 (2): 13–30.

Xu, Hui. 2007. "Sino-Africa Educational Cooperation under the FOCAC Framework." *Educational Development Research* 9: 1–7.

Xu, Tao. 2009a. "An analysis on Africans social relations and interaction logics in Guangzhou." *Youth Research*, 5:71–86.

Xu, Tao. 2009b. "African's Social Support in Guangzhou: Weakening, Fracture and Reconstruction." *South China Population* 24 (4): 34–44.

Xu, Tao. 2011. "Re-Analysis of the Relations of Social Contact of African Merchants in Guangzhou." *Journal of Zhejiang Normal University* 4: 10–15.

Xu, Tao. 2012. "Analysis of Characteristics of the Behavior of African Merchants in Guangzhou." *Journal of Zhejiang Normal University* 4: 55–63.

Xu, Tao. 2013. *The Social Adaptations of African Merchants in China*. Zhejiang People's Press.

Yang, Y. 2013. "African Traders in Guangzhou." In *Globalization from Below: The World's Other Economy*, edited by G. Mathews, G.L. Ribero and C.A. Vega. Routledge, Taylor and Francis.

Yan, Hairong and Barry Sautman. 2012. "Chasing Ghosts: Rumours and Representations of the Export of Chinese Convict Labour to Developing Countries." *The China Quarterly* 210: 398–418.

Yan, Hairong and Barry Sautman. 2017. *China in Africa: Discourses and Reality*. Beijing: Social Sciences Academic Press (China).

Yang, Mengjie, et al. 2016. "China-Hand's Dream of Mandarin." *China news*, November 28, 2016. China. http://www.chinanews.com/sh/2016/11-28/8076600.shtml. Accessed June 1, 2017.

Ye, Shuai. 2011. "A Comparative Analysis of the Cross-Cultural Communication Based on the Somali Students and the Chinese Students on Time and Family Concepts." *Kexue Wenhui* 11: 30–31.

Yi, Pei and Xiong Lijun. 2013. "An Empirical Study of Intercultural Adaptation of African Students in China." *Journal of Shenyang University (social science)* 15 (3): 364–368.

Zeleza, Paul Tiyambe. 2005. "Rewriting the African Diaspora: Beyond the Black Atlantic." *African Affairs* 104 (414, January): 35–68.

Zeleza, Paul Tiyambe. 2008. "The Challenges of Studying African Diasporas." *African Sociological Review* 12 (2): 4–21.

Zhang, Tieshan. 1999. *Friendship Road: A Report on the Construction of Tanzania-Zambia Railway*. Beijing: China Economic and Foreign Trade Press.

Zheng, Jianghua. 2012. "Research on Safety Management of African Students on University Campus." *Journal of Tianjin University of Technology and Education* 22 (4): 72–74.

Zheng, Jianghua, et al. 2013a. "Construction of Community Management System for Foreign Students in Universities." *Vocational and Technical Education* 34 (23): 66–68.

Zheng, Jianghua. 2013b. "Exploration of Compound Applied Talents Training Mode on African Students." *Journal of Tianjin University of Technology and Education* 23 (4): 64–70.

CHAPTER 22

China–Africa Cooperation in Poverty Reduction: The Evolution of Concepts and Characteristics of Practices

Chunying An

Today's China–Africa relations have reached a crucial stage at which to begin building on past achievements and forging ahead. African countries look forward to strengthening their means of practical cooperation, sharing opportunities, and, together with China, meeting the challenges ahead. China's achievements in poverty reduction have provided precious experience for African countries, who are eager to integrate China's experience—a global public good—into their own development practices. In this context, China should not settle for mere material considerations about trade and investment ties in its relations with Africa, but should further the collaborative efforts of poverty reduction to build a China–Africa community with a shared future on a steady foundation,

C. An (✉)
Institute of West-Asian and African Studies, Chinese Academy of Social Sciences (CASS), Beijing, China
e-mail: gaoj@ssap.cn

X. Li and C. An (eds.), *The Changing World and Africa*,
https://doi.org/10.1007/978-981-16-4983-7_22

respond to the aspirations of African countries, and advance the traditional China–Africa friendship to an even higher level characterized by a new momentum.

1 Concepts of Poverty Reduction in China's Africa Policy

China's Africa policy is mainly codified in its government declarations and agreements signed with the African side. In addition, relevant speeches, opinions, and views given by Chinese leaders also serve as guidelines for China's continued cooperation with Africa. The concepts outlining China's poverty reduction measures in its Africa policy can be derived from all of the above sources. Generally speaking, since the founding of the People's Republic of China, its notion of poverty reduction in Africa has gone through three stages, as reflected in the means and contents of corresponding China–Africa cooperation.

Poverty Reduction through Assistance

Between the 1950s and the 1970s, national liberation movements on the African continent reached a climax, with sovereign states established one after another. It was of utmost urgency for African countries at that time to maintain and consolidate their sovereign independence, achieve their own national economic development, and shake off poverty and backwardness. Recognizing the strength demonstrated by the rise of African nation-states, Mao Zedong and other Chinese leaders regarded African countries as China's reliable friends and put forward a momentous decision to provide African countries with international assistance for national independence and state-building.

During this period, China's poverty reduction scheme for Africa was part of its assistance policy system, whose main contents included the following. First and foremost was establishing the principle of equality and mutual benefit. At the Bandung Conference in April 1955, Premier Zhou Enlai clearly and comprehensively put forward the Five Principles of Peaceful Coexistence guiding China's foreign relations. From December 1963 to February 1964, during his visit to ten African countries, Premier Zhou, on behalf of the Chinese government, proposed the "five principles" concerning China's relations with African and Arab countries and the "eight principles" governing China's economic and technological aid to other countries,[1] marking the formal formation of China's policy on aid to Africa. Second, advocating charitable internationalism in China's relations with Africa. In 1963, when meeting with visiting African friends, Mao Zedong made it clear that "it is our internationalist duty that we, as the people who have won the revolutionary victory, help the people who are still fighting for liberation."[2] In June 1975, when Deng Xiaoping met with a delegation of Thai parliamentarians headed by Prasit Kanchanawat, he pointed out that: "It is for the people of Asia to run the affairs of Asia, so it is for Africa. Although China is an underdeveloped Third World country, we still have to fulfill our internationalist duty and provide some help to many countries in the Third World within our capacity. It's not much help, but it's our obligation. And it comes with no political strings attached. This is one of our principles."[3] Third, carrying out international development cooperation based on national conditions, with priority given to self-reliance. In May 1960, when meeting with social and peace activists, trade unions, youth, and students from twelve African countries and regions, Mao Zedong said: "Some of China's experience may serve as your reference, including experience of revolution and nation-building. However, I would like to remind my friends that China has its historical conditions and you have your own. China's experience can only serve as your reference."[4] In August 1975, while meeting a delegation of the Seychelles People's United Party led by France-Albert René, Deng Xiaoping pointed out that the key for Seychelles to develop was to "focus on domestic development and rely on itself, solve the problems of food, clothing, housing and transportation for its people, and develop friendly relations with the Third World."[5]

[1] "Eight Principles for Foreign Economic and Technological Assistance," *Selected Diplomatic Works of Zhou Enlai*, CCCPC Party Literature Publishing House, 1990, pp. 388–389.

[2] "Talks in Meeting with African Friends," *Mao Zedong's Manuscripts Since the Founding of the People's Republic of China*, Vol. 10, CCCPC Party Literature Publishing House, 1996, p. 340.

[3] *Deng Xiaoping's Chronicle 1975–1997 (I)*, CCCPC Party Literature Publishing House, 2004, p. 58.

[4] "Imperialism Is Not Horrible," *Selected Diplomatic Works of Mao Zedong*, CCCPC Party Literature Publishing House & World Affairs Press, 1994, p. 413.

[5] *Deng Xiaoping's Chronicle 1975–1997 (I)*, p. 79.

During this interval, China–Africa cooperation in poverty reduction was initially conceptualized under a framework of China's assistance to Africa. Although expressions such as "shake off poverty" and "people's livelihood" did not appear in China's Africa policy, given the different historical tasks and international environments facing the two sides, China's policy on assistance to Africa focused on helping African countries maintain and consolidate their national independence, develop their economies, and eliminate poverty, so as to lay the political cornerstone for their national prosperity and strength, as well as a foundation for poverty reduction. It should be noted that the eight principles governing China's aid to other countries were a pioneering policy declaration of China in its relations with Africa, providing the ideological basis for all subsequent cooperative measures in poverty reduction. The core conceptions of equality, mutual benefit, and an emphasis on helping recipient countries improve their own development capacity have been upheld to this day.

Poverty Reduction through Economic Growth

In the 1980s and 1990s, China adopted the policy of focusing on economic construction and subsequently carried out reform and opening-up measures, while African countries began to implement their own economic adjustments. The development of national economies became the focus of work in both China and African countries. Against this background, China–Africa development cooperation took on new features.

First, while giving priority to the development needs of African countries, China's Africa policy would, at the same time, appropriately reflect China's own development agenda. In January 1983, China announced the "four principles" of economic and technological cooperation with African countries, namely "equality and mutual benefit, emphasis on practical results, diversity in form, and pursuit of common development." As reflected by these principles, China and Africa not only adhered to the principle of equality and mutual benefit in development cooperation, but also stressed the results of cooperation and the diversity of cooperation forms, with the ultimate goal of achieving common economic and social development on both sides. In the 1990s, in the course of implementing the strategies of "market diversification" and "two markets (domestic and foreign) and two resources," China adjusted its mode of foreign aid, focusing on joint ventures and government loans with discounted interest rates. During his visit to Africa in May 1996, President Jiang Zemin proposed consolidating and developing long-term, stable, and comprehensive cooperative relations with African countries for the twenty-first century, emphasizing mutual benefit and common development. "We will encourage cooperation between businesses of the two sides, and particularly encourage those Chinese companies with certain strength to carry out mutually beneficial cooperation in different sizes, fields and forms in Africa," he said.[6] Second, the Chinese government continued to uphold the concepts of sincere friendship and assistance that were within its capacity. When Deng Xiaoping met with Mali's President, Moussa Traoré, in June 1986, he forecast that by the end of the twentieth century, when China would lift itself out of poverty and establish a well-off society, that "China still belongs to the Third World, which is the basis of our policy ... Even if China successfully develops itself, it will still regard itself as a Third World country. We must not forget any of our poor friends in the Third World and must help bring them out of poverty. When China becomes a medium-level developed country, we will still extend a helping hand. Only in this way can we truly demonstrate that China is pursuing socialism."[7]

During this period, what was changed and what was not in China's Africa policy were always in line with the evolving circumstances of social development on both sides. Regarding China's policy on poverty reduction in Africa, it was the fulcrum of cooperation that China took advantage of its vigorous economic growth to directly or indirectly improve the livelihood of African people and achieve poverty reduction.

Poverty Reduction through Development

Since the beginning of the twenty-first century, conditions in China and Africa—as well as China–Africa relations themselves—have evolved into a new stage. As for China, the rapid growth of its economy, the continuous improvement of its people's living standards, and the bolstering of its comprehensive national strength have attracted much attention from

[6] "Creating a New Historical Monument for China-Africa Friendship: Speech at the Organization of African Unity," *Selected Works of Jiang Zemin*, Vol. 1, People's Publishing House, 2006, p. 529.

[7] *Deng Xiaoping's Chronicle 1975–1997 (II)*, CCCPC Party Literature Publishing House, 2004, p. 1124.

the international community. Although African countries enjoy sound economic development, most of them have not been able to overcome the fundamental burden of poverty. The dual issues of poverty reduction and development are therefore urgently needed by African countries. In terms of China–Africa relations, mutual ties have intensified at a rapid rate since 2000, and the comprehensive strategic partnership between the two sides has been advancing steadily. In particular, the establishment of the Forum on China–Africa Cooperation (FOCAC) is a milestone in the history of China–Africa relations, one which further institutionalizes and systematically elaborates their cooperation on poverty reduction. During this period, China and Africa have expanded and deepened their strategies and policy connotations of poverty reduction on an unprecedented scale.

First, the two sides uphold justice while pursuing shared interests, and view their cooperation in poverty reduction measures from the strategic height of building a community with a shared future for humankind. Both China and Africa belong to the developing world, and both shoulder the mission of developing their countries and improving people's livelihood. In China's view, continued poverty on the African continent is not in the best interests of the world at large, and China will therefore help African countries achieve sustainable economic and social development. In a speech delivered in Tanzania in March 2013, President Xi Jinping stated that the history of China–Africa relations demonstrated that "China and Africa have always been a community with a shared future." He also fully expounded upon the connotations of "sincerity, real results, affinity and good faith" in China's Africa policy.[8] Based on this new concept informing its Africa policy, China has repeatedly stressed the importance of "upholding justice while pursuing interests" in its relationship with Africa. As China's *Africa Policy Paper*, released in December 2015, points out, "While valuing friendship and justice as well as shared interests, China places more importance on the former. The core principle is to connect assistance to developing countries, including those in Africa, for their independent and sustainable development with China's own development, achieve win–win cooperation and common development, and promote more balanced, inclusive and sustainable development of the

[8] "Xi Jinping's Speech at Nyerere International Conference Center, Tanzania," in *Xi Jinping: The Governance of China*, Foreign Languages Press, 2014, pp. 306–310.

world at large."[9] In his speech at the opening ceremony of the FOCAC Beijing summit in 2018, Xi Jinping articulated something similar: "China pursues common interests and puts friendship first in pursuing cooperation ... it is for China to complement Africa's development through its own growth, and it is for both China and Africa to pursue win–win cooperation and common development. In doing so, China follows the principle of giving more and taking less, giving before taking and giving without asking for return ... With top priority given to the interests of Chinese and African peoples, China advances its cooperation with Africa to improve the well-being of Chinese and African peoples and deliver more benefits to them."[10] The principle of "putting friendship first" in China–Africa cooperation for poverty reduction[11] reflects China's renewed understanding of the international situation and the development status of both itself and African countries. It embodies the value of promoting justice and integrating interests in contemporary Chinese diplomacy, as well as demonstrates China's strong sense of mission and responsibility as a major developing country. China–Africa cooperation not only requires "momentum," but also needs to produce "tangible results," thus closely integrating the interests of the Chinese people with those of the African people. It can be concluded from this that poverty reduction has been raised to a higher level on the priority list of China–Africa cooperation.

Second, China seeks to promote poverty reduction through development, and as such a more comprehensive range of measures has been undertaken for the continued cooperation between the two regions in poverty reduction. With a deeper understanding by the international community of what really constitutes poverty and the clarification of China's own poverty reduction concepts, China recognizes that the root causes of poverty in Africa are multifaceted and mainly associated with

[9] "Full Text: China's Second Africa Policy Paper," *China Daily*, December 5, 2015, https://www.chinadaily.com.cn/world/XiattendsParisclimateconference/2015-12/05/content_22632874.htm.

[10] "Full text of Chinese President Xi Jinping's Speech at Opening Ceremony of 2018 FOCAC Beijing Summit," *Xinhua*, September 3, 2018, http://www.xinhuanet.com/english/2018-09/03/c_129946189.htm.

[11] Luo Jianbo, "Upholding Justice while Pursuing Shared Interests and China's Diplomacy with Developing Countries," *West Asia and Africa*, No. 5, 2018, p. 11.

inadequate development. In the words of President Xi Jinping: "Development holds the master key to solving all problems."[12] Based on this, the policy on poverty reduction in Africa issued by the Chinese government covers a wide variety of issues concerning Africa's development. They are designed to guide the two sides to carry out poverty reduction cooperation in broader and farther-reaching areas, thus injecting new contents into China–Africa poverty reduction cooperation. The Johannesburg Action Plan (2016–2018) and the Beijing Action Plan (2019–2021), adopted by the two FOCAC summits in 2015 and 2018, put forward a multi-disciplinary cooperation plan covering Africa's industrialization, agricultural modernization, infrastructure, finance, green development, trade and investment facilitation, poverty alleviation, people's well-being, public health, culture, peace, and security. It highlights the new dimensions in China's poverty reduction policy toward Africa.

2 Development and Changes in the China–Africa Cooperation Model on Poverty Reduction

Over the past 70 years, from 1949 to 2019, China–Africa cooperation in poverty reduction has not only evolved at the conceptual level, but also served as a guide to action for deepening cooperative measures toward transformation, acceleration, and upgrading.

From Government-Led Assistance to Multi-Stakeholder Participation

Regarding stakeholders in China–Africa cooperative efforts in poverty reduction, before the 1990s, intergovernmental mechanisms dominated cooperative efforts in economic and social development between China and Africa, as during this period there were few actors in China involved in the cooperation. After the mid-1990s however, Chinese companies began to operate in Africa in search of market-oriented investment opportunities. At the same time, a growing number of NGOs in China were expanding public-interest cooperation in poverty alleviation at home and abroad. Joining together with the relevant government departments,

[12] Xi Jinping, "Open a New Era of China-Africa Win–Win Cooperation and Common Development: Address at the Opening Ceremony of the Johannesburg Summit of the Forum on China-Africa Cooperation," *People.com.cn*, December 4, 2015, http://politics.people.com.cn/n/2015/1204/c1001-27892314.html.

these two actors increased their participation in the China–Africa poverty reduction agenda.

First of all, the government plays a major role in executing poverty reduction programs through planned and sustained large-scale collaborative projects. Both in the early phases of the PRC and throughout the twenty-first century, the Chinese government has always been involved in poverty reduction cooperation in Africa, playing an important leading role in terms of the scope, scale, and intensity of such cooperation. As the main provider of mandatory institutional supplies and as the manager and user of public resources, the government can effectively advance the process and effect of China–Africa poverty reduction cooperation through policy tools. For example, in response to the lack of medical care and the suffering of African people, since 1963, when the first Chinese medical team was sent to Algeria, the Chinese government has continued to send medical personnel to Africa through top-down institutional arrangements. Second, the government, using its own administrative resources, has implemented China's poverty reduction plans in Africa in a systematic way through relevant departments such as the Ministry of Foreign Affairs, Ministry of Commerce, Ministry of Agriculture, and the Ministry of Education, thus integrating aid with poverty reduction. For example, the Ministry of Commerce has vigorously encouraged Chinese state-owned enterprises and private companies to carry out mutually beneficial forms of cooperation in terms of trade, investment, and employment in Africa. The Ministry of Agriculture has undertaken China's assistance and cooperation in agriculture-related projects, sending agrotechnicians to African countries to set up demonstration centers for agricultural technologies and carry out training programs for agricultural professionals and technical workers. In this framework of governmental assistance, the Ministry of Education is responsible for coordinating and arranging the reception of African students with the relevant domestic colleges and universities, while dispatching volunteer Chinese students to African countries in accordance with agreements in order to provide services across many fields, such as Chinese language teaching, medical and health care, physical education, information technology, and international rescue. The Ministry of Culture and Tourism has worked with relevant African countries in implementing a number of activities to promote cultural and people-to-people exchanges and mutual learning between China and Africa. The National Health Commission is responsible for sending medical teams to Africa, providing assistance in epidemic prevention and control, and donating medical supplies. China Development Bank and the Export–Import Bank of China, as two major national policy banks, offer effective financial services to support the investment and assistance projects of Chinese enterprises in Africa.

Second, the economic and social benefits generated by the investment activities of the business community have either directly or indirectly contributed to poverty reduction in Africa. Chinese enterprises started investing in Africa on a modest scale in the 1980s, but it was not until the 1990s that they began to regard Africa as an important destination for their international operations and overseas investment cooperation. By the end of 2017, about 3,413 Chinese enterprises had invested in Africa.[13] With their investment and business activities in Africa, Chinese enterprises have helped Africa in their poverty reduction and development ambitions in the following ways. First, Chinese enterprises provide financial support for African countries to implement economic and social development plans by creating output value through investment and increasing the tax revenues of host countries. By the end of 2017, Chinese enterprises had invested in 25 special economic zones in Africa, generating an output value of US$18.9 billion, and had paid $900 million in taxes.[14] This has contributed to Africa's economic development and, to some extent, helped host countries implement national poverty reduction strategies. Second, Chinese enterprises seek to create jobs for the local people, so that they can participate in economic activities to broaden their sources of income and improve the income level of the poor. The Zambia-China Economic and Trade Cooperation Zone, for example, increased the number of jobs created in the country from 2,647 to 7,251 between 2007 and 2017.[15] Third, Chinese enterprises in Africa focus on technology transfer in their investment projects. They train skilled technical personnel in host countries through vocational training and short-term technical courses, so as to improve the quality of local human

[13] Ministry of Commerce, State Statistical Bureau, and State Administration of Foreign Exchange of China, "Statistical Bulletin of China's Overseas Direct Investment in 2017," http://www.mofcom.gov.cn/article/tongjiziliao/dgzz/201809/20180902791492.shtm.

[14] "China's Investment Stock in Africa Exceeds US$100 Billion," *Guangming*, August 29, 2018, http://economy.gmw.cn/2018-08/29/content_30846138.htm.

[15] Materials provided by China Nonferrous Group, organizer of the Zambia-China Economic and Trade Cooperation Zone.

resources and help African countries upgrade their industries. Within four years of construction of the Maputo–Katembe Bridge, the largest suspension cross-sea bridge in Africa, the project has helped train more than 5,000 local welders, lathe workers, steel bar workers, drivers, mechanical operators, and other technical personnel. This project is regarded as an educational site for training local industrial workers in Mozambique.[16] Fourth, by donating funds, building roads, bridges, wells, hospitals, and schools, Chinese enterprises take the initiative to fulfill their corporate social responsibilities and partake in the public welfare undertakings of regional communities, so as to benefit local people and improve their livelihoods. For example, during the operation of CNPC's refinery project in Niger, off-site water pipelines were used to build water supply zones for nearby villagers, and more than 160 wells were drilled deep into the Sahara Desert.[17]

Third, Chinese NGOs have become a new force in China's poverty reduction engagement with Africa by carrying out charity and poverty relief projects in Africa. NGOs are the product of a market economy at a certain stage and are one of the organizational carriers of social works in public affairs. Set apart from the government's use of political and economic resources to implement national poverty reduction policies, and from the business community's fulfillment of their social responsibilities to promote poverty reduction, NGOs mainly engage in raising social awareness through interactions intended to mobilize all social actors to participate in poverty reduction. According to historical data, China's NGOs have grown gradually since 1978, whereas their participation in poverty reduction cooperation with Africa started at the beginning of the twenty-first century. The China NGO Network for International Exchanges, China Foundation for Poverty Alleviation, China Youth Development Foundation, and other domestic NGOs associated with the agenda of reducing poverty have jointly explored cooperation with African countries in poverty reduction. The "smiling children program," which was launched by the China Foundation for Poverty Alleviation, is a typical case in point. The program provides free breakfasts to hungry

[16] Materials provided by China Communications Construction, operator of the Maputo–Katembe Bridge project.

[17] Zhang Jun and Wang Hongyi, "Achievements of China Petroleum's Business in Africa and Suggestions on Policy Improvement," *Africa Development Report (2015–2016)*, Social Sciences Academic Press (China), 2016, p. 55.

children in recipient countries and has been implemented in Sudan and Ethiopia. In Sudan, the program has benefited thousands of children in primary schools since it was officially launched in August 2015.

In general, the government, enterprises, and NGOs each play a unique role in poverty reduction cooperation with Africa. The government takes livelihood assistance as its priority. The projects implemented by the government are sustainably organized, planned, and funded, and have become the most important method of support for poverty reduction cooperation with Africa. Although enterprises and NGOs have been involved in the cause of African poverty alleviation for only a little more than 20 years, their participation has gradually expanded and deepened with the increased momentum of China–Africa common development, which has allowed them to play a powerful role in complementing intergovernmental poverty reduction cooperation.

From Cooperation in Traditional Fields to Human Resources Development

With the establishment of the International Poverty Reduction Center in China in 2005, China's human resources training as it is related to poverty reduction has entered a period of systematic planning and institutional structuring. Simultaneously, the content of China–Africa poverty reduction cooperation has changed from the more conventional "survival-oriented poverty alleviation" to a "development-oriented poverty alleviation" aimed at improving the ability of Africa's poor population to lift themselves out of poverty.

China, using its experience in and capacity for economic and social development, has already had a long-term engagement in poverty reduction cooperation aimed at satisfying the basic needs of the poor in Africa since the independence of African countries. This is mainly evidenced in the following aspects. First, China supports the development of public infrastructure in Africa for the purposes of improving the working and living conditions for the poor and in order to directly benefit the local communities. These infrastructure projects include roads, bridges, housing, canals, etc. For example, the China-assisted Kambia Bridge project in Sierra Leone has effectively improved the land transportation conditions in the northwest of the country, lowered transaction costs, and benefited local producers. Second, China has been carrying out various forms of collaboration to increase the level of food security in poor areas.

Main practices in this regard include: helping with agricultural infrastructure, such as the Poilao Dam in Cape Verde; building farms, like those in Uganda, Ghana, Niger, and Rwanda; providing agricultural material assistance and a certain amount of funds in emergency humanitarian food aid to African countries; sending agrotechnical experts to teach breeding, seed selection, sowing, field management, and other farming techniques, and more. Third, medical aid to Africa has been the most long-standing poverty reduction project under the umbrella of China–Africa cooperation, involving the largest number of countries and achieving the most prominent results. By the end of 2016, China had sent more than 20,000 medical teams to 48 African countries.[18] It is fair to say that in the second half of the twentieth century, China–Africa poverty reduction cooperation mainly focused on poverty alleviation through cooperation in agriculture, medical care, infrastructure-building, and other fields immediately related to reducing poverty. Assistance services were provided to guarantee the basic survival needs of Africa's poor population.

Since the beginning of the twenty-first century, China's large-scale poverty reduction efforts have aroused great interest on the part of those African countries that are also a part of the developing world. Investigations were made into China's philosophy of poverty reduction and how it had provided and could continue to provide for the successful alleviation of poverty. At the same time, China has become aware of the importance of education in its poverty reduction cooperation with Africa. Therefore, China and Africa have begun to pay more attention to the exchange of experience in poverty reduction and development. The sharing and exchange of knowledge on poverty reduction have become institutionalized and more frequent than in previous stages. First, numerous forums and seminars on the experience of poverty reduction have been held. Since its first meeting in November 2010, the Africa–China Poverty Reduction and Development Conference[19] has become a core platform for the two sides' poverty reduction exchanges. This annually held event has been included in the FOCAC framework and is now known as the FOCAC Africa–China Poverty Reduction and Development Conference.

[18] Institute of International Trade and Economic Cooperation of the Ministry of Commerce, ed., *Report on Economic and Trade Relations between China and Africa 2017*, p. 18.

[19] The conference was initially sponsored by the International Poverty Reduction Center in China and the United Nations Development Program.

Second, China is now holding short-term training courses for African countries. The China Agricultural University, the Chinese Academy of Social Sciences, and the Academy for International Business Officials of the Ministry of Commerce often hold seminars on issues related to poverty reduction and development in Africa. The International Poverty Reduction Center in China is the most active in organizing international poverty reduction seminars for African countries. Today, the center has established a comprehensive training system for international efforts to lift populations out of poverty. From 2005 to the end of 2018, the center held 139 training courses on overseas poverty reduction, 106 of which were for Africa, and trained 2,219 participants from 52 African countries in a bid to share China's experience in its own battle against mass poverty.[20] Through theoretical explanations, case studies, field visits, and participatory discussions, the students were able to improve their perceptual and rational understanding of poverty reduction efforts in China. Third, through reciprocal interactions between "going global" and "bringing in," China carries out medium- to long-term cooperation in human resources development. In addition to sending volunteers to Africa and providing government scholarships for incoming young students from African countries, China established the Institute of South–South Cooperation and Development at Peking University in 2016, with master and doctoral degree programs and non-degree training programs. With a wide range of courses, the institute hopes to cultivate talents for the modernization of national governance system and capacity in developing countries. The first class of students in this program came from Ethiopia, Burundi, Eritrea, and Zimbabwe, among others. The above-mentioned cooperation in training and human resources management is conducive to enhancing the independent capacity for poverty reduction and development of recipient countries.

In summary, China–Africa cooperation in poverty reduction has covered a wide range of areas, including food security, industrial development, infrastructure construction, medical and health care, the sharing of development experience and capacity building, debt relief, and humanitarian assistance. It reflects the fact that the two sides have adopted a holistic approach to their ongoing cooperation in poverty reduction. The contents of their cooperation have been increasingly extended, and

[20] According to information on the website of the International Poverty Reduction Center in China, http://www.iprcc.org.cn.

have shifted from a predominantly philanthropic purpose in the past to fulfilling the goal of improving the working and living conditions of poor people in Africa today, enhancing their capacity for self-development and self-reliant poverty reduction, and sharing experience in poverty reduction and development. This demonstrates that the quality and extent of China–Africa poverty reduction cooperation is reaching to an even higher level.

From Bilateral Cooperation to Interactions at Both the Bilateral and Multilateral Levels

Since the establishment of diplomatic relations between China and African countries, the two sides have, in terms of formats, implemented intergovernmental poverty reduction and development cooperation projects through bilateral committees, strategic dialogues, political consultations between foreign ministries, and joint economic and trade commissions. Since the mid-1990s, as the African continent has become increasingly united and independent, the integration process at the continental and regional levels has entered a new stage. The African Union replaced the Organization of African Unity and set forth the New Partnership for Africa's Development (NEPAD) and Agenda 2063. Since the launch of reform and opening-up, China has accelerated its integration into the global system and increased its cooperation with international organizations. China–Africa relations have developed in an all-round and multifaceted way. The Forum on China–Africa Cooperation, as a one-to-many multilateral mechanism for international cooperation, thus emerged. It is precisely based on these changes that the approach to China–Africa cooperation in poverty reduction has been upgraded from a purely bilateral channel to a new stage of equal emphasis on both bilateral and multilateral interactions.

First, the two sides determine the direction and focus of poverty reduction cooperation, based on the FOCAC and the Outline on Strengthening Poverty Reduction Cooperation established between China and the African Union. After 18 years of development, the FOCAC has become an effective mechanism for collective exchanges and dialogues between China and African countries and an important platform for deepening practical collaboration. It is also a typical example of the type of international multilateral diplomacy that features an approach of "X+Africa."[21] The FOCAC ministerial meeting is usually held every three years, with representatives of the Chinese government and the African Union, as well as political leaders of African countries attending, to jointly discuss and formulate the general direction and outline for cooperation in the coming three years in the form of declarations and action plans. After the conclusion of the meeting, proposals made by the two sides will be followed up by the senior officials' meeting, the follow-up committee, and other coordinating bodies responsible for promoting implementation. For example, the FOCAC Beijing Summit in 2018 identified the essence of building a China–Africa community with a shared future, featuring joint responsibility, win–win cooperation, happiness for all, cultural prosperity, common security, and harmony between humanity and nature. The summit also outlined initiatives to strengthen cooperation in industrial promotion, infrastructure connectivity, trade facilitation, green development, capacity building, health care, people-to-people exchange, and peace and security. The FOCAC has served as a leading multilateral institutional platform for Africa's poverty reduction efforts and the common development of China and Africa. In particular, the above-mentioned Outline on Strengthening Poverty Reduction Cooperation, reached by China and the African Union when Premier Li Keqiang visited the AU headquarters in May 2014, is a policy document specifically focusing on poverty reduction cooperation. The 15-point Outline explains the necessity and urgency of China–Africa cooperation in poverty reduction, and calls for "upholding the principle of inclusiveness and diversity." According to the Outline, the two sides should explore additional cooperative approaches in industrialization, infrastructure, food security, youth development, vocational and technical training, and mutual learning on poverty reduction.[22] Both the FOCAC and the Outline constitute a general framework of poverty reduction cooperation agreed upon by China and Africa through multilateral

[21] Examples of the "X+Africa" diplomatic mechanism also include the US–Africa Leaders Summit, the Tokyo International Conference on African Development, the India–Africa Forum Summit, and the Turkey-Africa Partnership Summit. See Wang Tao and Bao Jiazheng, "An Analysis of the EU-Africa Summit: From the Perspective of 'Multilateral to Multilateral' Mechanism," *West-Asia and Africa*, No. 4, 2018, p. 136.

[22] "Outline on Strengthening Poverty Reduction Cooperation between China and the African Union," *Central People's Government of China*, May 6, 2014, http://www.gov.cn/xinwen/2014-05/06/content_2672503.htm.

channels, whereas specific projects must be implemented through bilateral channels. Therefore, China–Africa cooperation in poverty reduction is a combination of bilateral and multilateral approaches.

Second, China participates in poverty alleviation projects in Africa under the framework of South–South cooperation with various international organizations or third parties. In 1996, the United Nations Food and Agriculture Organization (FAO) launched the Special Program for Food Security under the South–South cooperation framework. China has been actively involved in the program since its inception and in 2006 signed a memorandum of cooperation with the FAO, becoming the first country to establish a strategic partnership for South–South cooperation with the FAO. Since then, cooperation between the two sides has been deepening in the following aspects. First, in terms of providing financial support. In 2005 and 2015, China donated US$30 million and $50 million, respectively, to the FAO trust fund for South–South cooperation in order to support agricultural and rural development projects in developing countries, including African ones. Second, sending agrotechnical experts to conduct policy research, field demonstrations of agricultural production techniques, and personnel training, covering fields such as aquaculture, crop production, irrigation, livestock and poultry breeding, agroforestry, and pest control. By the end of 2018, China had sent nearly 1,100 agrotechnical experts to twelve African countries, namely, Ethiopia, Gabon, Ghana, Liberia, Namibia, Malawi, Mali, Mauritania, Nigeria, Senegal, Sierra Leone, and Uganda, establishing China as the main leader and supporter of South–South cooperation.[23] In addition, China has conducted third-party cooperation with the US-based Gates Foundation at agrotechnical demonstration centers in Mozambique and Zambia. This shows the openness of China–Africa cooperation in poverty reduction. Together with other countries in the world, China has contributed its wisdom and solutions to the sustainable development goals of developing countries.

China–Africa poverty reduction cooperation conforms to the developments among and changes in national conditions on both sides, with more social forces and actors involved. In light of the new themes and trends of international cooperation, it has also added sustainable development and capacity building in pursuit of innovation on the basis of

[23] "South-South Cooperation," *Food and Agriculture Organization of the United Nations*, http://www.fao.org/partnerships/south-south-cooperation/en.

inheritance. Cooperation platforms have also been diversified and the number of involved partners has increased, thus highlighting the characteristics of diversity, innovation, and sustainability. Today, cooperation enjoys a firmer foundation, greater resources, and stronger development than ever before. China has become a major external force for promoting poverty reduction and development in Africa.

3 Focuses of Deepening China–Africa Poverty Reduction Cooperation

At present, China's Africa policy is guided by the principles of "sincerity, real results, affinity, and good faith" and "upholding justice while pursuing shared interests." Its goal is to build a more closely connected China–Africa community with a shared future. With the implementation of the "eight major initiatives," China–Africa cooperation in poverty reduction has entered a new era, or the "3.0 version," characterized by steady and sustained progress. The new positioning of China–Africa relations calls for accelerated improvement and a higher quality of their cooperation in poverty reduction, with the objective of improving African people's livelihoods and forging a new landmark for a China–Africa community with a shared future. To better advance the upgraded China–Africa poverty reduction in cooperative measures, we need to clarify the following issues.

Recognizing Changes in the Environment of Poverty Reduction Cooperation

At the international level, the role of South–South cooperation in the global development system has undergone major changes, moving from the periphery to a central position in international development cooperation alongside North–South cooperation. Regarding international actors engaging in poverty reduction in Africa, developed countries such as the US, the UK, France, and international organizations such as the European Union and the World Bank have maintained their traditional influence on development cooperation with Africa, based on historical ties and strategic considerations. In recent decades, however, emerging economies such as China, Brazil, India, and Turkey have also become important players in developmental cooperation with Africa as a consequence of their strong economic growth. The focus of South–South cooperation has also shifted from political issues to the economic arena. The status of developing countries on the stage of international development cooperation has thus witnessed changes, as they now stand in the "spotlight" alongside the countries of the global North.

Currently, the 2030 Sustainable Development Goals (SDGs) have replaced the earlier Millennium Development Goals (MDGs) as the bearer of international development cooperation. The SDGs, including 17 goals and 169 targets, have something in common with the MDGs, but also feature new innovations, such as the addition of targets concerning the system and conditions of poverty reduction. This indicates the international community's deeper understanding of poverty and anti-poverty issues and that relevant cooperation among countries should be better synergized to the new mechanism of international development cooperation. Given China's role as a substantial contributor to global poverty reduction, the international community also expects China to assume more responsibilities in this area.

Defining Leadership in China–Africa Poverty Reduction Cooperation

China and the African nations are all developing countries and face many similar development-related problems in their drive for modernization. Therefore, China–Africa poverty reduction cooperation essentially consists of mutual support between developing countries and is an integral part of South–South cooperation. Moreover, such cooperation in poverty reduction has gone beyond mere assistance. It is a decisive driving force behind China and Africa being able to realize the African Dream and the Chinese Dream of national rejuvenation, and also reflects their joint pursuit of sustainable economic and social development. As far as bilateral cooperation is concerned, China needs to respect Africa's strategic autonomy in the field of poverty reduction and development, which means that African countries can independently choose a suitable path for their poverty reduction and development strategies. In December 2015, for example, President Xi Jinping pointed out in his speech at the opening ceremony of the FOCAC Johannesburg summit: "China strongly believes that Africa belongs to the African people and that African affairs should be decided by the African people."[24] In his speech at the FOCAC Beijing Summit in September 2018, President Xi reiterated China's commitment to pursuing a "five-no" approach in cooperation with Africa, a policy that includes "no interference in African countries' pursuit of development paths that fit their national conditions" and "no imposition of China's own will on African countries."[25] The key points of the "ten cooperation plans" and "eight major initiatives" proposed during the two FOCAC summits are also in line with Africa's Agenda 2063. This reflects China's value orientation of pursuing international fairness, equality, harmony, and cooperation. China has taken the initiative to coordinate and synergize with Africa's strategic plan on poverty reduction, while underlining that cooperation with African nations should follow the principle of being "proposed, agreed, and led by Africa."

The Balanced Participation of Chinese Actors in Poverty Reduction Cooperation with Africa

In terms of the actors involved in poverty reduction cooperation between China and Africa, the role of government-level assistance has been predominant throughout history. Since the 1990s, Chinese enterprises have stepped up the pace of "going global," which is in line with the new ideas and policies for the development of South–South cooperation. Enterprises have gradually shouldered their social responsibilities for local development, and now take charity as a social investment with valuable economic benefits. By building a people-oriented corporate culture and engaging in development projects in local African communities, enterprises are able to establish a positive public image. At the same time, China's domestic NGOs for poverty reduction and development have gradually broadened their portfolio of services, expanded their international horizons, and launched a successive number of poverty reduction cooperation projects. China–Africa poverty reduction cooperation has thus formed a model led by the government and participated in by multiple parties. In the near future, the government will continue to play

[24] Xi Jinping, "Open a New Era of China-Africa Win–Win Cooperation and Common Development: Address at the Opening Ceremony of the Johannesburg Summit of the Forum on China-Africa Cooperation."

[25] "Full text of Chinese President Xi Jinping's Speech at Opening Ceremony of 2018 FOCAC Beijing Summit."

a major role in China–Africa poverty reduction cooperation, especially given its vast capacity to mobilize political and economic resources and its ability to implement a top-down execution. Meanwhile, the participation of enterprises and NGOs should be further expanded. In mobilizing NGOs to engage in the cause of poverty reduction, China can draw useful experience from the practices of developed countries and other developing economies in their poverty reduction cooperative measures and construct a pluralistic, yet differentiated, frame of reference to help supplement and improve China's own cooperation model. After all, the weak links in existing cooperation can be addressed by more innovative thinking.

Issues of Differentiation and Adaptability in China–Africa Poverty Reduction Cooperation

China's poverty reduction strategy has brought a new frame of reference and experience to Africa, mainly based on similarities in the experience of poverty on both sides. At the same time, Africa is a continent of 54 countries with both major and minor differences among them. From the perspective of political stability, there are countries facing severe challenges to their peace and development, such as South Sudan and Somalia; there are also stable states, such as Botswana and Namibia. In terms of basic industries of the national economy, there are countries with agricultural resources such as Tanzania and Senegal; there are those with mining resources such as South Africa, Angola, and Zambia; there are also those with tourism resources, such as Seychelles and Mauritius. With regard to the level of economic development, there are countries with an established industrial base, such as South Africa and Kenya; there are also those without any industrial foundation, such as Cape Verde and Djibouti. As for employment, there are countries with high youth employment rates, such as Ethiopia and Rwanda; there are also those with serious youth unemployment, such as Gabon and Tunisia. Regarding infrastructure, there are countries with crisscrossed highway networks, such as South Africa; there are also those with underdeveloped roads, such as Madagascar and Malawi. Similar instances are legion. Therefore, as the 54 African countries differ in developmental stages, development level, and national conditions, China and Africa need to embrace inclusiveness and diversity in poverty reduction cooperation. To implement specific poverty reduction programs, China should take into account the individual characteristics of each African country, take a tailor-made approach to each, and carry out cooperation in a targeted way. Only by doing so can we increase the effectiveness of poverty reduction cooperation.

Objectively, China's poverty reduction experience stems from its unique national conditions, history, and culture, from the continuous exploration of new paths since the founding of the PRC 70 years ago, from the prudent practices of Chinese leaders and the Chinese people, and from the innovative thinking about poverty reduction that builds on the past and keeps pace with the times. The Chinese government has played a firm leading role in the implementation of poverty reduction policies, which is in accordance with the political culture of a "strong government" in China. Relying on the political, economic, and cultural resources at its disposal, this kind of potent government has the powerful ability to mobilize and intervene in national economic construction. Additionally, the Chinese government is able to formulate medium- to long-term national development goals and long-term strategic plans, as well as maintain the continuation and sustainability of national poverty reduction and development plans regardless of changes in government leadership. Therefore, the political culture of "strong government" provides a similarly strong guarantee for the implementation of China's poverty reduction policies. On the contrary, most African countries have "transplanted," or followed certain Western models in the building of their own political systems, while various ethnic and religious conflicts and party rivalries in African countries, be it concealed or evident, dissolve or erode the power of the state to a certain extent, leading to the phenomenon of a "weak government and strong society." The lack of national authority makes it impossible for the state to mobilize and effectively integrate various resources to promote national plans for poverty reduction. The present multiparty system in Africa is also a result of external intervention. It is not rooted in African society, nor does it transcend narrow tribal or regional interests. When it holds state power, the ruling party usually pursues its own short-term interests, rather than setting the country's long-term development goals. After a change of government, it is often difficult to sustain the original national strategy for economic and social development. Therefore, when conducting cooperation with African countries in poverty reduction, especially in the process of sharing China's experience, China cannot simply impose or transplant its own recommended solution onto the local conditions. Rather, China's plan should serve as reference for African nations to learn from. Most importantly, China should create a favorable external environment for African countries to explore their own endogenous poverty reduction models suitable for their respective national conditions.

4 Conclusion

In today's world, human society has become a closely connected community with a shared future, one with tightly integrated interests and a high degree of interdependence. Under the China–Africa cooperative framework, poverty reduction cooperation will boost African development, consolidate the foundation of China–Africa relations, and optimize public opinion of China–Africa cooperation. Down-to-earth projects that benefit people's livelihoods will bring the people of China and Africa closer to each other, as well as help cement the foundation of China–Africa friendship. China–Africa cooperation in poverty reduction can also serve as a reference for further South–South cooperation and global poverty governance. The measures China has adopted in its cooperation, such as strengthening infrastructure construction, agrotechnical support, and human resources training for the poverty reduction sector (especially sharing China's poverty reduction experience), have enriched and further enhanced the agenda of global poverty governance. In addition to the North–South cooperation model, global poverty governance is witnessing the rise and concurrent development of the South–South cooperation model.

Correction to: The Changing World and Africa

Xinfeng Li and Chunying An

Correction to:
X. Li and C. An (eds.), *The Changing World and Africa*, https://doi.org/10.1007/978-981-16-4983-7

The original version of this book was inadvertently published with incorrect affiliation details for authors of Chapters 4, 8, 16, and 20, which have now been updated. The correction to the book has been updated with the changes.

The updated version of these chapters can be found at
https://doi.org/10.1007/978-981-16-4983-7_4
https://doi.org/10.1007/978-981-16-4983-7_8
https://doi.org/10.1007/978-981-16-4983-7_16
https://doi.org/10.1007/978-981-16-4983-7_20
https://doi.org/10.1007/978-981-16-4983-7

X. Li and C. An (eds.), *The Changing World and Africa*,
https://doi.org/10.1007/978-981-16-4983-7_23

The manufacturer's authorised representative in the EU is Springer Nature Customer Service Centre GmbH, Europaplatz 3, 69115 Heidelberg, Germany. If you have any concerns regarding our products, please contact ProductSafety@springernature.com

Printed and bound by CPI Group (UK) Ltd, Croydon, CR0 4YY
30/09/2026
02225286-0016